T0172533

Teaching Secondary School
MATHEMATICS
Research and practice for the 21st century

2nd Edition

Teaching Secondary School
MATHEMATICS
Research and practice for the 21st century

**Merrilyn Goos, Colleen Vale and Gloria Stillman
with Katie Makar, Sandra Herbert and Vince Geiger**

Routledge
Taylor & Francis Group

LONDON AND NEW YORK

First published 2017 by Allen & Unwin

Published 2020 by Routledge
2 Park Square, Milton Park, Abingdon, Oxon OX14 4RN
605 Third Avenue, New York, NY 10017

Printed and bound by CPI Group (UK) Ltd, Croydon, CR0 4YY
Routledge is an imprint of the Taylor & Francis Group, an informa business

Cataloguing-in-Publication details are available
from the National Library of Australia
www.trove.nla.gov.au

Index by Puddingburn
Set in 12/17 pt Spectrum MT by Newgen Knowledge Works (P) Ltd, Chennai, India

ISBN-13: 9781743315934 (pbk)

CONTENTS

Figures and Tables   vii

List of Abbreviations   xiii

Acknowledgements   xiv

About the Authors   xv

Preface   xix

Part 1: Introduction   1

   1. Doing, teaching and learning mathematics   3

Part 2: Mathematics pedagogy, curriculum and assessment   21

   2. Developing mathematical understanding   23

   3. Developing mathematical connections   46

   4. Effective use of technologies in mathematics education   75

   5. Mathematics curriculum models   110

   6. Assessing mathematics learning   137

Part 3: Teaching and learning mathematical content   167

   7. Teaching and learning number   169

   8. Teaching and learning algebra   199

   9. Teaching and learning geometry and spatial concepts   227

10. Teaching and learning measurement   257

11. Teaching and learning chance and data   274

12. Teaching and learning calculus   302

Part 4: Equity and diversity in mathematics education   **339**

13. Teaching students with diverse mathematical learning needs   341

14. Equity and social justice in mathematics teaching and learning   369

15. Gender equity and justice in mathematics teaching and learning   395

Part 5: Professional and community engagement   **415**

16. Working with parents and communities   417

17. Continuing professional learning   435

References   459

Index   525

FIGURES AND TABLES

Figures

1.1	Raymond's model of the relationships between teachers' mathematical beliefs and their teaching practice	5
1.2	Secondary school students' drawing of a typical mathematics teacher	11
1.3	Domains of Mathematical Knowledge for Teaching	15
2.1	Levels of understanding in the Pirie–Kieren theory	26
2.2	A student's attempt to solve simultaneous equations	27
2.3	Categories of complex mathematical thinking	35
2.4	Representation of process aspects of mathematics in curriculum materials	37
2.5	Factors contributing to successful problem solving	39
2.6	Teacher scaffolding questions during problem solving	40
2.7	First six stages of constructing the Koch snowflake curve	43
2.8	Characteristics of a classroom community of mathematical inquiry	44
3.1	Task 1 Lessons 1 and 2	50
3.2	Context as border	61
3.3	Context as wrapper	63
3.4	The modelling process	70
3.5	Mathematical modelling framework incorporating regulatory mechanism	71
4.1	Spreadsheet for investigating compound interest	77
4.2	Symbolic, graphical and numerical representations of the family of functions represented by $y = x^2 \pm c$	80
4.3	Excel spreadsheet with scrollbars for investigating properties of trigonometric functions	81
4.4	Dynamic geometry software screen display for investigating the angle sum of triangles	82
4.5	Exploring rotation	83
4.6	A spreadsheet simulation of a spinner	83

4.7	CensusAtSchool archive	85
4.8a	Modelling cooling with an exponential function	85
4.8b	Modelling periodic motion with a trigonometric function	85
4.9	Solving a quadratic equation	89
4.10a	Finding the turning point of a quadratic function	93
4.10b	Solving the equation $5000 \times (0.758)^x = 1$	93
4.11	Matrix transformation task	96
4.12	Different student programs for finding angle between vectors	98
4.13	Spreadsheet method for solving equation	99
5.1	Sources of influence on mathematics curriculum content	115
5.2	Superordinate and subordinate functions in curriculum development and implementation	117
5.3	Elaborations of the four proficiency strands in the current *Australian Curriculum: Mathematics*	127
5.4	Model of numeracy	129
5.5	Age distribution of the Australian population, 1960 and 2010	131
6.1	Purposes of assessment	140
6.2	Verbal descriptors of standards exemplar, Mathematics Stage 4	144
6.3	KWL chart	149
6.4	*NAPLAN* Year 9 multiple-choice item	154
6.5	PISA (15-yr-old) Quantity and Employ test item	154
6.6	TIMSS Year 8 Number and Reasoning multiple-choice test item	154
6.7	Fraction and Decimal interview item	155
6.8	Flowchart for preparing an assessment task for students	158
6.9	Recording feedback and assessment using scores	164
6.10	Teacher tracking document	164
7.1	A fraction wall showing $\frac{2}{3} = \frac{4}{6} = \frac{6}{9}$	171
7.2	Place value chart	174
7.3	Two examples of student decimal misconceptions	175
7.4	Three solutions for $5 \div 6$	178
7.5	Field laws	181
7.6	Arrays and area diagrams illustrating the distributive law for multiplication	183
7.7	Paper folding into sixths	184
7.8	$0.4 \times 0.6 = 0.24$	185
7.9	A double number line	189
7.10	A ratio table	189
7.11	Procedures for finding the percentage of amounts	191
7.12	Graphic organiser for index numbers	192
7.13	18×19 illustrating multiplying negative integers	194

7.14	Arithmetic calculations on CAS calculator	195
7.15	Right-angled isosceles triangle construction for a logarithmic spiral	196
7.16	Learning cycle	197
8.1	Spontaneous use of 'w'	203
8.2	Generalising from geometric patterns	207
8.3	Solving equations in a computer algebra environment	208
8.4	Algebraic insight needed to interpret unfamiliar form of solution	210
8.5	Equivalence tasks	212
8.6	Tasks investigating the notion of variable	214
8.7	Multiple views of the mathematical concept of rate	215
8.8	Function representations and the links to be developed between these	221
8.9	The family of graphs $y = ax + 1$ in the multiple representation environment (View^3) of the TI-SmartView	222
8.10	Possible shapes for graphs of cubic functions	223
8.11	Algebraic transformation with resulting numerical and graphical representations using a function grapher	224
8.12	Translation and dilation of function using a function as object manipulator	225
9.1	'Say what you see'	233
9.2	Many cubes	233
9.3	Federation Square, Melbourne	234
9.4	Slicing a cube	235
9.5	Verbal and visual word association diagram showing the concept of hypotenuse	236
9.6	Sum of exterior angles in a polygon	237
9.7	Isosceles triangles constructed using Geometer's Sketchpad	240
9.8	Leaning Tower of Pisa and experiment	244
9.9	Midpoint quadrilaterals formed by dragging vertex A of a quadrilateral	246
9.10	Constructing quadrilaterals from diagonals	247
9.11	Cutting folded paper to construct quadrilaterals	247
9.12	Rotation about the centre, the edge and a point not on the centre	249
9.13	Cairo tiling (pentagons)	250
9.14	Cylinder created by rotation of rectangle	251
9.15	Distance at the same latitude	253
9.16	Network diagram or graph	254
10.1	Incorrect response to scale item	263
10.2	Representing volume (volume = area × height)	266
10.3	Cans in a box	267
10.4	The freeway problem	269
10.5	Noticing constant ratio in similar triangles for cosine 30.11°	270
11.1	Estimating unusual areas	275

11.2	Comparing the quality of brands of peanut butter	280
11.3	A Punnett square	283
11.4	Hours of sleep of Years 7–10 students	286
11.5	Association between lengths of femur bones and heights in humans	288
11.6	Spring temperatures for Perth, disaggregated by month	289
11.7	The statistical investigation cycle	294
11.8	A comparison of the ages of 100 couples getting married	298
11.9	Random sample of 50 voters—22 in favour, 28 against	299
12.1	Use of GridPic to fit a quartic function to a Gothic arch	304
12.2	Graphic representations of $f(x) = x^2 - 2x + 1$ (left) and $g(x) = -x^2 + 2x + 1$ (right)	305
12.3	Numeric representations of $f(x) = x^2 - 2x + 1$ (left) and $g(x) = -x^2 + 2x + 1$ (right)	306
12.4	Students linking numerical and diagrammatic representations in a real-world task	307
12.5	Aeroplane speed models	309
12.6	Drawing a tangent to the curve at the point (–3.5, 0.9)	310
12.7	Drawing secants approaching the tangent to the curve at the point (–3.5, 0.9)	311
12.8	Constant rate of change of height of a stack of plastic cups, per cup	311
12.9	Varying rate of change of total distance run with station distance	312
12.10	Path and distance travelled by runner	312
12.11	Using Draw Tangent and a judicious choice of window	314
12.12	Use of numerical derivative to sketch gradient function of $f(x) = (x - 2)^2 (x + 5)$	314
12.13	Finding the limit informally using the symbolic facilities of a CAS calculator	317
12.14	Finding the limit informally using the graphing facility of a CAS calculator	317
12.15	Rate diagrams	318
12.16	Common textbook diagram for gradient at a point	320
12.17	Calculator notation for finding gradient at a point	321
12.18	Kendal and Stacey's (2003) concept map of differentiation in numerical, graphical and symbolic representations	325
12.19	Zandieh's (2000) framework	328
12.20	Developing a rule for the derivative of the sum or difference of two functions	330
12.21	Developing a rule for the derivative of the product of two functions	332
12.22	Use of slope diagrams to find anti-derivatives	333
12.23	Introduction to numerical integration	334
12.24	Demonstration of limiting process leading to formula	334

12.25	Using summation to find area bounded by a curve	335
12.26	Area bounded by a curve	336
12.27	Development of the Fundamental Theorem of Calculus	336
13.1	A rich mathematical task and an open-ended mathematical task	347
13.2	'Just right' self-assessment task	350
13.3	'Just right' tasks	351
13.4	Sample clinical interview item	355
13.5	Three students' drawings of $\frac{2}{3}$	356
14.1	Percentage of students for each level of mathematics literacy by socio-economic background from PISA 2012	370
14.2	Percentage of students for each level of mathematics literacy by geographic location from PISA 2012	370
14.3	Percentage of students for each level of mathematics literacy by Indigenous background from PISA 2012	371
14.4	Percentage of students for each level of mathematics literacy by immigrant background from PISA 2012	371
14.5	Cloze task	385
14.6	8 Aboriginal Ways of Learning	390
15.1.	Maryam Mirzakhani, the first female Fields Medallist	411
16.1	Network model for home–school–community partnerships	425
17.1	Reflection card	438
17.2	The IMPACT procedure	441
17.3	Relationships between the three 'zones of influence' in teacher learning	448
17.4	Adam's first year of teaching	451
17.5	Adam's second year of teaching	452

Tables

1.1	Relationship between beliefs about mathematics, teaching and learning	7
1.2.	Framework for analysing pedagogical content knowledge: clearly PCK	15
2.1	Evidence of understanding for secondary mathematics students (n = 329)	25
3.1	Japanese mathematics lesson sequence: linear equations	50
4.1	A typology of expected use of graphics calculators in assessment	103
4.2	Factors influencing technology use in mathematics education	105
5.1	Civilian population aged 15–69 years, labour force status—September 2010	134
6.1	SOLO taxonomy	145
6.2	Assessment tasks	147
6.3	Everyday rubric grading	159
6.4	Rubric for medicine doses	160
6.5	Rubric for communication category judging and scoring criteria for *Maths Talent Quest*	162

7.1	Division of fraction problem types	186
10.1	Common prefixes in the metric	262
11.1	Association between where UK citizens live and whether they smoke	288
12.1	Kendal and Stacey's (2003) Differentiation Competency Framework	324
12.2	Delos Santos and Thomas's (2003) Representational Framework of Knowing Derivative	327
13.1	Items from the Australian Mathematics Competition	363
14.1	Mathematics problems of personal or social relevance	392
15.1	Historical development of theories and practices for gender equity	407
17.1	Adam's practicum experience	450
17.2	Australian Professional Standards for Teachers	454
17.3	Domains of the AAMT Standards	455

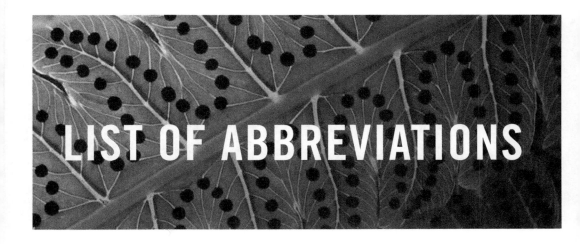

LIST OF ABBREVIATIONS

AAMT	Australian Association of Mathematics Teachers
ACARA	Australian Curriculum, Assessment and Reporting Authority
ACER	Australian Council for Educational Research
AITSL	Australian Institute for Teaching and School Leadership
CAS	Computer algebra systems
EDA	Exploratory Data Analysis
ICT	Information and Communication Technology
MCEETYA	Ministerial Council on Education, Employment, Training and Youth Affairs
MCTP	Mathematics Curriculum and Teaching Program
OECD	Organisation for Economic Co-operation and Development
NAPLAN	*National Assessment Program—Literacy and Numeracy*
NCTM	National Council of teachers of Mathematics
PCK	Pedagogical content knowledge
PISA	Program of International Student Assessment
RIME	Reality in Mathematics Education
SES	socio-economic status
SOLO	Structure of the Observed Learning Outcome
STEM	Science, technology, engineering and mathematics
TIMSS	Trends in International Mathematics and Science Study
ZPD	Zone of proximal development

ACKNOWLEDGEMENTS

Our goal in writing this book was to provide prospective and practising secondary mathematics teachers and university-based mathematics teacher educators with a research-based text that is also rich in practical teaching ideas. In this second edition we have updated each of the chapters in line with changes to the secondary mathematics curriculum in Australia and recent research on the teaching and learning of mathematics. We invited colleagues with particular expertise to assist with revising chapters in the text and have acknowledged their contribution as authors: Katie Makar, who wrote the original chapter on Teaching and Learning Chance and Data and revised this chapter for the second edition, Sandra Herbert, who revised the chapters on Teaching and Learning Algebra and Teaching and Learning Calculus, and Vince Geiger, who revised the chapter on Effective Use of Technologies in Mathematics Education.

We are grateful to the following for permission to reproduce material: *Computers in the Schools; Educational Studies in Mathematics; Journal for Research in Mathematics Education; Journal of Mathematics Teacher Education; Journal of Teacher Education; Mathematics Education Research Journal; Procedia Social and Behavioural Sciences; Teaching Mathematics and its Applications*; American Mathematical Society; Australian Association of Mathematics Teachers; Australian Bureau of Statistics; Australian Council for Educational Research; Australian Curriculum, Assessment and Reporting Authority; Australian Institute for Teaching and School Leadership; Department of Education and Training Victoria; Department of Education NSW; International Association for the Evaluation of Educational Achievement; Mathematical Association of Victoria; Mathematics Education Research Group of Australasia; National Council of Teachers of Mathematics; NSW Board of Studies, Teaching and Educational Standards; Sense Publishers; Taylor and Francis Group; Wolfram Alpha; Monica Baker; Helen Chick; Doug Clarke; Kaye Stacey; and Mike Thomas.

ABOUT THE AUTHORS

Merrilyn Goos is Professor and Head of the School of Education at The University of Queensland, where for ten years she coordinated mathematics curriculum studies for prospective secondary school teachers in the School's Graduate Diploma in Education and Bachelor of Education programs. She is an internationally recognised mathematics educator whose research is well known for its strong focus on classroom practice. She has led projects that investigated students' mathematical thinking, the impact of digital technologies on mathematics learning and teaching, the professional learning of mathematics teachers, and numeracy across the curriculum. She has served her professional and research communities as Chair of the Queensland Studies Authority's Mathematics Syllabus Advisory Committee, Vice-President of the Queensland Association of Mathematics Teachers, and President of the Mathematics Education Research Group of Australasia. In 2004 she won an Australian Award for University Teaching for her work as a mathematics teacher educator.

Colleen Vale is Associate Professor of Mathematics Education at Deakin University where she teaches in primary and secondary pre-service and associate teacher education programs. She works closely with schools and teachers to provide further teaching experiences for secondary pre-service teachers of mathematics and to provide professional learning for primary and secondary teachers. She has an international reputation for her research on gender equity and social justice in mathematics education. Her current research projects concern the identities and teaching practices of out-of-field and beginning secondary mathematics teachers, instructional leadership and professional learning for mathematical reasoning and inquiry-based teaching and learning. Colleen has been an active participant in the mathematics education and research field as a co-writer of the original VCE Mathematics course and held positions as the President of the Mathematics Association of Victoria, Vice-President of Mathematics Education Research Group of Australasia and editor of *Mathematics Teacher Education and Development*.

Gloria Stillman is an Associate Professor at the Australian Catholic University of Melbourne. She has had an extensive career as a lecturer teaching secondary mathematics methods at the University of Melbourne and in tertiary institutions in Queensland, where she was on the state panel for Mathematics C then A for the Board of Senior Secondary School Studies. Previously she was a secondary mathematics teacher in a Brisbane school. Gloria's research interests focus on mathematics education and, in particular, the teaching, learning and assessing of real-world applications and modelling. Significant projects of international regard include RITEMATHS, an Australian Research Council-funded Linkage project, and CCiSM, an ongoing research project into curriculum change in secondary mathematics in mathematical modelling. She has recently co-edited two books on international research about the teaching and learning of mathematical modelling and is President of ICTMA (International Community of Teachers of Mathematical Modelling and Applications). Gloria is a member of the Editorial Board of *ZDM—The International Journal of Mathematics Education.*

Katie Makar is Associate Professor and Deputy Head of the School of Education at the University of Queensland. She teaches the cohort of prospective secondary mathematics teachers, with an emphasis on taking an investigative approach to mathematics. A former secondary mathematics teacher for 15 years, she remains involved with schools through research projects that focus on supporting teachers to improve their pedagogy within their own classrooms. Her work with inquiry-based pedagogies over time, for example, has engaged with teachers in over 1500 lessons in the past ten years. Makar is known internationally for her research in statistical reasoning, particularly informal statistical inference in schools, and is the Co-Director of the International Collaborative for Research in Statistical Reasoning, Thinking and Literacy. She has won three awards for the quality and impact of her research as well as a commendation for her teaching of pre-service teachers.

Sandra Herbert is a Senior Lecturer in the Faculty of Arts and Education at Deakin University. She is B.Ed. (Primary) course co-ordinator at Warrnambool and teaches secondary mathematics associate teachers in a blended learning education program. These courses provide students from rural and regional low-SES backgrounds, who are typically under-represented in university courses and have lower retention rates and lower success rates, opportunities in teacher education, and prepare teachers to work and teach mathematics in these locations and communities. Sandra's research focuses on improved learning in mathematics and science in schools including mathematical reasoning, algebra and calculus. Her work has been published in high-quality international journals making a major methodological contribution to the application of phenomenography to educational research. With her prior experience in advanced mathematics and information technology, she brings knowledge and expertise to strengthen nationally recognised imperatives for education in mathematics to her primary and secondary pre-service and associate teachers. A particular focus of her teaching and research is fostering a more positive attitude to mathematics in school students.

Vince Geiger is an Associate Professor and Research Fellow within the Learning Sciences Institute Australia at Australian Catholic University. Prior to his current appointment he was heavily involved in teacher pre-service at ACU including the co-ordination of the School of Education's Graduate Diploma of Education, Master of Education and Bachelor of Education/Bachelor of Arts programs. Within this role he drew on his 22-year experience as a secondary teacher of mathematics. His research on teacher professional learning is internationally recognised, particularly in the areas of teachers as designers of students' mathematics learning, numeracy across the curriculum, mathematical modelling and applications, and the role of digital tools in enhancing mathematics instruction. He has contributed to education more broadly through various professional roles such as President of the Australian Association of Mathematics Teachers, Chair of the National Education Forum, Chair of the Queensland Studies Authority's State Assessment Review Panel for Mathematics B, and Secretary of the Mathematics Education Research Group of Australasia. His work with pre-service teachers was acknowledged in 2009 through an Australian Learning and Teaching Council citation for an outstanding contribution to student learning.

PREFACE

It has been ten years since we published the first edition of *Teaching Secondary School Mathematics: Research and Practice for the 21st Century*. While change, especially with regard to innovations in mathematics teaching practice, appears to have occurred slowly, the context for mathematics teachers of secondary school students has continually changed. Increasing globalisation has impacted the mathematics classroom in a number of ways. Classrooms are becoming more ethnically diverse due to the increasing diversity of people seeking refuge from ongoing war and atrocities, immigration of skilled people from a broad range of countries and the influx of international students into Australian secondary schools. Schools and teachers therefore need to develop new understandings of their students and build relationships with new and diverse local ethnic communities. Innovations in digital technology, including increased connectivity, bring new tools and resources to the mathematics classroom along with opportunities for online engagement and collaboration at a distance. These changes in context are an international phenomena. But there have also been some particular changes in the Australian context with regard to curriculum and education policy.

Australia now has a national curriculum for mathematics from Kindergarten to Year 10; indeed the second version (V8.1) was published in 2015 along with a national curriculum for senior secondary mathematics. Numeracy is one of the general capabilities in the national curriculum and is expected to be developed across the disciplines. A *National Assessment Program for Literacy and Numeracy* (NAPLAN) is implemented annually across the country for all students in Years 3, 5, 7 and 9. This test uses the standards in the national mathematics curriculum, rather than standards associated with numeracy as a general capability, so there are some inconsistencies in the meaning and assessment of numeracy in the Australian context. Alongside changes to the curriculum, a set of national professional standards for teachers at different stages of their career, including graduation, have been developed and used for accreditation of teacher education programs and assessment of graduating students.

The field of mathematics and its applications has continued to develop. There have been two highlights regarding mathematics achievement of note for this book. Australia has its first Fields Medalist, Prof. Terence Tao, born and educated in Adelaide, and the world has its first female Fields Medalist, Prof. Maryam Mirzakhani, an Iranian woman (see Chapters 14 and 15). Mathematics education research has continued to explore and deepen understanding of mathematical thinking and learning, effective mathematics teaching practice, and affordances of representations and resources. Mathematics education researchers have continued to work with teachers to investigate pedagogical approaches that address the most challenging issue for teachers—the diversity of learners and their learning needs in the classroom. Enabling mathematics engagement and success for Indigenous students and other students from socially and culturally disadvantaged communities continues to be an important focus of mathematics education research in Australasia. Approaches and findings from these studies reveal the importance of engagement of students in mathematical thinking for meaning making.

The second edition of *Teaching Secondary School Mathematics* is intended to be a reference for pre-service teachers aspiring and training to be mathematics teachers of secondary school students. It is also a professional learning resource for in-service teachers and out-of-field mathematics teachers, a reality in Australian schools, especially schools in rural and remote regions or low socio-economic communities.

The structure of the second edition remains the same. There are five sections: Part 1 is an introduction including one chapter; Part 2 includes four chapters that focus on mathematical proficiencies, curriculum, and pedagogical approaches including assessment; Part 3 includes six chapters, each focusing on the teaching and learning of particular mathematics content; Part 4 contains three chapters regarding diversity and social justice; Part 5 contains two chapters regarding professional engagement with colleagues and the community, now recognised as the third domain within the *Australian Professional Standards for Teachers*: professional engagement. We have updated each chapter with regard to the *Australian Curriculum: Mathematics*, significant findings from research, the application of new generation digital technologies, and changes in education policy and practices impacting secondary mathematics teaching. Each chapter also includes revised or new 'review and reflect' activities related to recently published research findings discussed in the chapter, or curriculum and policy documents and materials for teacher learning including online video recordings.

Updates to Chapter 1 include recently published frameworks of mathematics teachers' pedagogical content knowledge that inform professional learning and the evaluation of professional knowledge, and a revised section on the current challenges facing teachers of secondary mathematics. Chapters 2 and 3 focus on pedagogical approaches and the planning and enactment of lessons, especially for developing students' understanding, problem solving and modelling. Explicit connections to the mathematical proficiencies that now underpin the *Australian Curriculum: Mathematics*—understanding, fluency, problem solving and reasoning—have been included in these chapters along with relevant recently published studies. Significant changes made to Chapter 4 explain, illustrate and critique the affordances and use of the current generation of digital tools and resources for mathematics teaching. Chapter 5 about curriculum models also includes significant changes. It explains the development and structure of the *Australian Curriculum: Mathematics*, including elaborations of the proficiencies underpinning this curriculum. It also includes a new section about numeracy, a general capability of the *Australian Curriculum* which includes findings from recent studies that have researched the teaching and learning of numeracy within mathematics and across the curriculum. Substantial changes in Chapter 6 concern more explicit discussion of assessment *for* learning and assessment *as* learning, acknowledged as a component of effective mathematics teaching by researchers and increasingly emphasised in education policy and an expected practice of mathematics teachers. A critical review of the *National Assessment Program of Literacy and Numeracy* is also included in the second edition of this chapter.

Revisions to the chapters in Part 3 concerning teaching and learning of mathematics content report recent findings from research studies and make explicit connections to the *Australian Curriculum: Mathematics* and, where relevant, developments in Australian senior secondary mathematics curriculum. Formative assessment tools for fractions and decimals along with recent research findings concerning students' decimal misconceptions, mental computation strategies for multiplication and division, the source of student misconceptions regarding proportion, students' proportional reasoning and effective approaches for teaching integers are the updates included in Chapter 7. Recent research regarding students' learning of functions, including a new approach to introducing functions, are the main revisions to the algebra chapter, Chapter 8 in this second edition. Reference to recent studies and a more specific focus on geometric reasoning as included in the *Australian Curriculum: Mathematics* are the main revisions to Chapter 9 on geometry. New material included in Chapter 10 on measurement include approaches and strategies to ensure that learning about measurement is meaningful and connected to other disciplines and real-world problems, and research findings on students' skills in using measurement instruments and using and interpreting units of measurement. Recent research findings on the teaching and learning of statistics are cited and discussed in Chapter 11. Reference to and a review and reflect task about statistics content included in the different mathematics subjects in the recently published Australian Senior Secondary Mathematics Curriculum are other changes. Recent research on students' understanding of rate and rate of change and approaches to introducing integral calculus are the main revisions to Chapter 12 on calculus.

The chapters in Part 4 are concerned with equity and social justice and have been re-ordered and re-organised in this second edition, and findings from recent studies including further developments in theories informing equitable practice are the main revisions to these chapters. Chapter 13 on diversity in classrooms includes recent studies of effective approaches in teaching heterogeneous classes and studies that critique the practice of streaming. A new section on teaching differently abled students, that is, students with special learning needs, is included. Recent research regarding enrichment programs and intervention programs are also included. At the beginning of Chapter 14 readers are invited to analyse the findings from international studies to identify the impact of socio-cultural factors on student achievement and participation in mathematics. Geographic location is a new socio-economic factor included and discussed in this chapter. Recent studies providing revised theoretical perspectives are included to provoke thinking about the meaning of equity and social justice and its practice in the classroom. Recent models of socially and culturally response-able pedagogies are presented. The final chapter on gender equity and justice includes findings from recent research studies and also makes connections with emergent feminist theory to present pedagogical approaches representing a shift from inclusion to participation and gender response-able pedagogy.

Revisions to the final two chapters in the last section of the book acknowledge recent developments and changes in government policy along with references to recent studies in the field. These changes include recognition of the importance of establishing relations with parents and community for improving student participation and learning outcomes by governments and recent studies supporting this action. Recent literature concerning reflective practice and links to and discussion of the *Australian Professional Standards for Teachers* are the revisions made in the final chapter.

We have stressed the importance of reflective practice for ongoing professional learning and achievement of effective mathematics teaching in the initial and final chapter of this book. This second edition of *Teaching Secondary School Mathematics: Research and Practice for the 21st Century* represents our reflection on practice, and our reflection on research, both our own and others' research, that has deepened our understanding and practice of secondary mathematics teaching and teacher education. We hope that this book equips you with the pedagogical knowledge to take up the challenge of teaching secondary mathematics in your school and community context.

PART 1

Introduction

CHAPTER 1
Doing, teaching and learning mathematics

Excellent teachers of mathematics are purposeful in making a positive difference to the learning outcomes, both cognitive and affective, of the students they teach. They are sensitive and responsive to all aspects of the context in which they teach. This is reflected in the learning environments they establish, the lessons they plan, their uses of technologies and other resources, their teaching practices, and the ways in which they assess and report on student learning. (Australian Association of Mathematics Teachers (AAMT), 2006)

This statement appears in a professional standards framework that describes the unique knowledge and skills needed to teach mathematics well. It reflects findings from a multitude of research studies that show how students' mathematics learning and their dispositions towards mathematics are influenced—for better or for worse—by the teaching that they experience at school (see Mewborn, 2003). While it is sometimes difficult for researchers to untangle the complex relationships that exist between teaching practices, teacher characteristics and student achievement, it is clear that teachers do make a difference to student learning.

This chapter discusses what it means to be a teacher of secondary school mathematics and the requirements and challenges such a career choice entails. We first consider the mathematical beliefs of teachers and students, as well as students' perceptions of mathematics teachers, reflecting on how teachers communicate powerful messages about the nature of mathematics and mathematics learning to the students they teach. Next, we turn our attention to the secondary school mathematics classroom by examining recent research on mathematics teaching practices, and identify the types of knowledge needed for effective teaching of mathematics. Finally, we review some of the challenges for mathematics curriculum and teaching arising from this and other research.

Mathematical beliefs

Whether we are aware of it or not, all of us have our own beliefs about what mathematics is and why it is important. In summing up findings from research in this area, Barkatsas and Malone (2005) conclude that 'mathematics teachers' beliefs have an impact on their classroom practice, on the ways they perceive teaching, learning, and assessment, and on the ways they perceive students' potential, abilities, dispositions, and capabilities' (p. 71). However, the relationship between beliefs and practices is not quite so straightforward as this, and many researchers would agree that rather than being related in a linearly causal way in either direction, beliefs and practice influence one another and develop together (Beswick, 2007). Context also contributes to the relationship between beliefs and practice. Raymond's (1997) model of the relationships between beliefs and practices and the factors influencing them is informative in this regard (see Figure 1.1). The model suggests some of the complexity involved in understanding how beliefs shape, and are shaped by, teaching practices, and why inconsistencies sometimes exist between the beliefs that teachers might espouse and those they enact through their practice. Beswick (2012) also identified inconsistencies between mathematics teachers' beliefs of mathematics as a discipline and as a school subject, to further confirm the complexity of the relationship between beliefs and practice.

Beliefs about the nature of mathematics

Given that teachers communicate their beliefs about mathematics through their classroom practices, it is important to be aware of one's beliefs and how they are formed.

REVIEW AND REFLECT : In your own words, write down what you think mathematics is and why it is important for students to learn mathematics at school. Compare your thoughts with a fellow student and try to explain why you think this way.

Look up the definition or description of mathematics provided in the *Australian Curriculum: Mathematics* (Australian Curriculum, Assessment and Reporting Authority (ACARA), 2015) and by notable mathematicians or mathematics educators (Gullberg, 1997; Hogan, 2002; Kline, 1979). Compare these with your own ideas.

Discuss with a partner some of the possible influences on the formation of your beliefs, using Raymond's (1997) model as a guide (see Figure 1.1).

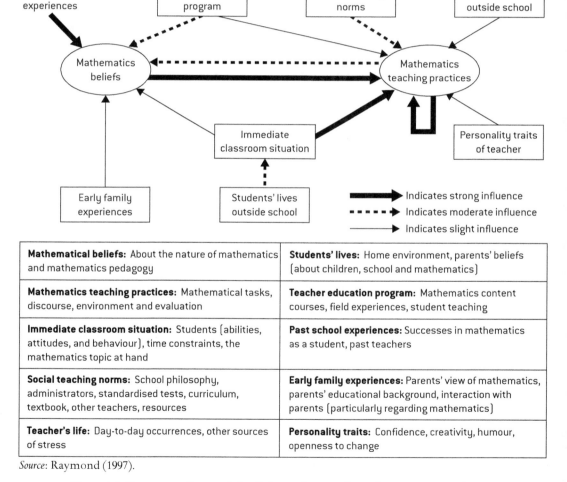

Mathematical beliefs: About the nature of mathematics and mathematics pedagogy	Students' lives: Home environment, parents' beliefs (about children, school and mathematics)
Mathematics teaching practices: Mathematical tasks, discourse, environment and evaluation	Teacher education program: Mathematics content courses, field experiences, student teaching
Immediate classroom situation: Students (abilities, attitudes, and behaviour), time constraints, the mathematics topic at hand	Past school experiences: Successes in mathematics as a student, past teachers
Social teaching norms: School philosophy, administrators, standardised tests, curriculum, textbook, other teachers, resources	Early family experiences: Parents' view of mathematics, parents' educational background, interaction with parents (particularly regarding mathematics)
Teacher's life: Day-to-day occurrences, other sources of stress	Personality traits: Confidence, creativity, humour, openness to change

Source: Raymond (1997).

Figure 1.1 **Raymond's model of the relationships between teachers' mathematical beliefs and their teaching practice**

Mathematics and numeracy

In recent years, the idea of numeracy has gained prominence in discussions about the essential knowledge and competencies to be developed by school students for participation in contemporary society. Indeed, it is included in the *Australian Curriculum: Mathematics* (ACARA, 2014) as a general capability, meaning that it applies to and can be developed across the curriculum, and is not confined to mathematics or the sole responsibility of mathematics teachers. Thus, it is important for mathematics teachers to have a clear understanding of the nature of numeracy and its relationship with mathematics (Queensland Board of Teacher Registration, 2005). The term 'numeracy' is common in Australia, New Zealand, Canada and the United Kingdom (UK), but in other parts of the world, the terms 'quantitative literacy', 'mathematical literacy' or 'statistical literacy' are used. These different names convey different meanings that may not be interpreted in the same way by all people. For example, some definitions of quantitative literacy focus on the ability to use quantitative tools for everyday practical purposes, while mathematical literacy is understood more broadly as the capacity to engage with mathematics in order to act in the world as an informed and critical citizen (Organisation for Economic Cooperation and Development [OECD], 2000). Steen (2001) offers the following distinction between mathematics and numeracy:

> Mathematics climbs the ladder of abstraction to see, from sufficient height, common patterns in seemingly different things. Abstraction is what gives mathematics its power; it is what enables methods derived in one context to be applied in others. But abstraction is not the focus of numeracy. Instead, numeracy clings to specifics, marshalling all relevant aspects of setting and context to reach conclusions … Numeracy is driven by issues that are important to people in their lives and work, not by future needs of the few who may make professional use of mathematics or statistics. (pp. 17–18)

These definitions suggest that numeracy is broader than, and different from, the way that mathematics traditionally has been viewed by schools and society. These meanings and approaches to teaching numeracy are discussed in Chapter 5.

REVIEW AND REFLECT: Consider your beliefs about numeracy and revisit your beliefs about the nature of mathematics and compare these with the distinction between mathematics and numeracy proposed by Steen (2001).

How is numeracy described in the *Australian Curriculum*? To what extent does this description incorporate ideas about mathematics being used for practical purposes, in real-world contexts and for developing critical citizenship?

Beliefs about mathematics teaching and learning

Just as important as mathematics teachers' beliefs about the nature of mathematics are their beliefs about mathematics teaching and learning. Beswick (2005, 2007, 2012) shows the connections between these types of beliefs by drawing on categories developed by Ernest (1989) and van Zoest et al. (1994), as shown in Table 1.1.

REVIEW AND REFLECT : Researchers usually obtain information about teachers' mathematical beliefs via questionnaires (e.g., Barkatsas & Malone, 2005; Beswick, 2005; Frid, 2000a; Perry et al., 1999). Obtain a copy of one of these beliefs questionnaires and record your responses. Discuss your answers with a partner in the light of the classifications in Table 1.1.

Table 1.1 Relationship between beliefs about mathematics, teaching and learning

Beliefs about the nature of mathematics (Ernest, 1989)	Beliefs about mathematics teaching (Van Zoest et al., 1994)	Beliefs about mathematics learning (Ernest, 1989)
Instrumentalist: mathematics as a tool kit of facts, rules, skills	Content-focused with an emphasis on performance	Skill mastery, passive reception of knowledge
Platonist: mathematics as a static body of absolute and certain knowledge comprising abstract entities	Content-focused with an emphasis on understanding	Active construction of understanding
Problem solving: mathematics as a dynamic and expanding field of human creation	Learner-focused	Autonomous exploration of own interests

Student beliefs about mathematics

Thus far, we have given our attention to teachers' mathematical beliefs, but what do students believe about the nature of mathematics? A subtler way to investigate this than to ask a direct question involves using metaphors for mathematics, such as:

> If mathematics were a food, what kind of food would it be? If mathematics were a colour, what colour would it be? If mathematics were music, what kind of music would it be? (See Frid, 2001; Ocean & Miller-Reilly, 1997 for more ways of using metaphors for mathematics.)

Pre-service teachers who tried this activity with their junior secondary students during a practice teaching session were surprised, and somewhat disturbed, by the results. If mathematics were a food, most students agreed that it would be a green vegetable such as broccoli, brussels sprouts or zucchini. According to them, these vegetables taste terrible but we have to eat them because they are good for us, thus implying that mathematics is a necessary but unpleasant part of their school diet. Others who were more favourably disposed towards mathematics compared it with bread (a staple food), fruit salad (because it contains a variety of ingredients) or lasagne (different layers are revealed as you eat it). These responses perhaps suggest that students had varying perceptions of mathematical knowledge as either necessary, diverse or sequenced in layers of complexity. Students thought that if mathematics were a colour it would be either black (depressing, evil), red (the colour of anger and pain) or brown (boring). The few who admitted to liking mathematics often said it would be blue because this colour is associated with intelligence or feelings of calm and peacefulness. There was more variety in metaphors for mathematics as music. Many students said that mathematics was like classical music because they found it difficult to understand; some likened it to heavy metal music because 'it hurts your brain'; while one responded that it was like the theme from the movie *Jaws*—because 'it creeps up on you'. Writing in her practice teaching journal, one pre-service teacher lamented, 'There was not one person in the class who admitted to liking maths and compared it with McDonald's or Guy Sebastian!'

REVIEW AND REFLECT : Try the mathematical metaphors activity with some school-aged children and some adults (if possible, with mathematics teachers, non-mathematics teachers and non-teachers). Analyse the results and compare them with those of a partner.

Investigating students' views about mathematics and comparing these with teachers' beliefs might lead us to reflect on the role of teachers in enriching or limiting students' perspectives on the nature and value of mathematics, and to consider how students' dispositions towards mathematics might be shaped by their experiences in school mathematics classrooms. The important message here is that *through their words and actions, teachers communicate their beliefs about what mathematics is to the students they teach.*

Perceptions of mathematics teachers

Through their daily experiences in classrooms, students develop perceptions of mathematics teaching and long-lasting attitudes to mathematics. Some of these perceptions involve memories of particular teachers, such as those below, while others are more stereotypical, arising from students' experiences over time in many different mathematics classrooms:

> I was never very good at maths in primary school. I particularly remember a teacher who shamed me and ridiculed me in front of the class. That was a shattering experience, and every time I was asked to answer a mental maths question after that I'd just freeze. Things did improve, and in my last year of primary school I actually did quite well. That led to the next problem, when I got to secondary school and was put in the A stream class with all the students who were really good at maths, and I constantly felt like I was swimming against the tide to keep my head up. I didn't take maths for my A-levels. I really regret not doing more maths at school, as I still have a big confidence problem with maths and I hate being put on the spot—working behind a till, to give change—my primary school days come back to haunt me and I still get a bit panicky.

I always excelled at maths during primary school. I enjoyed recall activities used to teach the times tables, especially when the teacher timed us with a stopwatch and the quickest student received a prize. (I used to win a lot!) I couldn't understand why other students didn't feel the same way—but I get it now! Despite this I struggled with maths in the junior secondary years. I didn't like my maths teacher very much because he was intimidating, boring and hard to approach. By Year 10 I was finding maths easier again so I decided to take senior maths subjects. I had a fantastic teacher and found maths easier than other subjects as there was more of a focus on understanding and application than on memorising content. I graduated from Year 12 with an A for maths, largely thanks to great maths teaching.

A number of studies have shown the impact of teacher–student relationships on students' attitudes to mathematics. Students entering school develop more negative attitudes to mathematics when they have difficulties forming positive relationships with their secondary mathematics teacher (Attard, 2010). Conversely, a caring approach and willingness to adapt teaching to meet the needs of students was received positively by students (Averill & Clark, 2012).

> **REVIEW AND REFLECT :** Write your personal mathematical life history, describing your experiences of learning mathematics at home, at school and at university, and recalling the influence of different teachers and other people you may have encountered. What are your earliest memories of doing mathematics? What have been your 'highs' and 'lows'? Compare your mathematical life history with a partner. Together, compile a list of qualities of the best mathematics teachers in your experience.

To emphasise further the key role that teachers play in influencing students' dispositions towards mathematics, we can also explore school students' perceptions by inviting them to draw a typical mathematics teacher. A pre-service teacher tackled this task by drawing a stick figure on the whiteboard and asking the class to give her instructions on what additional features to include. The finished drawing, complete with annotations provided by the class, is reproduced in Figure 1.2.

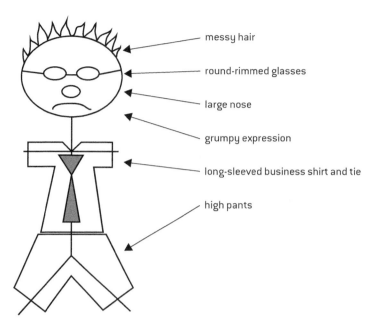

Figure 1.2 **Secondary school students' drawing of a typical mathematics teacher**

The school students also commented on aspects of a typical mathematics teacher's personality, using words such as 'boring', 'old', 'depressing', 'cranky' and 'ugly'. Other pre-service teachers found that their students produced very similar drawings and described mathematics teachers in much the same way. Likewise, local and international studies of students' images of mathematicians have identified stereotypes such as *the foolish mathematician* (lacking common sense or fashion sense), *the mathematician who can't teach* (doesn't know the material or can't control the classroom) or even *mathematics as coercion* (mathematicians as teachers who use intimidation or threats) (Picker & Berry, 2001). While you may not recognise yourself in these drawings or descriptions, the clear message here is that *teachers have the power to engage or alienate students in ways they will remember for the rest of their lives.*

Mathematics teaching practices: perspectives on Year 8 classrooms

The preceding discussion has touched on relationships between teachers' mathematical beliefs and their mathematics teaching practices, and students' perceptions of the way mathematics is taught. Observational research such as that undertaken in the Trends in International Mathematics and Science Study (TIMSS) 1999 Video Study (Hollingsworth,

Lokan & McCrae, 2003) gives a clearer picture of what actually happens in mathematics classrooms. This study was funded by the US Department of Education to describe and compare teaching practices across seven countries—Australia, the Czech Republic, Hong Kong SAR, Japan, the Netherlands, Switzerland and the USA. These were among the 40 countries that participated in the 1995 TIMSS an international comparative study of student achievement in mathematics and science. In the 1995 TIMSS mathematics assessment, students in the USA were outperformed by students in each of the other six countries, while Japanese students recorded the highest scores.

The TIMSS 1999 Video Study collected data from 638 Year 8 mathematics lessons across the seven countries listed above. The Australian sample comprised 87 schools randomly selected to be proportionally representative of all states, territories and school systems, as well as metropolitan and country areas. One teacher was randomly selected from each of these schools and was filmed for one complete Year 8 mathematics lesson. The coding and analysis of the videotapes was rigorous and comprehensive, and the study provides us with detailed information about the distinctive features of Year 8 mathematics lessons in each country. Some of the main conclusions about the average Year 8 mathematics lesson in Australia are summarised below.

One type of analysis considered the *procedural complexity* of the mathematical problems presented to students. Low-complexity problems required few decisions or steps by students, while high-complexity problems required many decisions and contained two or more sub-problems. In common with all countries except Japan, most problems (77 per cent) presented in Australian lessons were of low procedural complexity and few (8 per cent) were of high complexity. Another type of analysis considered the *relationships among problems* as a measure of the mathematical coherence of the lesson. Four types of relationships were classified: repetition, mathematically related, thematically related and unrelated. Three-quarters of problems presented in Australian lessons were repetitions of preceding problems (a higher proportion than for any other country), and only 13 per cent were mathematically related in that they extended or elaborated a preceding problem (a lower proportion than for any other country). Problem statements were also analysed to determine the *mathematical processes* to be engaged for their solution. 'Using procedures' problems could be solved by applying a procedure or set of procedures involving, for example, number operations or manipulation of algebraic symbols. 'Stating concepts' problems called on mathematical conventions or examples of mathematical concepts (e.g., 'draw the net of an open rectangular box'). 'Making connections' problems required students to construct relationships between mathematical ideas, and often to engage in the mathematical reasoning processes of conjecturing, generalising and verifying. The analysis found that in Australia, and in all other countries except Japan, most of the problems presented in lessons (62 per cent) focused on using procedures. Only 15 per cent of problems in Australian Year 8 lessons involved making connections (compared with 54 per cent in Japan).

Although Australian students perform reasonably well in international comparative studies of mathematics achievement, the findings of the TIMSS Video Study point to areas in which different methods of teaching might lead to higher achievement. In particular, it seems that the average Year 8 mathematics lesson in Australia displays 'a cluster of features that together constitute a syndrome of shallow teaching, where students are asked to follow procedures without reasons' (Stacey, 2003, p. 119). An analysis of Australian Year 8 textbooks also found a predominance of low-level procedural tasks with examples used to prepare students for practice exercises rather than to develop thinking and reasoning tools for solving more challenging problems, justifying or connecting and abstracting mathematical ideas (Stacey & Vincent, 2009b). These studies suggest that when mathematics teachers make choices about pedagogy, lesson organisation, content and how the content is presented, they need to provide students with:

- less exposure to repetitive problems
- more exposure to higher-order problems
- more opportunities to appreciate connections between mathematical ideas
- more opportunities to understand the mathematics behind the problems they are working on.

Increasing the emphasis on challenge, connections and understanding may go some way towards addressing the sense of alienation experienced by many students in secondary school mathematics classrooms.

Knowledge needed for teaching mathematics

Statements of the professional requirements for successful teaching of mathematics, such as the *Standards for Excellence in Teaching Mathematics in Australian Schools* (AAMT, 2006), usually identify three domains that structure the professional work of mathematics teaching: knowledge, practices and attributes. These domains align with the *Australian Professional Standards for Teachers* (Australian Institute for Teaching and School Leadership [AITSL], 2014): professional knowledge, professional practice and professional engagement. The statement that opened this chapter described the professional *practice* of mathematics teachers who make a positive difference to their students' learning in terms of an inclusive and supportive learning environment, coherent planning for learning, teaching approaches that challenge students' thinking, and timely and informative assessment and reporting. The *attributes* of excellent teachers encompass a belief that all students can learn mathematics, commitment to lifelong professional learning, and constructive interaction with a range of communities relevant to their professional work. These teachers also have a strong *knowledge* base that includes knowledge of students and their social and cultural contexts, knowledge of the mathematics appropriate to the level of students they teach, and knowledge of how students learn mathematics. Numerous models of knowledge for mathematics teaching have been developed. In this section, we take a closer look at one aspect of the knowledge base needed for effective teaching of mathematics.

Research on mathematics teachers' knowledge has largely been concerned with identifying their *pedagogical content knowledge* (PCK), defined by Shulman (1986) as 'the blending of content and pedagogy into an understanding of how particular topics, problems, or issues are organised, represented, and adapted to the diverse interests and abilities of learners, and presented for instruction' (p. 8). The attributes of PCK have been elaborated in various models of knowledge. For example, the domains of Mathematical Knowledge for Teaching (Ball et al., 2008) include specialised content knowledge, that is, mathematical knowledge and skill unique to teaching, and three sub-categories of PCK: knowledge of content and students, knowledge of content and teaching, and knowledge of content and curriculum (see Figure 1.3).

The PCK framework developed by Chick et al. (2006) also included three sub-categories: clearly PCK, content knowledge in a pedagogical context and pedagogical knowledge in a content context. The elements of clearly PCK are reproduced in Table 1.2.

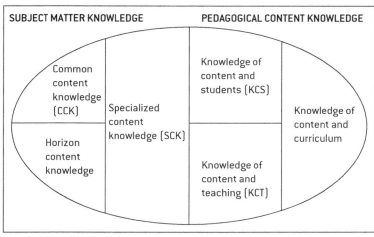

Source: Ball et al. (2008, p. 3).

Figure 1.3 **Domains of Mathematical Knowledge for Teaching**

Table 1.2 Framework for analysing pedagogical content knowledge: clearly PCK

PCK category	Evident when the teacher . . .
Teaching strategies	Discusses or uses strategies or approaches for teaching a mathematical concept
Student thinking	Discusses or addresses student ways of thinking about a concept or typical levels of understanding
Student thinking—misconceptions	Discusses or addresses student misconceptions about a concept
Cognitive demands of task	Identifies aspects of the task that affect its complexity
Appropriate and detailed representations of concepts	Describes or demonstrates ways to model or illustrate a concept (can include materials or diagrams)
Knowledge of resources	Discusses/uses resources available to support teaching
Curriculum knowledge	Discusses how topics fit into the curriculum
Purpose of content knowledge	Discusses reasons for content being included in the curriculum or how it might be used

Source: Based on Chick et al. (2006).

Other frameworks of PCK that have been developed and used to analyse teachers' knowledge in the process of planning and teaching also focus on connecting knowledge of mathematics and knowledge of students and their learning. For example, Roche and Clarke (2011) identified these aspects of PCK:

- *Pathways:* understanding possible pathways or learning trajectories within or across mathematical domains, including identifying key ideas in a particular mathematical domain.
- *Selecting:* planning or selecting appropriate teaching/learning materials, examples or methods for representing particular mathematical ideas including evaluating the instructional advantages and disadvantages of representations or definitions used to teach a particular topic, concept or skill.
- *Interpreting:* interpreting, evaluating and anticipating students' mathematical solutions, arguments or representations (verbal or written, novel or typical), including misconceptions.
- *Demand:* understanding the relative cognitive demand of tasks/activities.
- *Adapting:* adapting a task for different student needs or to enable its use with a wider range of students.

Rowland et al. (2009) developed the Knowledge Quartet to analyse teachers' pedagogical knowledge as they enact their teaching. The four quadrants included *foundation, transformation, connection* and *contingency. Foundation* refers to content knowledge but also indicates that teachers tend to stick to the textbook when planning and teaching. *Transformation* is about selecting appropriate tasks and representations, and being able to model and explain procedures and concepts, and thus is similar to Roche and Clarke's (2011) *selecting, demand and adapting.* Demonstration of knowledge of *connections* when teaching involves anticipating complexity and sequencing activities to support student learning and enabling them to make connections between concepts and between procedures. The final category, *contingency,* is not evident in other frameworks, though it is related to anticipating student solutions and explanations. It involves being able to 'think on your feet', that is, respond to students' responses and ideas, take up opportunities for learning as they arise and alter the lesson in response to these events.

Current challenges for mathematics teaching

Secondary school mathematics teachers in the 21st century face at least two significant challenges. The first was foreshadowed in 1956 by Ken Cunningham (2006), the director of the Australian Council for Educational Research (ACER), who argued that the mathematics curriculum needed to be more relevant because many secondary students felt alienated in mathematics classrooms. The post-war secondary mathematics curriculum was designed for the very small number of students who would study mathematics in Year 12 rather than taking into account the needs of all students, irrespective of their likely career paths and employment prospects. He made a very strong case for what we now call numeracy, or mathematical literacy, in the secondary mathematics curriculum, and especially quantitative literacy—or what we now call statistical literacy.

As mentioned earlier in this chapter, numeracy is now a general capability in the *Australian Curriculum*. Even so, students continue to seek relevance in mathematics and there is a continuing expectation that mathematics teachers will focus on developing students' capacity to apply mathematical knowledge, solve problems and conduct investigations of mathematical phenomena in their world. This includes being able to use and interpret the ever-increasing amount of information and data available for active citizenship and decision making, not only in the workplace, but also in all aspects of life. However, teachers find this a challenging aspect of their work—although it can be very rewarding when successful. We pay particular attention to this issue in Chapter 3, on making mathematical connections, as well as in other chapters throughout this text.

With the much higher retention rates to the end of secondary school, and an expectation to retain and improve retention rates, especially for students from disadvantaged communities, mathematics curricula in the senior secondary years must exhibit relevance and make connections with real-world phenomena for all students. Education systems in each state have provided a breadth of senior secondary mathematics subjects and, in 2015, ACARA published a national senior secondary curriculum. It includes four mathematics subjects taught across Years 11 and 12 that cater for the diverse learning trajectories and career pathways of students. One of the major developments in this suite of senior secondary mathematics subjects is the inclusion of statistical literacy in all four of the senior mathematics subjects in the Australian national curriculum. This is expected to be a challenge for many teachers of mathematics, as statistics has not been included in all state-based senior secondary subjects designed. The nature and depth of statistical literacy content and proficiencies across the secondary years are discussed in Chapter 11.

A second challenge for mathematics teachers, presented by the Office of the Chief Scientist, relates to Australia's potential to use science, technology, engineering and mathematics (STEM) skills to innovate and compete on the global stage. The Office identified a set of so-called 'future-focused' skills including critical thinking, complex problem solving, creative problem solving and design thinking for each of the STEM disciplines (Prinsley & Baranyai, 2015). The issue for mathematics teaching and learning is to help learners to make sense of mathematics. It is the flexibility, depth and diversity in thinking mathematically that comes from making sense of situations and mathematical abstractions that will be important for working mathematically to meet the social, economic and environmental challenges of the 21st century. This challenge—to promote mathematical understanding and reasoning—is taken up in the next chapter. It continues to be evident throughout this book as we discuss the use of particular tools and teaching strategies, and examine how students develop the facility to use the various concepts and skills that constitute the field of mathematics.

The structure of this book

The remainder of this book addresses the professional knowledge, professional attributes and professional practice of secondary school mathematics teachers. It is divided into four sections. Part II deals with issues around mathematics pedagogy, curriculum and assessment, as well as the role and influence of technologies in mathematics education. Part III analyses relevant research on students' learning of specific mathematical content (number, algebra, geometry and spatial concepts, measurement, chance and data, and calculus) and identifies implications for effective teaching approaches. Part IV considers equity and diversity in mathematics education in terms of social and cultural issues, gender and teaching mathematics to students with diverse learning needs. Part V discusses responsibilities of secondary mathematics teachers with regard to community engagement and ongoing professional development.

Recommended reading

Australian Association of Mathematics Teachers. (2006). *Standards for excellence in teaching mathematics in Australian schools.* Retrieved from www.aamt.edu.au/standards/

Frid, S. (2001). Food for thought. *The Australian Mathematics Teacher,* 57(1), 12–16.

Hogan, J. (2002). Mathematics and numeracy: Is there a difference? *The Australian Mathematics Teacher,* 58(4), 14–15.

Hollingsworth, H., Lokan, J. & McCrae, B. (2003). *Teaching mathematics in Australia: Results from the TIMSS 1999 Video Study.* Melbourne: Australian Council for Educational Research.

Sullivan, P. (2008). Knowledge of teaching mathematics: Introduction. In P. Sullivan & T. Wood (Eds), *Knowledge and beliefs in mathematics teaching and teaching development* (pp. 1–9). Rotterdam: Sense.

PART 2

Mathematics Pedagogy,
Curriculum and Assessment

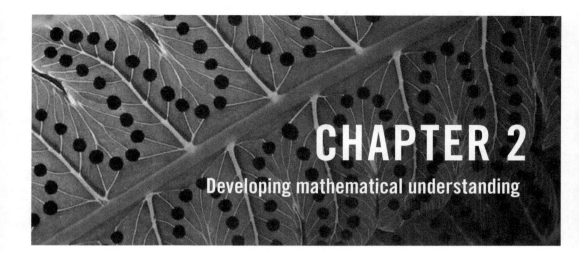

CHAPTER 2
Developing mathematical understanding

The desire to make sense of what we see, hear and learn is driven by a need to understand, and knowing whether one understands something is essential for learning. Research in mathematics classrooms has shown that mathematical thinking and reasoning are important for the development of conceptual understanding in mathematics. This chapter looks at the nature of mathematical understanding and what teachers can do to develop deep understanding of mathematical concepts in their students. We compare two general theories of learning and consider what they might be able to tell us about mathematics learning and teaching. This is followed by a discussion of mathematical thinking, with specific attention given to mathematical reasoning and problem solving. The final part of the chapter considers the role of the teacher in creating a classroom culture of mathematical inquiry.

What does it mean to understand something in mathematics?

Mathematics syllabuses and curriculum documents in many countries place great emphasis on building students' understanding (e.g., National Council of Teachers of Mathematics (NCTM), 2000; National Curriculum Board, 2009). Most mathematics teachers would claim that they value learning with understanding, but what exactly does this mean?

> **REVIEW AND REFLECT :** How do you know when you understand something in mathematics?
>
>   Discuss your responses to this question with a peer or in a small group. Compare with other groups—are there any similarities or differences in your responses? What kind of answers do you think secondary school students would give to this question?

One of the authors of this book asked the question 'How do you know when you understand something in mathematics' of over 300 Queensland secondary school mathematics students in Years 10, 11 and 12. Their responses were grouped into the categories shown in Table 2.1. The majority of students considered they understood something in mathematics if they could complete the associated problems and obtain the correct answer. A few described understanding in affective terms; that is, understanding was accompanied by feelings of increased confidence or enjoyment or excitement. Only a small proportion of students associated understanding with knowing why something worked or made sense, and even fewer referred to the ability to apply their knowledge to unfamiliar problems as evidence of understanding. Perhaps the most sophisticated kind of response came from students who knew they understood something when they could explain it to someone else. Our observations of many secondary mathematics classrooms and interviews with students suggest that explaining actually does more than allow students to assess their understanding—it is also a process through which understanding is clarified and refined.

Researchers often describe mathematical understanding in terms of the structure of an individual's internal knowledge representations. For example, Hiebert and Carpenter (1992) define understanding as 'making connections between ideas, facts, or procedures' (p. 67), where the extent of understanding is directly related to the characteristics of the connections. (This definition is much like Category III responses in Table 2.1.)

It is also helpful to distinguish between different kinds of mathematical understanding, and these are often expressed in the form *knowing*-[preposition]. For example, Skemp (1987) described instrumental understanding as knowing-*what* to do in order to complete a mathematical task, and contrasted this with relational understanding as both knowing-*what* to

Table 2.1 Evidence of understanding for secondary mathematics students (n = 329)

Response category	Sample responses	Frequency	Proportion
I Correct answer	When I get it right. You can do heaps of them without mistakes.	234	0.71
II Affective response	I get interested. I feel confident when doing it.	35	0.11
III Makes sense	It fits in with my previous knowledge. You realise why you use the formula, what reasons.	52	0.16
IV Application/transfer	When I can apply it to something else outside school. When I can understand a complex problem and do all the related problems.	27	0.08
V Explain to others	When I can explain it to other people without confusing myself. I can explain theory to other students.	24	0.07

do and knowing-*why* the particular piece of mathematics works. The actions of students with instrumental understanding are driven by the goal of obtaining the correct answer (see Category I responses in Table 2.1). Students who learn mathematics as a set of fixed, minimally connected rules whose applicability is limited to a specific range of tasks cannot adapt their mental structures to solve novel or non-routine problems. Conversely, students who have relational understanding construct richly connected conceptual networks that enable them to apply general mathematical concepts to unfamiliar problem situations (see Category IV in Table 2.1).

Others have expanded this framework in ways that provide insightful contrasts. Mason and Spence (1998) identify differences between:

- knowing-*that*, as in stating something (e.g., the sum of interior angles of a triangle is 180 degrees)
- knowing-*how*, as in doing something (e.g., finding the area of a triangle)
- knowing-*why*, as in explaining something (e.g., why the algorithm to divide one fraction by another involves inverting and multiplying)

- knowing-*to*, as in seizing the opportunity to use a strategy that comes to mind in the moment of working on a problem.

They argue that it is possible for students to get into situations in which they have understanding in the forms of knowing-that, knowing-how and even knowing-why, but the relevant knowledge does not come to mind (knowing-to) when it is needed.

It is becoming more commonly acknowledged that mathematical understanding is not an acquisition or a product, as implied by Hiebert and Carpenter's (1992) definition, but a continuing process of negotiating meaning, or of attempting to make sense of what one is learning. Pirie and her colleagues (Pirie & Kieren, 1994; Pirie & Martin, 2000) have attempted to represent the dynamic and recursive nature of this process by conceptualising growth in understanding as movement back and forth through a series of nested layers, or levels, each of which illustrates a particular mode of understanding for a specified person and a specified topic. Figure 2.1 provides a diagrammatic representation of these levels of understanding.

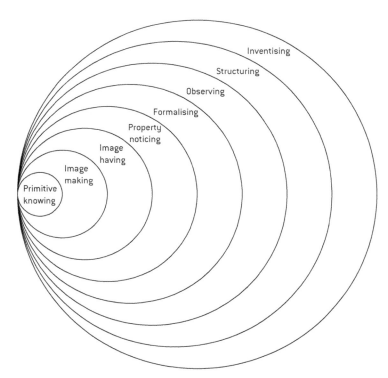

Figure 2.1 **Levels of understanding in the Pirie-Kieren theory**

$$2x + y = 10 \quad [1]$$
$$x - y = 2 \quad [2]$$

$$x = 8$$

Substitute into [2]: $8 - y = 2$

$$y = 6$$

Figure 2.2 **A student's attempt to solve simultaneous equations**

Primitive knowing describes the starting place for growth of any particular mathematical understanding. At the second level, image making, learners use previous knowledge in new ways. Image having involves using a mental construct about the topic without having to do the activities that brought it about. Property noticing occurs when learners can combine aspects of images to construct relevant properties. In formalising, the learner abstracts a quality from the previous image, while observing is a process of reflecting on and coordinating this formal activity and expressing such coordinations as theorems. Structuring occurs when learners attempt to think about their formal observations as a theory. Inventising presents the possibility of breaking away from existing understanding and creating new questions.

School mathematics and inquiry mathematics

When we look at the kind of mathematical errors that students make, it is natural to assume that these arise because of a lack of understanding. However, more often 'errors are based on systematic *rules* which are usually distortions of sound procedures' (Perso, 1992, p. 12). A classroom example will help to illustrate this point. Figure 2.2 shows the work of a Year 11 student who was attempting to solve a pair of simultaneous equations as a homework exercise. Instead of adding the two equations, she subtracted equation (2) from equation (1) and assumed that this would eliminate y. She then checked her answer (x = 8, y = 6) by substituting into equation (1), and was puzzled to find that she obtained a different result.

When it was suggested to her that it would be more appropriate to add the equations instead of subtracting them, she insisted 'but this is what the teacher did—we were taught always to *subtract* them!' It is unlikely that the teacher really said these words, but this is how the student had interpreted, and probably over-generalised, the teacher's advice about handling simultaneous equations. This example demonstrates that if students simply aim to reproduce the procedures demonstrated by the teacher without understanding when and why these procedures work, they will create their own reasons and rules in a systematic, if flawed, effort to achieve understanding.

In many mathematics classrooms, learning is assumed to involve mastering a predetermined body of knowledge and procedures. Likewise, it is assumed that the teacher's job is to present the subject matter in small, easily manageable pieces and to demonstrate the correct technique or algorithm, after which students work individually on practice exercises. Richards (1991) has described this as the 'school mathematics' culture, where teaching and learning are structured as information transfer or *transmission*. However, as the example in Figure 2.2 shows, knowledge cannot be transferred directly from teacher to learner; instead, learners reinterpret and transform the teacher's words and actions. In contrast, in an inquiry mathematics culture, students learn to speak and act mathematically by asking questions, proposing conjectures and solving new or unfamiliar problems.

In recent years, researchers have been interested in studying the characteristics of these contrasting classroom cultures and identifying consequences for students' mathematical achievement. Boaler (1998, 2000, 2002a, 2002b) conducted a landmark study of students' mathematics learning in two English secondary schools, which she called Amber Hill and Phoenix Park. She chose these schools because their students were similar in terms of socio-economic and cultural background but their teaching methods were very different. Mathematics teachers at Amber Hill used traditional teaching methods, consistent with the 'school mathematics' culture described above, and assessment was based solely on written tests in preparation for the national examinations at the end of Year 11. Classrooms here were quiet and orderly and students appeared to be motivated and hardworking. Yet, when interviewed, the students revealed their dislike of mathematics, which for them was a boring and difficult subject. In addition, despite their diligence in listening to the teacher and working through textbook exercises, this passive approach to their work left them unable to apply their mathematical knowledge to unfamiliar tasks. Boaler said this was because the students had developed 'inert knowledge', which she attributed to their belief that learning mathematics required memorising set rules, equations and formulae.

Teaching approaches at Phoenix Park were more progressive and mathematics lessons often quite unstructured. Students worked on open-ended projects for most of the time, and there was a strong emphasis on meaning and explaining one's thinking. The philosophy here was that students should encounter mathematics in a context that was realistic and meaningful, and teachers taught new mathematical content when the need arose as students worked on their projects. Most of the students enjoyed this inquiry mathematics approach and found mathematics interesting because it involved thinking and solving problems.

One of the most important findings of this study concerns the evidence Boaler found of differences in students' learning between the two schools. At the beginning of the study, when students were in Year 9, there were no significant differences in their mathematical performance as measured by standardised tests and other questions Boaler devised to assess aspects of number work. During the study, Boaler used various methods to assess students' mathematics learning. One of these was an applied investigation task that required students to interpret and calculate from the scale plan and model of a house. Phoenix Park students performed substantially better than Amber Hill students on this task, which is perhaps not surprising given the former school's project-based teaching approach. More significantly, Phoenix Park students achieved just as well as, and sometimes better than, Amber Hill students on conventional written tests assessing mathematical content knowledge. Amber Hill students' greater experience in working textbook exercises did not help them in formal test situations if the questions required them to do more than simply reproduce a learned rule or procedure. By comparison, Phoenix Park students were flexible and adaptive mathematical thinkers who could apply their knowledge to unfamiliar tasks.

Boaler and Staples (2008) conducted a similar study in the USA that compared students' mathematics achievement and attitudes in three secondary schools. In two of these schools, the mathematics teaching was similar to that in Amber Hill, while at the other school, Railside High, teachers adopted an inquiry-oriented approach characterised by high cognitive demand, an emphasis on achievement as the product of effort rather than ability, and communication of clear expectations about the learning practices that would help students to achieve understanding. Compared with the other two schools, Railside students demonstrated higher achievement on curriculum-aligned assessments, enjoyed mathematics more, and had a stronger belief in their capacity to use their own ideas and validate their own reasoning. This result is especially noteworthy because Railside had students from more socially and economically disadvantaged backgrounds than the other schools.

These two studies provide compelling evidence that a school mathematics approach produces only instrumental understanding—knowing-*that* and knowing-*how*—while an inquiry mathematics approach can generate relational understanding—knowing-*why* and knowing-*to*.

Theories of learning

For teaching to be effective, it must be grounded in what we know about how students learn. This section outlines the two theories of learning that are currently most influential in mathematics education, constructivism and socio-cultural perspectives. Both of these theoretical positions have something to say about relationships between social processes and individuals' learning (Cobb, 1994), and the theories have intermingled in recent years (Confrey & Kazak, 2006). Constructivism gives priority to individual construction of mathematical understanding, and sees social interaction as a source of cognitive conflict that brings about learning through reorganisation of mental structures. This position contrasts with the socio-cultural perspective, which views learning as a collective process of enculturation into the practices of the mathematical community (e.g., Lerman, 1996, 2001).

Constructivism

The central claim of constructivism is that learners actively construct knowledge and personal meanings by connecting their prior knowledge with new knowledge gained from their own interaction with the world (Davis, 1990). The emergence of constructivism was significantly influenced by the work of Swiss psychologist Piaget (1954) on theories of cognitive development. Piaget realised that babies, children, adolescents and adults think in ways that are qualitatively different; that is, adults do not simply know more than children do, they know *differently*. From his observations of children, Piaget concluded that intellectual development proceeded through a series of stages as children matured. The age ranges he attached to these stages are no longer accepted; for example, he underestimated the reasoning capacity of younger children and overestimated that of adolescents. We also now know that learners may exhibit different types of thinking in different contexts and for different topics. Nevertheless, Piaget's ideas about stages of development in children's thinking help us to think about the kinds of things learners may be able to do that are limited by their age rather than by their skill.

In mathematics education, constructivism 'attends to how actions, observations, patterns and informal experiences can be transformed into stronger and more predictive explanatory idea through encounters with challenging tasks' (Confrey & Kazak, 2006, p. 316). In such encounters, cognitive change begins when students experience conflict with their previous ways of knowing and take action to resolve this perturbation. Recording and communicating their thinking allows students to reflect on their actions and the adequacy of their new understanding. In order to participate successfully in a constructivist classroom, students and teachers have to renegotiate classroom norms that regulate patterns of interaction and discourse. These include *social* norms, such as expectations that students should explain and justify their reasoning, as well as *socio-mathematical* (or discipline-specific) norms, for example, about what counts as an acceptable, efficient or elegant mathematical solution to a problem (see McClain & Cobb, 2001; Wood et al., 2006).

Socio-cultural perspectives

The term *socio-cultural* is used to describe a family of theories whose origins can be traced to the work of the Russian psychologist Lev Vygotsky in the early twentieth century. Vygotsky's work was virtually unknown to the Western world until the 1970s when English translations became available (e.g., Vygotsky, 1978). Since then, his ideas have been explored and extended by many other researchers (see Forman, 2003; Sfard, Forman & Kieran, 2001 for reviews of socio-cultural research in mathematics education).

Vygotsky claimed that individual cognition has its origins in social interaction; that is, memory, concepts and reasoning appear first between people, as social processes, and then within an individual, as internal mental processes. He also claimed that mental processes are mediated by cultural tools—such as language, writing, systems for counting, algebraic symbol systems, diagrams, drawing tools and physical models—and that this mediation transforms people's thinking by changing the way they formulate and solve problems. In connection with these ideas, Vygotsky introduced the notion of the *zone of proximal development* (ZPD) to explain how a child's interaction with an adult or more capable peer might awaken mental functions that have not yet matured. He defined the ZPD as the distance between a child's problem-solving capacity when working alone and with the assistance of a more advanced partner, such as a teacher. The metaphor of *scaffolding* (introduced by Wood, Bruner & Ross, 1976) became associated with interactions in the ZPD to describe how a teacher structures tasks to allow students to participate in joint activities that would otherwise be beyond their reach, and then withdraws or *fades* support as students begin to perform more independently.

Vygotsky also drew on his observations of how children learned by playing together without adult intervention to explain the ZPD in terms of more equal status partnerships. From an educational perspective, there is learning potential in collaborative group work in which students have incomplete but relatively equal expertise, each partner possessing some knowledge and skill but requiring the others' contributions to make progress. This approach has informed research on collaborative ZPDs in small group problem solving in mathematics education (e.g., Goos, Galbraith & Renshaw, 2002).

Contemporary socio-cultural theory views learning as increasing participation in a community of practice (Wenger, 1998). In mathematics classrooms, this means that the teacher is responsible for initiating students into a culture of mathematical inquiry in which discussion and collaboration are valued in building a climate of intellectual challenge. Van Oers (2001) proposes that this process begins with the teacher's demonstration of a mathematical attitude, that is, a willingness to deal with mathematical concepts and to engage in mathematical reasoning according to the accepted values in the community, and, consequently, from the teacher's mathematical expectations about the learners' activity. Learners appropriate this mathematical attitude by participating in shared practice structured by the teacher's expectations and actions.

A classroom scenario

During his first practicum session, Damien (a pre-service teacher) was assigned to teach a very challenging Year 10 mathematics class. The students were unmotivated and had a history of poor achievement in mathematics, and there was a great deal of disruptive behaviour. Damien decided that the best approach was to use whole-class exposition and questioning in order to maintain order and control, and he offered very simple tasks to give students some experience of success. This approach was not successful, as shown by the post-lesson debriefing notes Damien recorded with the help of his university supervisor, who had observed the lesson.

Teacher expectations	Teacher actions	Student actions
• Listen to teacher, copy example, practise similar tasks. • Low-ability class, I will accept performance before understanding.	• Whole-class work, blackboard example, worksheet for individual practice. • Choose simple exercises.	• Confused, few finished task. • Unwilling to work, used delaying tactics, off task, restless. • Rude to each other and to teacher.

During the second practicum session, Damien was teaching the same class. In one lesson, he decided to take a different approach by trying a practical activity. He wanted students to work out for themselves some properties of equilateral and isosceles triangles; for example, when triangles have equal side lengths, their angles are also equal. The students had to use rulers and compasses to draw triangles with given side lengths, measure the angles, tabulate their results (as in the example below), and draw conclusions.

REVIEW AND REFLECT : Draw the equilateral triangles with side lengths shown below (AB = BC = AC). Measure and record the size the angles for each triangle.

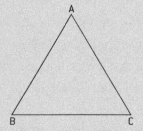

Side length (cm)	Angle (degrees)		
	A	B	C
5			
6.5			
9			

Not only did this activity allow students to discover mathematical properties for themselves, but it also had the unplanned effect of stimulating peer interaction and discussion—something that Damien had previously discouraged because he thought the class would become unruly and difficult to manage. It also gave him opportunities to listen to students' conversations and ask questions that moved their thinking towards generalising relationships between angles and sides. Below are the debriefing notes for the lesson recorded by Damien and his university supervisor.

Teacher expectations	Teacher actions	Student actions
• Work things out for themselves. • Learn by doing, 'hands on'. • Raised expectations, wanted understanding.	• Provided investigation task. • Toured room, asked questions of individuals and groups.	• Shared results, explained to each other. • Asked me questions, wanted to show me their work. • Cooperative, on task, excited.

REVIEW AND REFLECT : Consider the above two lessons and explain in terms of constructivist theory why the second lesson was more successful than the first. (Think about challenging tasks, cognitive conflict, recording of and reflection on new understanding.)

Now use socio-cultural perspectives to explain why the second lesson was more successful than the first. (Think about cultural tools, scaffolding, peer interaction.)

Mathematical thinking

Williams (2002) has developed a framework for describing students' mathematical thinking that may prove useful for teachers as they observe the nature of learning occurring in their classrooms. The framework is based on work of Dreyfus, Hershkowitz and Schwarz (2001), who view mathematical thinking as abstraction and generalisation. Three categories of student thinking are identified, at increasing levels of complexity:

- *recognising:* realising that a known mathematical procedure applies in a new situation
- *building-with:* using several previously known mathematical procedures to solve an unfamiliar problem

- *constructing:* selecting previously known strategies, mathematical ideas and concepts to integrate when solving an unfamiliar challenging problem.

Williams (2002) added further detail to create a set of nested categories showing how students move from analysis to synthesis to evaluation (see Figure 2.3).

Complexity of thinking	Examples of thinking
Evaluation (Constructing)	Continually check for inconsistencies as different aspects of the problem are explored. Look for how the new mathematical processes used may be applicable to other situations. Recognise new mathematical insights developed may be useful in solution pathways used to resolve subsequent spontaneous questions.
Synthesis (Constructing)	Select relevant strategies, ideas and concepts. Integrate mathematical ideas, concepts and strategies to produce a new mathematical insight. Develop a mathematical argument to explain *why*. Find and resolve a spontaneous question. Progressively resolve subsequent spontaneous questions. Develop a new process to add to problem-solving repertoire.
Evaluative- analysis (Building-with)	Use a quick method or estimate to check a finding by relating the mathematics to the context. Refer to the context to provide reasons why a pattern works.
Synthetic- analysis (Building-with)	Interconnect different solution pathways by considering why two or more solution pathways may be appropriate to solve the same problem. Consolidate insights developed through the problem by building-with them in the development of further insights.
Analysis (Building-with)	Search for patterns. Search for alternative solution pathways to solve the same problem. Work backwards where you have only been taught to solve the question forwards. For example, find the length of a rectangle when given the width and the area if previous questions always gave the length and width and required the student to find the area. Analyse a situation to find what known mathematical procedures to apply (analyse to recognise).

Figure 2.3. **Categories of complex mathematical thinking**

Consider the following task: 'Angles in polygons' (Williams, 2002, pp. 335–336).

Students were provided with an A3 page containing polygons with three to ten sides. Students were asked to work in pairs and were directed to:

- Use straight lines to divide each polygon into triangles.
- Cut out the triangles for one polygon (at a time).
- Tear off each angle of each triangle.
- Place the angles around a point and find the sum of all the angles from triangles for a particular polygon.
- Record their findings in a table with headings like 'number of sides of polygon' and 'total of angles'.

Students were shown how to piece the angles together and instruction was given about how to tell the size of the total angle by the number of rotations made. Students were asked to look for a pattern in the table they generated. Once the pattern was found, the teacher developed a rule, and then the students practised the rule on other polygons.

Williams explains that this task, when implemented as described above, promotes thinking only at the level of analysis (building-with) (see Figure 2.3). It requires students to find a pattern in the table that shows the number of sides and the sum of the angles of the polygon, and to use known information about 90-, 180- and 360-degree angles or about protractor use to work out what an unknown angle is.

REVIEW AND REFLECT : Work with a partner on the 'Angles in polygons' task by following the instructions given to students.

Verify that the task calls for thinking at the level of analysis (building-with).

How could the task be refined or its manner of implementation be changed so that it calls forth more complex thinking? (Use the examples of thinking in Figure 2.3 as a source of ideas.)

Mathematical thinking is represented in curriculum documents in a variety of ways that distinguish mathematical processes from mathematical content. Figure 2.4 summarises

Australian Curriculum (Australia)	Understanding Fluency Problem solving Reasoning
NCTM Principles and Standards (USA)	Problem solving Reasoning and proof Communication Connections Representation
National Curriculum (UK)	Develop fluency Reason mathematically Solve problems

Figure 2.4 **Representation of process aspects of mathematics in curriculum materials**

approaches to specifying process aspects of mathematics in curriculum materials provided for teachers in Australia (ACARA, 2013), the USA (NCTM, 2000) and the UK (Department for Education, 2013). Common to all of these curricula are the thinking processes of *reasoning* and *problem solving*.

Mathematical reasoning

Mathematical reasoning involves making, investigating and evaluating conjectures, and developing mathematical arguments to convince oneself and others that a conjecture is true. Yackel and Hanna (2003) assert that 'explanation and justification are key aspects of students' mathematical activity in classrooms in which mathematics is constituted as reasoning' (p. 229). Students learn to give explanations and justifications when teachers (a) provide tasks that require them to investigate mathematical relationships (as in the example from Williams, 2002, above), and (b) foster a classroom climate in which students are expected to listen to, discuss and question the claims made by others (NCTM, 2000).

REVIEW AND REFLECT : Write down your age. Add 5. Multiply the number you just got by 2. Add 10 to this number. Multiply this number by 5. Tell me the result.

I can tell you your age (by dropping the final zero from the number from your result and subtracting 10).

Investigate this task with a partner. Why does it work? Formulate a generalisation to explain why and present your argument to the class. Evaluate any alternative solutions presented by others.

Source: NCTM (2000, pp. 56–57).

Mathematical problem solving

Problem solving is considered integral to all mathematics learning (NCTM, 2000), and a great deal of research on mathematical problem solving was carried out during the 1980s and 1990s (see Schoenfeld, 1992 and Lesh & Zawojewski, 2007 for comprehensive international reviews; and Clarke et al., 2007 for a review of the evolution of problem-solving research, curricula and teaching approaches in Australia). Since this time, problem solving has been incorporated into the aims of mathematics syllabuses and other curriculum documents throughout Australia.

'Problem' and 'problem solving' have had many, often contradictory, meanings in the past (Schoenfeld, 1992). However, a commonly accepted definition is that a task is a problem if the person attempting it does not know the solution method in advance (NCTM, 2000). This means that a particular task could be a problem for one person but a routine exercise for another, because the 'problem' does not lie solely in the task but in the interaction between task and student.

Figure 2.5 identifies factors contributing to successful problem solving identified by research in the 1980s and 1990s. The mathematical *knowledge base* includes intuitive knowledge, facts and definitions, routine procedures and algorithms, and knowledge about the rules of mathematical reasoning. *Heuristics*, general strategies or 'rules of thumb' for making progress with non-routine tasks (e.g., work backwards, look for a pattern, try a simpler problem), are also an important strategic resource. *Metacognition* has two components: *awareness* of one's own mathematical strengths and weaknesses, task demands and factors affecting task difficulty; and *regulation* of one's thinking while working on mathematical tasks. Regulation involves such activities as planning an overall course of action,

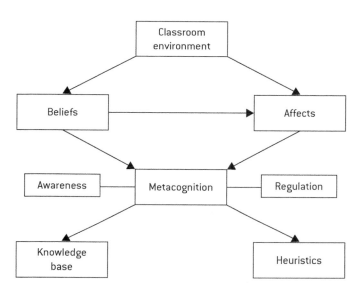

Figure 2.5. **Factors contributing to successful problem solving**

selecting specific strategies, monitoring progress, assessing results and revising plans and strategies if necessary. Without effective metacognitive awareness and regulation, students will not be able to recognise or use their knowledge to help them solve a problem when they get stuck, and they typically persist with inappropriate strategies that lead nowhere.

Beliefs and affects may enhance or interfere with metacognitive activity. Students' *beliefs* represent their mathematical worldview. Op't Eynde, de Corte and Verschaffel (2002) propose that 'students' mathematics-related beliefs are the implicitly or explicitly held subjective conceptions students hold to be true about mathematics education, about themselves as mathematicians, and about the mathematics class context' (p. 27). Their beliefs about themselves as doers of mathematics, and about particular topics, the nature of mathematics in general, and the mathematics classroom environment contribute to their metacognitive awareness and influence their metacognitive regulation (Schoenfeld, 1992). Self-beliefs also reinforce *affects*, in particular attitudinal traits such as motivation, confidence and willingness to take risks (McLeod, 1992). These affective reactions may in turn influence students' capacity to maintain task involvement. Variations in the *classroom environment* can trigger different types of affective responses, but the classroom context has an even more important role to play in shaping students' beliefs (Henningsen & Stein,

1997). We have seen that teaching that emphasises memorisation, formal procedures and correct answers at the expense of understanding can lead students to believe that they are meant to be passive participants, neither capable of, nor responsible for, proposing and defending ideas of their own (Boaler, 1998; Schoenfeld, 1988).

Holton and Clarke (2006) have discussed the idea of scaffolding, introduced earlier in connection with teacher–student interaction in the ZPD, as a means of developing students' metacognition. They propose that scaffolding questions posed by the teacher during three stages of problem solving (listed in Figure 2.6) can support students in becoming aware of and regulating their thinking, and that students can become self-scaffolding by asking themselves the same questions as the teacher's support is withdrawn (the teacher *scaffolds and fades*).

Getting started	While students are working	After students are finished
What are the important ideas here?	Tell me what you are doing.	Have you answered the problem?
Can you rephrase the problem in your own words?	Why did you think of that?	Have you considered all the cases?
What is this asking us to find?	Why are you doing this?	Have you checked your solution?
What information is given?	What are you going to do with the result once you have it?	Does it look reasonable?
What conditions apply?	Why do you think that that stage is reasonable?	Is there another solution?
Anyone want to guess the answer?	Why is that idea better than that one?	Could you explain your answer to the class?
Anyone seen a problem like this before?	You've been trying that idea for 5 minutes. Are you getting anywhere with it?	Is there another way to solve the problem?
What strategy could we use to get started?	Do you really understand what the problem is about?	Could you generalise the problem?
Which one of these ideas should we pursue?	Can you justify that step?	Can you extend the problem to cover different situations?
	Are you convinced that that bit is correct?	Can you make up another similar problem?
	Can you find a counter-example?	

Figure 2.6 **Teacher scaffolding questions during problem solving**

Try solving the problem below. Work on your own for a few minutes to get a feel for the problem, and then work with a partner or in a small group. Use the scaffolding questions in Figure 2.6 to monitor and regulate the group's problem-solving attempts.

REVIEW AND REFLECT : Divide $5 among 18 children such that each girl gets 2 cents less than each boy.

Most mathematically experienced students attempt an algebraic solution to this problem and are surprised to discover that other approaches can be more productive. (See Goos, Galbraith & Renshaw, 2000 for an analysis of secondary school students' solution methods for this task.)

Creating a classroom community of inquiry

It should be clear from the discussion we have presented throughout this chapter that teachers have a pivotal role to play in developing students' understanding of mathematics. We have also seen that an *inquiry mathematics* approach leads to deeper understanding and more flexible mathematical thinking in students. Much recent research based on constructivist and socio-cultural theories has asked how the mathematics classroom can become a *community of inquiry* and what teachers should do to engage students in mathematical thinking, reasoning and problem solving. While it is not feasible to give a list of prescriptive actions or recipes for teachers to follow, researchers have identified characteristics of classroom communities of mathematical inquiry and the teacher's role within such classrooms. For example, Anthony and Walshaw (2007) synthesised the findings from 660 research studies to present a strong evidence base for quality teaching in mathematics that improves student engagement and learning. From these studies, they concluded that creating classroom communities of inquiry:

- demands an ethic of care
- creates a space for the individual and the collective
- demands explicit instruction
- involves respectful exchange of ideas
- demands teacher content knowledge, knowledge of mathematics pedagogy, and reflecting-in-action.

A closer look at one of the Australian research studies cited in Anthony and Walshaw's (2007) synthesis provides greater insight into key elements of the teacher's role. In this senior secondary mathematics classroom, the teacher's role involved:

- modelling mathematical thinking
- asking questions that scaffolded students' thinking
- structuring students' social interactions
- connecting students' developing ideas to mathematical language and symbolism (Goos, 2004b).

An example from a Year 11 lesson on matrices early in the school year illustrates how the teacher placed explicit emphasis on the processes of mathematical inquiry. The aim of the lessons was to have the students discover for themselves the algorithm for finding the inverse of a 2 x 2 matrix $\begin{pmatrix} a & b \\ c & d \end{pmatrix}$. The teacher first chose a matrix $A = \begin{pmatrix} 3 & 1 \\ 5 & 2 \end{pmatrix}$ with a determinant of 1 and asked the students to find the inverse A^{-1} by using their existing knowledge of simultaneous equations to solve the matrix equation $AA^{-1} = I$. He then elicited students' conjectures about the general form of the inverse matrix, based on the specific case they had examined. Since the nature of the example ensured that students would offer $\begin{pmatrix} d & -b \\ -c & a \end{pmatrix}$ as the inverse, the teacher was able to provide a realistic context for students to test this initial conjecture. A counter-example, whose inverse was found to have the form $n\begin{pmatrix} d & -b \\ -c & a \end{pmatrix}$, allowed the students to find a formula for n as $\frac{1}{ad-bc}$, which only then was labelled by the teacher as the determinant. In this example, the teacher modelled and scaffolded mathematical thinking by presenting a specific problem for students to work on, and eliciting a series of conjectures that had to be tested by students in order to arrive at a valid generalisation.

As the school year progressed, the teacher continued to withdraw his support to pull his students forwards into their zones of proximal development, and students responded by taking increasing responsibility for their own mathematical thinking. The effect is illustrated by the following exchange, which occurred about three months before the end of the school year. Dylan, a student who had previously displayed a highly instrumental approach to understanding mathematics (Skemp, 1987), was struggling with a task that asked students to prove that there is a limit to the area of a Koch snowflake

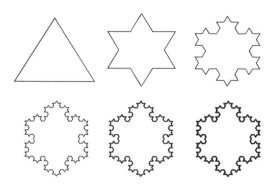

Figure 2.7 **First six stages of constructing the Koch snowflake curve**

curve (see Figure 2.7). The following dialogue occurred after Dylan had spent several minutes with his hand raised hoping to seek the teacher's assistance.

Dylan:	[Plaintively] I can't keep going! I want to know why!
Alex:	[Looks up, both laugh] Have you got my disease?
Dylan:	Yeah!
Alex:	Dylan, wanting to know why!
Dylan:	Me wanting to know why is a first, but I just want to. It's a proof—you need to understand it. [Alex resumes work, Dylan still has his hand raised.]

Dylan's new-found insistence on 'knowing why', rather than glossing over elements of a proof he did not understand, indicates that he was moving towards fuller participation in the practices of mathematical inquiry—in which being able to reason is essential to understanding.

The teacher regularly structured entire lessons around complex problems, a strategy that promoted collaborative peer interactions between students with incomplete but comparable expertise. The value of these interactions was memorably described by one student:

Adam helps me ... see things in different ways. Because, like, if you have two people who think differently and you both work on the same problem you both see different areas of it, and so it helps a lot more. More than having twice the brain, it's like having *ten* times the brain, having two people working on a problem.

Dimension	Change
Questioning	Shift from teacher as questioner of students to students and teacher as questioners. Student–student talk and questioning increase, especially 'why' questions.
Explaining mathematical thinking	As the teacher asks more probing questions to stimulate student thinking, students increasingly explain and articulate their mathematical ideas.
Source of mathematical ideas	Shift from teacher as source of all mathematical ideas to students' ideas also influencing the direction of lessons. The teacher uses student errors as opportunities for learning.
Responsibility for learning	Students increasingly take responsibility for learning and evaluation of others and self. Mathematical sense becomes the criterion for evaluation. The teacher encourages student responsibility by asking them questions about their own and each other's work.

Figure 2.8 **Characteristics of a classroom community of mathematical inquiry**

A similar study carried out in the USA (Hufferd-Ackles, Fuson & Sherin, 2004) summarised just how these teacher–student interactions change over time as a community of inquiry grows (see Figure 2.8).

We should not underestimate the challenges associated with implementing these teaching approaches in secondary mathematics classrooms. Often, it seems that content coverage and assessment pressures get in the way of developing understanding and mathematical thinking. This is especially the case when we are faced with a classroom full of students who have been socialised, over many years of schooling, into believing that teaching is telling and learning involves memorisation and practice. Research into inquiry pedagogies has shown that teachers need substantial time investment, sustained systemic support, and repeated opportunities to try out new practices (Makar, 2011). Frustration and disappointment are common experiences for teachers attempting innovative pedagogies, and shifts in practice are seldom linear or predictable. Yet, the evidence from research also convinces us that an inquiry mathematics approach offering worthwhile tasks and stimulating social interactions is the most effective way to foster learning with understanding.

Recommended reading

Anthony, G. & Walshaw, M. (**2007**). *Effective pedagogy in mathematics/pāngarau: Best evidence synthesis iteration [BES]*. Wellington: New Zealand Ministry of Education.

Boaler, J. (1998). Open and closed mathematics: Student experiences and understanding. *Journal for Research in Mathematics Education*, 29, 41–62.

Boaler, J. & Staples, M. (2008). Creating mathematical futures through an equitable teaching approach: The case of Railside School. *Teachers College Record*, 110, 608–645.

Goos, M. (2004). Learning mathematics in a classroom community of inquiry. *Journal for Research in Mathematics Education*, 35, 258–291.

Holton, D. & Clarke, D. (2006). Scaffolding and metacognition. *International Journal of Mathematical Education in Science and Technology*, 37(2), 127–143.

Wood, T., Williams, G. & McNeal, B. (2006). Children's mathematical thinking in different classroom cultures. *Journal for Research in Mathematics Education*, 37, 222–255.

CHAPTER 3
Developing mathematical connections

Being able to understand the world from a mathematical perspective is what we hope to facilitate for our students as secondary school teachers of mathematics. To this end, all students need to build a cohesive and comprehensive picture of mathematics by establishing and appreciating how:

- The mathematics they are doing in secondary school connects with and builds on the mathematical concepts and skills they started constructing and developing in the primary years.
- The mathematics they do in their mathematics classes connects across other curriculum areas they study, such as geography, science and health, and physical education.
- Various connections within mathematics itself are established.

Teachers of mathematics have many opportunities for forming and strengthening these connections through ensuring explicit connections are made:

- among content topics across lessons
- among content within lessons
- through mathematical applications
- through mathematical modelling to the world around them.

Connections across the middle years

The transition from primary to secondary school appears to be an ongoing problem, despite many transition programs being implemented over the years. For example, when baseline data for the Middle Years Numeracy Research Project (MYNRP) in Victoria were collected in 1999, researchers noted a significant drop in performance in numeracy tasks from Year 6 (the last year of primary school) to Year 7 (the first year of secondary school) in urban schools (Siemon et al., 2001). They also noted that teachers in Years 5–9 classes could expect a range of up to seven years in numeracy-related performance, echoing similar findings in international comparison studies. More recently, Vale et al. (2013) have identified transition issues from one year to the next, with a slide in student performance over the summer especially evident for junior secondary students from low socio-economic schools.

To overcome difficulties associated with transition into the lower secondary school, and between year levels, it was recommended that classroom teaching strategies involve the regular and systematic use of open-ended questions, mathematical games, authentic problems and extended investigation to enhance students' mathematics learning and capacity to apply what they know. In addition, teachers should focus on the connections within mathematics, across tasks and topics, and explicitly develop students' strategies for making connections. Teachers who model the making of such connections themselves are on the road to beginning this process.

Teachers need to actively engage all students in conversations and texts that encourage reflection on their learning, and explanation and justification of their thinking. Simply engaging students by doing does not ensure mathematical learning occurs: there must be a focus on meaning and ensuring that students are attending to what matters and how it matters mathematically (Mason, 2004). Highly atomised, topic-based approaches tend to mask the 'big ideas' of middle-years mathematics (i.e., place value and multiplicative thinking, see Chapter 7) and 'crowd the curriculum' (Siemon et al., 2001). Merely 'getting learners to do tasks in mathematics lessons is not sufficient to ensure that they make mathematical sense of what they are doing' (Mason, 2004, p. 79). Even supplementing this by discussion is not sufficient to ensure mathematical learning if learners are focusing their energies and attention elsewhere (Mason, 2004). Thus, learning activities need to be designed or chosen so they are appropriate to learners' needs and interests; however, the teacher also needs to work at actively gaining and focusing students' attention on the relevant aspects of the task.

The MYNRP researchers used data collected from student responses to rich assessment tasks to develop an emergent numeracy profile. This profile can be used to select learning materials and structured, numeracy-specific teaching across the middle years. The major discriminating factors distinguishing performance at different levels in the profile were:

- students' understanding of, and capacity to apply, rational number ideas
- metacognitive activity—namely, monitoring of cognitive goals (indicative of conceptual understanding of the situation/task) and monitoring cognitive actions (indicative of procedural management of the solution attempt)
- the extent to which students could deal with patterns.

Chapters 7 to 11 provide a brief overview of the extent of the relevant content areas covered in the primary school years to help teachers facilitate students' transition from primary school; however, it would also be worthwhile consulting the *Australian Curriculum: Mathematics* for the middle years in your state or territory.

Connections across curriculum areas

Until recently, formal cross-curricular cooperation in Australian secondary schools was limited to some attempts at the lower secondary levels. Use of mathematics in other subject areas does occur (e.g., in chemistry or geography), but when a mathematical problem is set in, say, a physical education context, the motivation to do this usually emanates from the mathematics classroom. Often there is no formal arrangement in which the physical education teacher and the mathematics teacher cooperate in writing a cross-curricular unit to serve as the basis for both the next mathematics unit and the next physical education unit. Thus, subject disciplines protect their time allocation in the curriculum and their territory.

At the lower secondary level, interdisciplinary projects are a possibility for formal cross-curricular cooperation (e.g. Goos & Askin, 2005; Peters et al., 2012). However, there has to be a payoff between the amount of mathematics applied or learned through these projects and the time devoted to them in an already decreasing time allocation to mathematics classes. With this in mind, mathematics needs to be the anchor subject to ensure that a significant amount and depth of mathematics is involved. (The meaning and models of numeracy and its application across the curriculum is discussed and illustrated further in Chapter 5.)

Connecting content across lessons

According to Lampert (2001):

> In school, students will connect what they learn in one lesson to what they learn in the next lesson in one way or another. The teacher can work to deliberately structure the making of connections to enable the study of substantial and productive relationships in the content. If the teacher can make the conceptual connections among lessons obvious, students will have the opportunity to study aspects of the content that are not easily contained in single lessons. They will be able to study the kinds of ideas that make a subject coherent across separate topics. (p. 179)

Lampert suggests that a suitable context be selected as a basis for generative problems that act as a thread connecting the mathematical content from one lesson to the next. This context can be a unifying real-world situation, such as back problems resulting from improper lifting techniques as a connecting thread for several lessons on dynamics at the senior secondary level, or it can be a mathematical context, such as an investigation into the patterns and algebra of square numbers as an over-arching context for teaching mathematical notions such as powers or Pythagoras' Theorem at the lower secondary level.

Connecting content across lessons is similar to the task of selecting problems that will allow connections across content within a single lesson, but it is a more substantial task, as the teacher needs to be able to anticipate and identify the mathematics within the situation that will allow the same context to be sustained across several lessons. The problems need to be developed or selected in a way that will allow the focusing of students' attention on significant mathematical ideas and the important relationships among the mathematical concepts contained within them. The problem of designing a roller-coaster ride, for example, can be used as the basis for a generative theme for the study of functions in Year 11.

Table 3.1 Japanese mathematics lesson sequence: linear equations

Number of lessons	Sub-units	Lesson activity
2	Changing variables and linear functions (1)	Task 1a, Staircases: find what varies and stays the same when more stairs are added. Task 1b, find perimeter of different diagrams, using algebra to generalise.
3	Changing variables and linear functions (2)	Task 2, Origami Folding: find what varies, what stays constant. Communicate relationship between perimeter and area. Communicate graphically. Reflect.
2	Change of values of Linear Functions and the graph	Draw graphs of linear functions: – given slope and intercept – given equation
3	Finding expressions for Linear Functions	
3	Applying a Linear Function	
2	Simultaneous Equations and Linear Functions	
1	Use of functions in daily life	

Source: Shimizu (2010).

The first three figures have been drawn for you. Draw the next two figures by stacking 1 cm sided squares on top of each other. What changes when the number of steps changes?

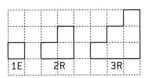

Source: Shimizu (2010).

Figure 3.1 **Task 1 Lessons 1 and 2**

Designing the ride using functions to model the track, and possibly other aspects of the ride, is the ultimate goal; however, along the way, students build an understanding of a variety of functions, their transformations, key features, gradient functions, and various mathematical and technological tools for working with functions. Table 3.1 provides a unit plan for a series of lessons regarding rate of change for linear functions (Shimizu, 2010). The sequence starts with a number of problem-solving lessons designed to draw students' attention to what stays the same and what changes (see Figure 3.1). The lessons later in the sequence focus on consolidating understanding of the key idea and developing fluency by practising procedures.

Making connections in lessons

The use of open questions in the mathematics classroom is often advocated to foster the development of students' higher-order thinking. However, Herbel-Eisenmann and Breyfogle (2005) point out that 'merely using open questions is not sufficient' (p. 484). This is especially true if the goal is to have students develop connections in a lesson that lead to an interconnected web of mathematical knowledge (Noss & Hoyles, 1996), rather than the students being 'funnelled' into the teacher's or text-book writer's way of thinking. Three ways in which teachers may assist students to develop connections in a lesson are explored in the box below in three different class-room scenarios about a task, *Shot on goal*. These methods include *triadic dialogue, funnelling* and *focusing*.

Shot on goal

You have become a strategy adviser to a group of new soccer recruits. Your task is to educate them about the positions on the field that maximise their chance of scoring. This means: when a player is dribbling the ball down the field, running parallel to the sideline, where is the position that allows this player to have the maximum amount of the goal exposed for a shot on the goal?

Initially, you will assume the player is running on the wing (i.e., close to the sideline) and is not running in the goal-to-goal corridor (i.e., running from one goalmouth to the other). Find the position for the maximum goal opening if the run line is a given distance from the near post. (Relevant field dimensions and a diagram of a soccer field as well as guiding questions are provided.)

Source: Adapted from Galbraith et al. (2007, p. 135).

Scenario 1: Triadic dialogue

The common use of *triadic dialogue* (Initiation–Response–Feedback) in many classrooms is well documented. The following exchange between a teacher and Year 9 students working on the *Shot on goal* task is typical of the type of questioning associated with triadic dialogue:

Teacher:	Now the question says, 'Find the angle of the shot on the run line 20 metres from the goal line.' See if you can find where that 20 metres should be put. [Pause] Has anyone worked out where that 20 metres belongs? [Initiation]
Mary:	It is AB. [Response]
Teacher:	Good. Thank you very much. [Feedback] Is that in agreement with everybody else? [Initiation]
Several students:	Uhuh. [Response]
Teacher:	Good. Now that's the first thing we need to do. [Feedback]

Source: Extract from RITEMATHS project classroom transcript (2005).

This extract also illustrates teachers' tendency to interpret the problem for students and thereby change the task from a problem-solving and modelling task to a routine, wrapper problem, in which students apply known procedures (Jackson & Cobb, 2010).

Scenario 2: Funnelling

In an effort to avoid such restricted patterns of interaction, but at the same time stay in control of the direction of the lesson, teachers commonly engage in *funnelling* (Herbel-Eisenmann & Breyfogle, 2005). This technique often gives the impression that students are making the connections in a lesson for themselves, but this is not in fact the case, as demonstrated by the following example of a different teacher using *Shot on goal* in the context of hockey with a Year 9 class.

This is the fourth lesson the class has spent on the task. The teacher drew the following diagram on the whiteboard before the students entered the room, and as a class, they decided the formula shown would find the shot angle.

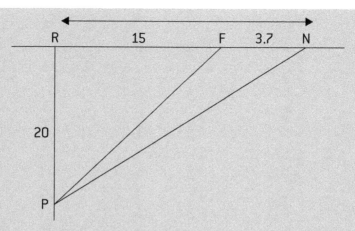

Shot angle = ∠ NPF
= ∠ RPN - ∠ RPF

Teacher:	Now, has anyone worked out where that [the formula] goes in terms of LISTs? Which LIST would that formula be in? [Referring to LISTs in the graphics calculator used for storing and manipulating data.]
Ben:	The last one.
Teacher:	The last one. We don't write it in that form but why is it in the last one?
Ben:	Because it is the last step?
Teacher:	Because this is the [pause] last step. This is 'the answer is' step, isn't it? [The teacher writes in large capital letters and circles ANSWER IS then draws a four-column table representing LISTs on the whiteboard with the second and third columns headed L2, L3.] So, could anyone suggest what should be in this one [second column] and that one [third column] if you are going to get there [fourth column]? [Spiro raises his hand.] What should be in LIST 2 and LIST 3 if you want LIST 4? [Waves at students] No, I am not going to ask you. My question was this: what do I put here and here if I want angle NPF there [pointing to the last column of the table he has drawn]?

Mat	RPF and RPN.
Teacher:	I want to put RPF there [writes on board with an arrow to column 2] and I want to put [writes RPN on board with an arrow to column 3], agree?
Mat:	Yeah.
Teacher:	How then do I calculate LIST 4?
Abdi:	You take LIST 2 from [stops].
Teacher:	I heard starters, someone?
Mat:	LIST 3 from LIST 2.
Teacher:	So LIST 4 is LIST 3 minus LIST 2, agree? [Doesn't detect the error.]
Mel:	Yes.
Sue:	But what numbers do you put in LIST 2?
Teacher:	Yes, that's an issue. Now we have to find out this question: how do I find the angle in LIST 2 [writes 'Find L2']. How do I calculate, this is, there's 4 [writes 'L4' above NPF]. That's in LIST 2 [writes '2' above RPF]. That's in LIST 3 [writes '3' above RPN]. How do I calculate angle RPF [writes angle below L2.]? Yes?
Spiro	Using tan.
Teacher:	Using tan, so what did you use? I will call this A [indicating ∠RPF.
Spiro:	Tan A equals 15 over 20. A equals 20^{-1}.
Teacher:	Angle A equals [waits]?
Spiro:	Tan^{-1}. No, no, 15^{-1}.
Teacher:	OK, I am going to ask this question: which is the number that changes all the time? Is it the 20 or the 15 that continues to change as the runner moves [drawing their attention to the shot spot on the diagram by writing '(Runner)' beside P]?
Sue and Spiro:	The 20.
Teacher:	The 20 moves. So sometimes this could be 20, next time it could be [waits]?
Sue and Spiro:	15.
Teacher:	And then?

Students:	10.
Teacher:	And then?
Students:	5.
Teacher:	Or it could go 1 and 2 and 3 and 4 and 5. Where are these—1, 2, 3, 4—coming from? Of course, positions along the line, aren't they [drawing various dots down the run line, RP, from R]?
Mel:	Yes.
Teacher:	So how can I get this formula [pointing to $A = \tan^{-1}\left(\frac{15}{20}\right)$], which is really good for one case, this case [pointing to the board], to a general formula so that it works for all these different points? Fatima?
Fatima:	I just thought of 15 divided by LIST 1.

[The transcript continues with the teacher 'funnelling' students towards developing a corresponding formula for the tan of $\angle$RPN.]

Source: Extract from RITEMATHS project classroom transcript, SOG lesson 4 (2005).

Funnelling results in the classroom discussion converging to the thinking pattern of the teacher. However, funnelling can also be used to scaffold students along a predetermined path of solution and connection making. In these circumstances, the teacher needs to make students aware of the metacognitive purpose of the questions asked and explicitly encourage students to start asking themselves these same questions. As students take responsibility for doing this, the teacher then fades the scaffolding. In the above exchange, the teacher's goal was to make sure all the students could 'code it up' (set up the problem) on their graphics calculators, and the teacher's funnelling was directed at bringing the class to this point.

Scenario 3: Focusing

Herbel-Eisenmann and Breyfogle (2005) suggest using a third questioning technique, *focusing*, to develop students' own connection making. Here, the teacher listens to students' responses and guides the discussion on the basis of the students' thinking, not on the basis of how the teacher would necessarily solve the task. The following transcript is from the same classroom introduced above; however, it is taken from the first lesson when students were beginning the hockey version of the *Shot on goal* task. The teacher facilitated the students' coming to know which angle was the focus of the task through focusing questions during a whole-class discussion when three students attempted to draw the relevant angle on the whiteboard at the front of the room.

They began with a diagram showing a goal box on the goal line and a run line perpendicular to this line.

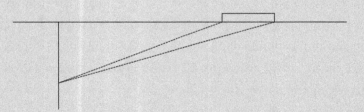

Teacher: Now, next part. Where do those lines go? Would someone like to come out and show where they think they have to hit? Dave?

Abdi: Is it like the best spot?

Teacher: Not the best spot but where he can hit the ball and still get it in. [Dave has drawn lines from a spot on the run line to the near post and the far post.]

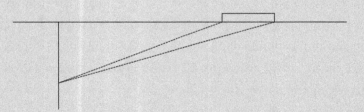

Teacher: Explain.

Dave: Ah, well, you can still hit that post [pointing to far goal post] and it will still go in and you can still hit the inside of that post [pointing to near goal post] and it will most likely go in.

Teacher: OK, is there a second person who would like to come out?

Spiro: There is no wall around the field, is there?

Ben: There is a goalie.

Teacher: Come on Spiro [holding out whiteboard pen]. First, debate was that a solution [pointing to Dave's diagram]. What are you going to do— agree with the first speaker or are you going to change it?

Spiro: I am not going to agree. Hit it off the wall. Is there another wall?

Teacher: What wall are you referring to? [Spiro draws a line down the side of the field diagram.]

Dave: There is no wall. [Spiro shades the diagram on the whiteboard to indicate a wall down the side of the field.]

Spiro: That wall there.

Dave: Spiro, this is a gutter—there is no wall.

Teacher: OK, now, so you are going to bounce it off the wall?

Mel: That was my idea.

Teacher: Can you show me the paths that would be allowed for you? [Spiro draws a trajectory from the shot spot to the wall then into the goal.]

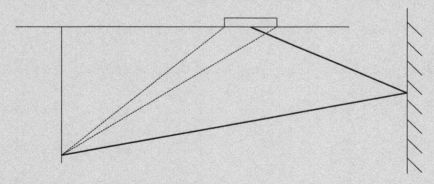

Teacher: Third speaker.

Mel: All right. OK, what if mine is the same as that one?

Teacher: OK, yours is the same as that one. I'll give you a different coloured pen.

Mel: What will I do? [Draws an angled trajectory showing the ball bouncing off the wall but lower down.] That's about right.

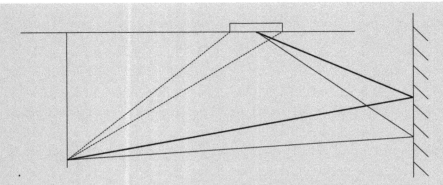

Teacher:	Now, where are my hockey players? Robyn? How many people play hockey? Robyn, as a judge of these things …
Mel:	[Interrupting] I didn't mean it to be that angle.
Teacher:	As a judge of these, which ones would you accept in a hockey game? Mel's? Spiro's? Dave's?
Mike:	Is there a wall there or not? [General laughter]
Teacher:	Is there a wall or is there not? OK, when you play hockey on a field, is there a wall there?
Fay:	No.
Teacher:	If there were a wall there, would it be inside the field or outside the field?
Cate:	Outside.
Teacher:	Therefore, if you hit the wall it would be … [waits]?
Jo:	It would be out.
Teacher:	It would be out so [walks to diagram on whiteboard] … Do you think the second one and third one are acceptable hits?
Students:	Nooo!
Spiro:	[Clapping] All right then. I agree with Dave.
Teacher:	Then you agree with Dave. Why do you agree with Dave?
Spiro:	Because it is a straight line hit until at least it is in.
Cate:	I have a better solution.

Teacher: OK, let's hear the better solution. [Hands pen to Cate who goes out to whiteboard.] Could you listen, please? Yeah, go for it. [Cate draws a large dot at each goal post then a thick line across the goalmouth.]

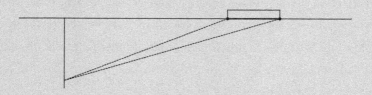

Teacher: Explain to me your better solution.

Cate: Ah, because Dave said you could hit the two poles but you can hit the ball between the poles, anywhere from this post to that post.

Abdi: [Almost disdainfully] But you have got a goalie!

Teacher: No, just forget that. They broke their leg over here. So what you are saying is, it's not just two lines? [Shades in between the two lines.]

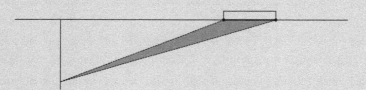

Dave: That's what I meant.

Mel: That was what I was going to ask.

Teacher: What were you going to ask?

Mel: Whether Dave just meant the lines or between the lines as well.

Teacher: OK, if it is that case, which angle are we being asked to find for the best angle? Where is the best angle? Where is this angle that is the shot on goal angle? Would you draw it quickly freehand in the exercise book?

Source: Extract from RITEMATHS project classroom transcript, SOG lesson 1 (2005).

In this approach, students have to articulate their thinking so others can understand what they mean. It also allows the teacher to see more clearly what students are thinking and what connection making they are doing for themselves. It is obvious from this transcript that the teacher values student thinking, and that students are encouraged to contribute. The teacher asks *clarifying questions* and *restates aspects of the solution* to keep attention focused on the *discriminating aspects* of the particular student's solution. However, for this to be used effectively, the teacher must be able to see the critical mathematical aspect of the task and, on a moment-by-moment basis, the essence of a task solution proffered by a student (Mason, 2010). Through this dialogue, the teacher encourages listening by other students as the clarifying questions provoke students to *explain* and *justify* and other students are invited to compare their thinking and solution with others.

REVIEW AND REFLECT : Reread the dialogue above. Note when the teacher is asking clarifying questions and restating aspects of a particular solution.

Reflect on the merits or otherwise of the teacher allowing the students this amount of time to come to an understanding of where the best angle might be for the shot on goal.

The dialogue above occurred near the beginning of the lesson, after the students had some time to think about the problem. In problem-solving and modelling lessons in which multiple solutions or expressions of solutions are expected, a whole-class discussion should also occur at the end of the lesson, with students (or groups of students) presenting, explaining and justifying their solutions. Teachers, after monitoring students' work on the problem, give careful consideration to the order in which they invite students to present (Hino, 2015; Inoue, 2011; Stein, Engle, Smith & Hughes, 2008). For example, the teacher may select the most common solution first, to ensure that students can access and contribute to the discussion. Alternatively, they may choose the most easily understood solution, and build to the more generalisable or useful solution. Another approach is to start with a common solution that includes a misconception or misinterpretation of the problem. Choosing the order in which to present solutions will depend on knowledge of your students' mathematical understandings and dispositions and the goal of the lesson, so that students may validate and develop their understanding through active listening, questioning, explaining and justifying.

REVIEW AND REFLECT : When observing classes, look at what happens *after* the first question is posed. What interaction follows? How does the teacher choose the order in which students present their solutions? How does the teacher support active listening? Reflect on how you can pose questions or ensure students pose questions for others so that all students engage in mathematical thinking and make connections for themselves.

Connections through mathematical applications

Curriculum documents often advocate making connections to the real world using mathematical applications for teaching and assessing (e.g., Board of Studies New South Wales, 2002; Queensland Board of Senior School Studies, 2001) as a means of motivating and engaging students, as well as illustrating the usefulness of mathematics to describe and analyse real-world situations. Galbraith (1987) categorises as applications 'the problem type questions typical of the teaching and examining tradition' (p. 6). He sees these as serving an important—if limited—function in 'requiring translation, interpretation, and the successful use of relevant mathematics' (p. 6). The limitation comes from the closed nature of the task: 'the situation is carefully described, relevant data are provided, and the student knows that each datum must be used in finding the solution. Assumptions needed to define the outcome … are explicitly provided' (p. 6).

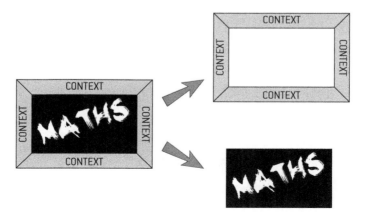

Figure 3.2 **Context as border**

The underlying assumption in selecting application examples for the classroom is that mathematical awareness and understanding are fostered through an active engagement in finding and applying solutions to real-life problems that fall within the sphere of students' personal understanding. Teachers' knowledge of students' interests is therefore crucial in bringing such tasks to the classroom, unless teachers allow students to pose their own problems or the teacher chooses open tasks in which the students have freedom to choose their own pathway and pose questions of interest to themselves in solving the task.

Unfortunately, for pragmatic reasons, application problems are often presented in the classroom with more reduced task contexts rather than highly life-like ones. There appear to be three levels of embedding of context that characterise these problems. These have been termed *border, wrapper* and *tapestry problems* (Stillman, 1998).

Border problems

For *border problems,* the context is merely a border around the mathematics (see Figure 3.2). Here, the supposed connection to the real world is superfluous. The mathematics and context are in fact entirely separate, and the mathematics can easily be disembedded once it is realised that this is all the task entails as the context does not obscure the mathematics at all.

However, these types of problems are deceptively difficult for students, particularly in the lower secondary years, as their form is not as transparent to many students as to the adult who wrote them. In the *Microwave ovens* problem in the box below, the connection of the mathematics to microwave ovens is merely window dressing for the mathematical task of equating the two equations and solving for *t*. There is no motivation for why you would need to do this in a real-life context. Giving an answer as a number with years written after it is also no indication that a student is working within the context and making connections to the real world outside the classroom.

Microwave ovens

The number of radioactive emissions from a certain faulty microwave oven is given by $N_1 = 64(0.5)t$ at t years from the time of use. The number of emissions from a second faulty microwave is given by $N_2 = 4^{15}(0.0625)^t$ at t years from the first time of use. Find out when both microwaves will emit the same number of radioactive emissions.

Source: Stillman (2002).

Wrapper problems

In *wrapper problems,* the mathematics is hidden within the context but the two can be separated by unwrapping the mathematics (see Figure 3.3). The context, in a sense, can be thrown away, as the mathematics is all that is needed to solve the problem. The more diligent problem solver may pick up the wrapper after a mathematical solution is reached to check that it makes sense. However, the presence of the context does make the unwrapping

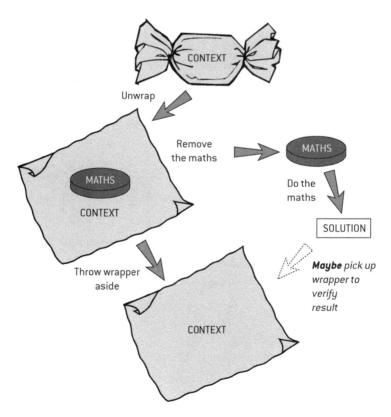

Figure 3.3 **Context as wrapper**

of such problems quite challenging for some students. The box below shows an example of a wrapper problem, *Road construction.*

Road construction

A new traffic lane (minimum width 6 metres) is to be added to a section of highway which passes through a cutting. To construct the new lane, engineers need to excavate an existing earth bank at the side of the roadway, which is inclined at 25° to the horizontal. This will make the inclination steeper. Local council regulations will not allow slopes greater than 40° due to the potential for erosion.

Decide whether the new traffic lane can be excavated without expensive resumption of properties at the top of the bank, which is 7 metres above the road surface.

Source: Galbraith and Stillman (2001, p. 306).

In Chapter 14, we discuss the influence of socio-cultural factors on students' interpretation of the context of such problems. In particular, there is evidence that students from low socio-economic, rural, Indigenous or other cultural communities rely on personal experience or cultural knowledge of contexts and find it difficult to 'unwrap the mathematics'.

> **REVIEW AND REFLECT :** Read the *Road construction* problem again. Solve it. Give possible reasons why it took some students just 6 to 8 minutes to solve while others took 18 to 25 minutes. It would be helpful to look at the different ways in which you could draw a diagram for the task.

Tapestry problems

In *tapestry problems,* the context and the mathematics are entwined or intermingled and the solution process proceeds by continually referring to the context to check you are on track. There is always the sense that the two are very much interrelated. The upper bound of these applications is actually true modelling problems such as the task in the box below, *Drying out.*

Drying out

There are many lakes in Australia that are dry for most of the time, only filling for short periods immediately after rain. Lake Eyre in South Australia is an example of one of these normally dry lakes. When a dry lake is filled with water, how rapidly will the lake empty?

Source: Henry and McAuliffe (1994, pp. 41–48).

Over-use of border and wrapper problems leads to a rather impoverished view of applications of mathematics, and certainly does not foster the modes of thought associated with the modelling of real-world situations. The teacher whose interview transcript appears below elaborates on this issue.

Teacher: [Kids] believe they will get questions that may not make sense to them but they know it has to be mathematical.

Interviewer: Where would they develop that idea from?

Teacher: Going through school, right through from junior [high school]. Right through … Context is just a coat that they wear.

Source: Stillman (2002).

REVIEW AND REFLECT : Watch the presentation 'Math Curriculum Makeover' by Dan Meyer (2010) on YouTube www.youtube.com/watch?v=BlvKWEvKSi8.

In small groups, examine the application tasks in a mathematics textbook. Categorise them as border, wrapper or tapestry problems. Compare your findings to the claim made by the teacher in the box above and Dan Meyer. Report to your peers. Identify an application task that could be developed into a modelling task.

Applications problems in the secondary setting can provide a useful bridge between the contextualised practice problems of the past and full-blown modelling tasks. In the busy classroom, it is frequently not possible to explore and model real-life situations. Often, the situation is simplified to make it more approachable with regard to the skill level of students, so the 'real situation' that is being modelled is 'pseudo-real' at best. However, reducing the realism may prevent students from posing and asking questions, making simplifying assumptions, generating and selecting variables, and formulating the mathematical model themselves. Stillman and Ng (2013) discuss industry contexts for modelling tasks, and Stillman et al. (2013) explain how to turn ideas and application problems into modelling tasks.

REVIEW AND REFLECT : Reread the *Drying out* problem above. White (n.d.) restated the problem as, 'Find how long it takes the lake to become dry again. We expect the answer to be in days.'

In groups, brainstorm the problem. Generate a list of variables involved. Select from these those that will be considered in an initial model of the situation.

Read the article by White, discussing the suggested classroom techniques for carrying out this modelling task.

Connections through mathematical modelling

As Kaiser and Sriraman (2006) show, there is wide variation in what is meant by the term *modelling* in mathematics education. This causes difficulties for teachers, students and assessors when modelling is expected to form a significant part of the curriculum. One interpretation sees a role for mathematical modelling primarily for the purpose of motivating, developing and illustrating the relevance of particular mathematical content (e.g., Zbiek & Conner, 2006). Alternatively, mathematical modelling can be seen as an approach in which the modelling process is driven by the desire to obtain a mathematically productive outcome for a problem with genuine real-world motivation (e.g., Galbraith & Stillman, 2006; Pollak, 1997; Stillman, 2006). At times, this is directly feasible, while at other times, the descriptor 'life-like' is more accurate. The point is that the solution to such a problem must take seriously the context outside the mathematics classroom, within which the problem is located, in evaluating its appropriateness and value (see Galbraith et al., 2006). The learning goal is to support the students' development of modelling competencies as modelling itself is considered content.

In the latter view, mathematical modelling involves more than just application problems. It is a process through which students work, and the techniques and metaknowledge about applying mathematics gained in this process are as important as the eventual solution. Many secondary textbooks take the approach of mathematical theory followed by examples, purely mathematical practice exercises and then the mathematics just covered embedded in contextualised problems. According to Blum and Kaiser (1984), 'when we start with situations which have already been idealized, the resulting mathematisation appears almost compulsory, which is practically a falsification of any genuine process of model building' (p. 213). Applications problems of this nature largely circumvent the modelling process. A purely applications approach does not engender the cultivation of the metaknowledge about applying mathematical processes that accompanies the modes of thought associated with modelling real-world situations. Two motivations for adopting a modelling approach at the senior secondary level are the application of technology such as graphics and computer algebra systems (CAS) calculators (discussed in Chapter 4) and the opportunity modelling affords of making mathematics relevant through connections to the real world. This approach to modelling has been adopted in curriculum documents in some states, as the interview below with a Queensland teacher indicates. Stillman et al. (2013) have investigated senior secondary curricula in other states to identify opportunities for social critique through mathematical modelling tasks.

Interviewer:	Why was the introduction of applications and modelling considered to be a valuable initiative at the senior secondary level?
Teacher:	One of the big things between the old syllabuses … and the new ones … the kids were always asking: 'Why do we do this? What's the point of all this? Where am I ever going to use this?' … The old syllabus was very much content driven whereas the new syllabus and, even down the track and as I understand it more and more, it's more about understanding concepts and how the concepts relate to each other. And mathematical modelling allows

you to build the understanding of the concepts. The content is important and you need the content but it is the application of that content in a way that is meaningful and the kids can make sense of that. So there is a purpose for it. There is a relevance to the mathematics that they are doing and the kids can see that. Nowadays, unless the kids can see a reason for doing something they just park up [refuse to budge]. You will get the small handful of kids who will just jump through any hoop that you give them but the vast majority of kids, unless there is a reason for doing it, they just won't engage.

Source: Extract from CCiSM project interview (2005).

REVIEW AND REFLECT : Investigate the senior secondary subject descriptions and textbooks in your state and the Australian Senior Curriculum. Is mathematical modelling an expected learning outcome in one or more of these subjects? How is it assessed? Present an example of tasks to your peers.

The modelling process

Although application problems can require the problem to be translated into a suitable representation, a mathematical model formulated and relevant mathematics used successfully in solving the problem and validating the solution, there are important aspects of modelling that are omitted in this process. These are crucial to the modelling task and vital for students to develop their own repertoire of modelling techniques. Treilibs et al. (1980) and Klaoudatous (1994) have argued that modelling ability is different from conventional mathematics ability, and their research showed that facility with conventional mathematics was no indication of students' modelling ability. Students who are very successful at conventional mathematics are often the most resistant to change from traditional pedagogy to a modelling approach (Clatworthy & Galbraith, 1989). Clatworthy comments in an interview that 'you become so locked into being successful in the standard mode of mathematics that you become too scared and too tight mentally to look at the processes that are involved in modelling' (Clatworthy & Galbraith, 1989, p. 102).

With most modelling situations (e.g., the *Mad Cow disease* example below), the task is not so constrained that only a limited number of possible mathematical models can be used in the solution, as is the case with applications problems. The process of defining the real-world problem in such a way as to allow it to be explored and investigated is crucial to modelling. Questions need to be asked about what has to be known and what tools are available to try to solve the problem. Specific assumptions made in reducing the problem to a manageable mathematical model must be identified. Decisions have to be made about which variables are important and which can initially be ignored in simplifying the problem. The modellers make these decisions, they are not made for them, as happens in applications problems. In addition, when the modeller has solved the particular problem, the validation process requires testing with a different data set from that used in the model development. The model may need modification if there are inconsistencies with empirical test data or if it is felt that the simplifying assumptions have produced a model that is too unrealistic or undesirable to be useful. The modeller is thus required to explore the strengths and weaknesses of the model, clearly communicating any limitations to possible users.

Mad Cow disease

Creutzfeldt-Jakob disease (CJD) is a fatal brain disease of humans first classified in the 1920s. In 1996, doctors in the UK reported a variant of the disease, vCJD. Research since suggests that vCJD is the result of exposure to the agent that causes the cattle disease commonly called Mad Cow disease. Parts of the brain are destroyed—hence the name. This is caused by abnormal versions of agents called prions, which build up over time because they are resistant to the body's normal mechanisms for breaking them down. They can join together into sheets (or fibrils) that destroy the nervous tissue around them. It is also possible that the presence of abnormal prions increases the probability that further ones will be produced. The height of Mad Cow disease in the UK was the mid-1980s. The first cases of vCJD occurred in the mid-1990s. Once symptoms are displayed, the sufferer generally survives around two years.

By 4 August 2006, there had been 156 definite and probable recorded deaths from vCJD in the UK. In 2001, Lawson and Tabor (2001) reported that the Government Chief Medical Officer was still saying that the final toll was likely to be between 'hundreds and hundreds of thousands'. Is it possible that there really is this number of sleepers out there in the general population who are infected with vCJD but who are yet to display the symptoms?

Source: Adapted from Lawson and Tabor (2001).

Although several simple sequential cycle frameworks of the modelling process (e.g., White, n.d.; see Figure 3.4) have been developed, when the process is studied holistically, it really is quite messy (Clements, 1989). In reality, following a sequential modelling cycle from analysis of the real-world situation through several iterations of the cycle to develop a refined model of the situation may be a long way from what really happens. The phases of the process are all interconnected and a particular mode of attack can lead directly from one phase to any other phase. *Some sense of this complexity is necessary if students are to develop the metaknowledge associated with modelling.*

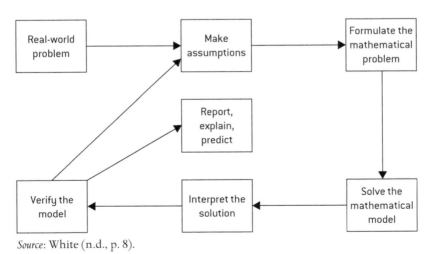

Source: White (n.d., p. 8).

Source: Stillman (1998, p. 245).

Figure 3.4 **The modelling process**

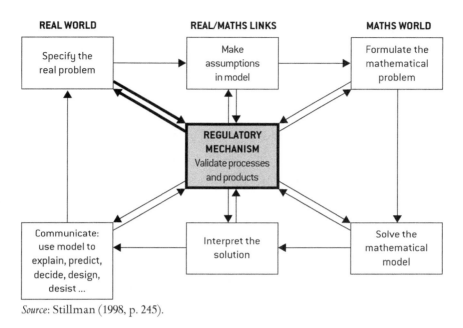

Source: Stillman (1998, p. 245).

Figure 3.5 **Mathematical modelling framework incorporating regulatory mechanism**

> **REVIEW AND REFLECT :** Work in small groups on the *Mad Cow disease* modelling problem presented above. Statistics on CJD in the UK are available from www.dh.gov.uk/PublicationsAndStatistics/Statistics/fs/en. Reflect on the processes you used in the light of the modelling cycle developed by White (n.d.). How adequately does it describe the complexity of your thinking?

Modellers need to be continually aware of the connections between the situation being modelled and the mathematical activity taking place, with some validation of the partially complete model occurring at each stage. In essence, validation then becomes a regulatory or controlling activity (i.e., it is metacognitive; see Chapter 2), affecting the whole modelling process. Recent research (e.g., Galbraith & Stillman, 2006; Maaß, 2006) confirms the presence of this metacognitive activity, even when students are beginning modellers in Year 9.

Figure 3.5 shows how this regulatory mechanism monitors the modelling processes throughout the solution. It is still possible to follow the cycle sequentially as before, but now deviating or backtracking via the regulatory mechanism and bringing into play a continual process of reconsidering possible assumptions, strategies, actions and their results in terms of other components in the modelling activity are highlighted. The links between the problem context and the regulatory mechanism are thicker to show the importance of carrying out this reconsideration in terms of the original problem specification. Connections between the real world and the mathematical world are continually being made to lessen the sense of separation present in other frameworks. This separation often reinforces the notion that context is at best a wrapper for the mathematics, leading to failure by many students to make the connections and use the context as a means of monitoring the solution process throughout (Stillman, 2002).

Modelling sub-skills

As modelling ability differs from conventional mathematical ability, it is imperative that the component skills that characterise modelling competency (Maaß, 2006) be fostered in the classroom, both by being the focus of particular instruction and through application in modelling tasks. The sub-skills include:

- formulating the specific question to be answered mathematically
- specifying assumptions associated with mathematical concepts or the modelling context
- identifying important variables or factors
- modelling different aspects of objects
- modelling different aspects of situations
- generating relationships
- selecting relationships
- making estimates
- validating results
- interpreting results.

Using a test to assess these sub-skills, Grunewald (2013) showed that working on a modelling project enabled Year 9 students to develop these sub-skills. The specific sub-skills can be developed from discussions of solutions to modelling tasks such as *Drying out* (see p. 65) (White, n.d.) or by constraining the assumptions when designing the problem.

Galbraith (1996), in 'Modelling comparative Olympic performance' provides examples of four distinct types of assumptions. They are those associated with:

- mathematical concepts (e.g., a shot in shot put can be treated as a free projectile)
- mathematical detail (e.g., the range of a projectile varies only slowly with the angle of projection)
- the modelling context (e.g., 172 centimetres is a representative height for a female hurdler)
- major leaps in the solution process (e.g., 10 per cent is a reasonable estimate to compensate for air resistance).

Often the same object can be modelled by a variety of models, depending on the situation being investigated. For example, a person could be modelled by a cylinder with a particular diameter and height if we wanted to know how much space had already been taken up in space, or alternatively by a rectangle if we were considering legal specifications for the size of entrance and exit doors that may or may not be rectangular. Here is an example from Stacey (1991) of a task to develop students' formulation skills.

Developing formulation skills

Match the models with the appropriate situations. Construct a similar task in which a variety of different models is appropriate for something (e.g., a person) when different situations are being investigated (e.g., an opinion poll or hurdling).

Models of the Earth	Situations being investigated
A. Stationary point	a. Eclipse of the moon
B. Stationary straight line	b. Driving with a road map around Melbourne
C. Stationary plane	c. Propagation of earthquakes
D. Point revolving around sun	d. A ball thrown through the air
E. Circle (or disc)	e. Eclipse of the sun
F. Sphere filled with layers of material	f. Calculating how far you can look out to sea from a cliff
	g. Sending a probe to intercept Halley's Comet

REVIEW AND REFLECT : As a beginning teacher, it is useful for you to both be able to create resources suitable for your students and other teachers to use, and to familiarise yourself with the applications of mathematics in different workplaces.

Gather information or data about the use of mathematics in the workplace by visiting a commercial or industrial site. From this information, develop a resource designed by you for students in Year 9 or Year 10 addressing topics in the *Australian Curriculum: Mathematics*. Incorporate technology use into your resource where practical and appropriate. Provide teacher notes and solutions to student tasks.

Conclusion

In this chapter, we have endeavoured to highlight ways to facilitate students' ability to understand the world from a mathematical perspective. In particular, we have looked at developing connections across the transition from the primary to lower secondary years, from the mathematics classroom to doing mathematics in other curriculum areas, and within mathematics itself. In later chapters, these ideas will be explored further.

Recommended reading

Galbraith, P. (1996). Modelling comparative performance: Some Olympic examples. *Teaching Mathematics and Its Applications*, 15(2), 67–77.

Galbraith, P. L. & Stillman, G. A. (2006). A framework for identifying blockages during transitions in the modelling process. *ZDM*, 38(2), 143–162.

Galbraith, P., Stillman, G. & Brown, J. (2010). Turning ideas into modelling problems. In R. Lesh, P. Galbraith, C. Haines & A. Hurford (Eds), *Modeling students' mathematical modeling competencies* (ICTMA 13, pp. 133–144). Rotterdam, The Netherlands: Springer.

Herbel-Eisenmann, B. A. & Breyfogle, M. L. (2005). Questioning our patterns of questioning. *Mathematics Teaching in the Middle School*, 10(9), 484–489.

Stein, M., Engle, R., Smith, M. & Hughes, E. (2008). Orchestrating productive mathematical discussions: Five practices for helping teachers move beyond show and tell. *Mathematical Thinking and Learning*, 10(4), 313–340.

Stillman, G., Kaiser, G., Blum, W. & Brown, J. P. (Eds). (2013). *Teaching mathematical modelling: Connecting to research and practice*. Dordrecht, The Netherlands: Springer.

CHAPTER 4
Effective use of technologies in mathematics education

In 2008, the state, territory and Australian government ministers of education agreed on a set of goals for schooling in Australia in the 21st century. The *Melbourne Declaration on Educational Goals for Young Australians* (Ministerial Council on Education, Employment, Training and Youth Affairs [MCEETYA], 2008) specified two educational goals related to young Australians: Australian schooling to promote equity and excellence; and that all young Australians become successful learners, confident and creative individuals, and active and informed citizens. As a foundation for success in all learning areas (Goal 2), it was seen that students must 'have the essential skills in literacy and numeracy and are creative and productive users of technology, especially ICT' (MCEETYA, 2008, p. 8). Consistent with this aim, the *Australian Curriculum* (ACARA, 2015) includes Information and Communication Technology (ICT) as a General Capability to be developed across all learning areas. Specific to mathematics, the *Australian Curriculum* states that students develop 'ICT capability when they investigate, create and communicate mathematical ideas and concepts using fast, automated, interactive and multimodal technologies. They use their ICT capability to perform calculations; draw graphs; collect, manage, analyse and interpret data; share and exchange information and ideas; and investigate and model concepts and relationships.' Thus, it is an expectation within all Australian states and territories that students make use of digital technologies when learning and doing mathematics.

Mathematics curriculum policy in Australia has promoted the use of technology to aid students' learning and understanding of mathematics for more than two decades (e.g., Australian Education Council, 1991). This agenda is reflected in the various state and territory mathematics curriculum statements and syllabuses that permit, encourage or expect the use of digital tools such as computers, tablets, graphics calculators or calculators with computer albegra systems (CAS), as well as the internet. The mathematics teaching profession in Australia also recognises that teachers need knowledge of a range of technologies, and that 'teachers at all levels are actively involved in exploring ways to take full advantage of the potential of technology for mathematics learning within the total curriculum' (AAMT, 2014).

Technologies currently used in secondary school mathematics education include mathematics-specific hardware (e.g., handheld graphics calculators and CAS calculators), generic digital tools (e.g., computers and tablets), mathematics-specific software (e.g., graphing, statistics and dynamic geometry software), general-purpose software (e.g., spreadsheets) and increasingly, multimodal and Web 2.0 technologies. Other technologies, such as 3-D printers and robots, are also being considered for their potential to enhance mathematics teaching and learning. These, and future developments on the horizon, make this an exciting time to be a mathematics teacher. However, to become *effective* users of technology in mathematics education, teachers need to make informed decisions about how and why to integrate different types of technology into their classroom practice in order to support students' learning of mathematics. We begin this chapter by identifying and illustrating ways in which technology can enhance student learning, and then discuss some of the implications of technology use for mathematics curriculum, pedagogy and assessment. The chapter concludes by considering some problems and challenges in creating technology-rich classroom learning environments.

REVIEW AND REFLECT : Before reading any further, record your own views about learning mathematics with technology. Do you think secondary students should be allowed to use scientific and graphics calculators when learning mathematics? Are computers a useful learning tool? What types of apps should students be allowed to use? What might be some benefits and disadvantages in teaching and learning mathematics with technology? Then ask other adults (teachers and non-teachers, if possible) and some school students what they think. Discuss your views and findings with other pre-service teachers in your class.

Potential benefits of learning mathematics with technology

Education researchers recognise the potential for mathematics learning to be transformed by the availability of technology resources such as computers, iPads, graphics calculators and the internet (see Bartolini, Bussi & Borba, 2010; Geiger, Forgasz, Calder, Tan & Hill, 2012; Hoyles & Lagrange, 2010; Larkin & Calder, 2015; Trouche & Drijvers, 2010 for reviews of recent research). In Australia and internationally, teacher organisations encourage the use of technologies as natural media for mathematics learning. For example, the AAMT's (2014) *Position Paper on Digital Learning in School Mathematics* recommends that 'all students have ready access to appropriate technology and the associated virtual learning environments and communities that support and extend their mathematics learning experiences' (see NCTM, 2011 for a US perspective). Let us explore some of the specific ways in which technology affords learning opportunities in mathematics.

Learning from instant feedback

Learning is assisted by the instant feedback that technology can provide. Figure 4.1 shows how using a spreadsheet to investigate the effect of compound interest on savings allows students to experiment with different interest rates, initial investments and monthly

C4				fx =0.05/12*B4		
	A	**B**	**C**	**D**	**E**
1	$100 monthly savings, 5% interest compounding monthly				
2					
3	Month	Principal	Interest	Savings	Amount
4	1	5000.00	20.83	100.00	5120.83
5	2	5120.83	21.34	100.00	5242.17
6	3	5242.17	21.84	100.00	5364.01
7	4	5364.01	22.35	100.00	5486.36
8	5	5486.36	22.86	100.00	5609.22
9	6	5609.22	23.37	100.00	5732.59
10	7	5732.59	23.89	100.00	5856.48
11	8	5856.48	24.40	100.00	5980.88
12	9	5980.88	24.92	100.00	6105.80
13	10	6105.80	25.44	100.00	6231.24
14	11	6231.24	25.96	100.00	6357.21
15	12	6357.21	26.49	100.00	6483.70

Figure 4.1 **Spreadsheet for investigating compound interest**

deposits by changing these amounts or formulae and observing how the dependent values are modified.

Using a spreadsheet

Use Excel software to reproduce the spreadsheet shown in Figure 4.1. The spreadsheet shows the monthly balance in a savings account in which a person has invested $5000 at 5 per cent interest, compounding monthly, and adds $100 per month to the account. Design a task with instructions for students that will lead them through an investigation of the effects of changing the principal, interest rate or amount of monthly savings. Include activities that would help students develop knowledge of compound interest needed for the spreadsheet investigation.

Make use of the spreadsheet to investigate the returns on investment from a number of different financial institutions in Australia by considering their savings account interest rates (available via the internet).

Spreadsheets are an important tool for investigating many topics within financial mathematics, such as budgeting, investing, borrowing money and calculating income tax. Identify the financial mathematics topics in the *Australian Curriculum: Mathematics* and state-based senior secondary curriculum documents. Select one topic and devise an application or modelling task (see Chapter 3) that makes use of spreadsheets.

Observing patterns

Technology can also help students to understand patterns, such as those related to linear functions in which there is a constant rate of change. For example, when asked to compare mobile phone company charge rates (see the box opposite), students might make a table of values for minutes and costs for each company, enter these into their graphics calculator or a relevant app for their computer or tablet and plot the points. They might then describe the two patterns verbally, write equations for the costs, and check these equations by plotting them on the same axes as the points originally graphed.

Comparison of mobile phone company charge rates

Phones-R-Us is a new mobile phone company that offers phone services for $45 per month plus $23 for each gigabyte (GB) of additional data required above the 1 GB that comes with the monthly charge. Long-established market leader Telecorp has a $15 monthly fee but charges $35 per GB of additional data above the 1 GB of initial data. Both companies charge similar rates for phone calls and SMS messages. Compare the two companies' total charges according to the total data used per month. If you are currently a Telecorp subscriber, should you change to Phones-R-Us? Why or why not?

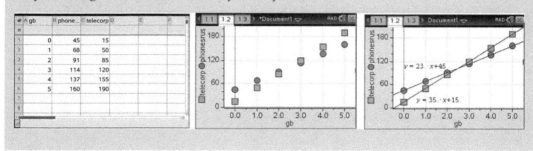

REVIEW AND REFLECT : Devise a graphics calculator task similar to the mobile phone investigation that will allow students to explore linear functions as models of real-world situations. (See Geiger, Makar, Faragher & Goos, 2012; Geiger, et al., 1997, 1999; and Goos, 2002 for examples.)

Making connections between multiple representations

Technology makes it possible for students to see connections between multiple representations of a concept and gain insights into abstract entities such as functions. Graphing software or graphics calculators can be used to explore families of functions represented symbolically, graphically and numerically (in tables) much more quickly, and with much less chance of error, than if this task was done by hand (as in Figure 4.2).

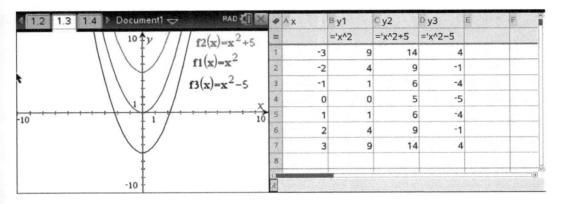

Figure 4.2 **Symbolic, graphical and numerical representations of the family of functions represented by** $y = x^2 \pm c$

While the linking of different representations offers clear benefits for students' deeper understanding of mathematical ideas, to do so effectively requires careful lesson planning. Pierce and Stacey (2008) identified four key principles for designing lessons that are intended to link multiple representations:

- Focus on the main goal of that lesson (despite the possibilities offered by technology).
- Identify different purposes for using different representations to maintain engagement.
- Establish naming protocols for variables that are treated differently when working with pen and paper and working with a digital tool.
- Reduce extraneous cognitive load.

In summary, it is very important to be clear about the specific goals of a lesson and how these goals will be achieved. Further examples are discussed in Chapter 8.

Working with dynamic images

Technology supports inductive thinking by allowing students to quickly generate and explore a large number of examples, and make conjectures about patterns and

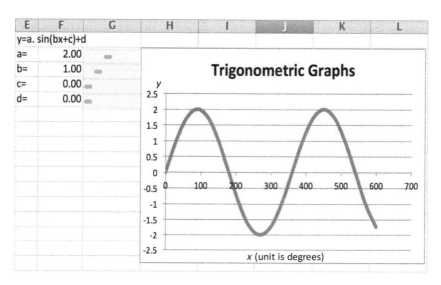

Figure 4.3 **Excel spreadsheet with scrollbars for investigating properties of trigonometric functions**

relationships. This process is greatly enhanced when students work with dynamic images or use interactive tools that can change the appearance of the mathematical objects on screen. Figure 4.3 illustrates how scrollbars inserted in an Excel spreadsheet can be used to vary the parameters in equations describing trigonometric functions and draw conclusions about related changes in amplitude, period, phase shift and vertical displacement of the related graphs.

Dynamic geometry software provides tools for constructing geometric objects in two or three dimensions and then manipulating these objects—for example, by 'dragging' vertices—to identify invariant properties. An example is provided in Figure 4.4, which shows the results of generating many examples of triangles and calculating the sum of their internal angles. The internet also offers many useful sites on which students can work with dynamic images of various kinds. Figure 4.5 shows a segment of a virtual manipulative website that enables students to interact with visual objects to develop an understanding of symmetry, transformations and other spatial concepts. See Chapter 9 for further discussion of the use of dynamic geometry software.

REVIEW AND REFLECT : Prepare an investigation of facial symmetry using digital photographs and image manipulation software such as Adobe Photoshop, Paint Shop Pro, scanning software, or good-quality drawing software (see Todd Edwards, 2004 for an example of an investigation). Discuss any issues you may need to consider in using photographs of teachers, students or celebrities whose images can be found on the internet.

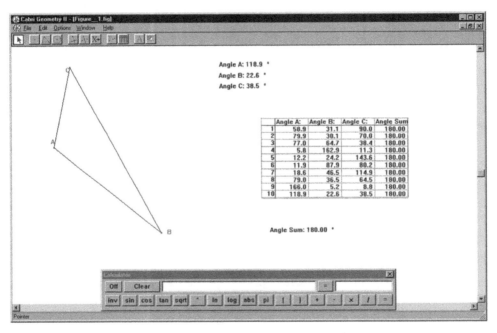

Source: Created using Cabri Geometry II

Figure 4.4 **Dynamic geometry software screen display for investigating the angle sum of triangles**

Exploring simulated or authentic data

With access to technology, students are no longer limited to working with simple data sets contrived by the teacher or textbook to make calculation easier. Computer software or graphics calculator programs can be used to simulate random phenomena, such as tossing coins and rolling dice, or to construct randomising devices like a spinner with sectors of the same (or different) area (see Figure 4.6).

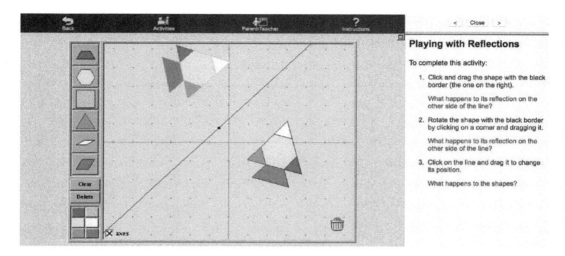

Source: National Library of Virtual Manipulatives, <nlvm.usu.edu>.

Figure 4.5 **Exploring rotation**

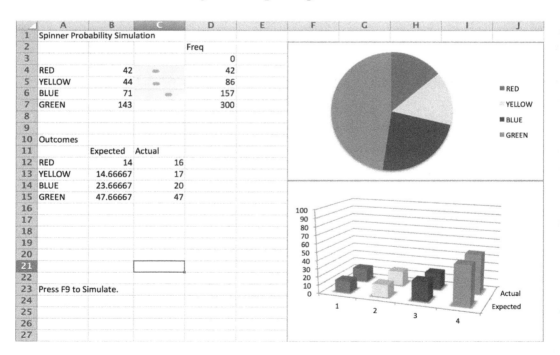

Figure 4.6 **A spreadsheet simulation of a spinner**

> **REVIEW AND REFLECT** : Find out how to generate random numbers using a spreadsheet, graphics calculator or app and investigate ways of using these functions to generate random data for different situations.

An enormous range of authentic data sets is now available from internet sites. The Australian Bureau of Statistics website at www.abs.gov.au offers data sets, including census data and other education resources suitable for secondary mathematics classrooms. This website provides access to data related to different characteristics of school students collected between 2006 and 2014 via the CensusAtSchool project. This project encouraged students to respond to questions of interest about themselves. The data are still available for students to investigate samples of the response data from the total population of responses received Australia wide. Figure 4.7 shows the opening page of the CensusAtSchool archive. Each heading is linked to a page displaying the corresponding information or data.

Data collection becomes even more realistic when students use data-logging equipment, such as motion detectors and probes that measure temperature, light intensity, pH, dissolved oxygen, heart rate and the frequency of sound waves, to investigate physical phenomena and the mathematical relationships that describe them. Using these technologies, students can discover that the rate at which hot water cools can be modelled by an exponential function and the motion of a pendulum by a trigonometric function (see Figures 4.8a and 4.8b). The instruction booklets that come with data loggers usually provide examples of classroom activities, worksheets for students, sample solutions to the questions asked and suggestions for making connections with other curriculum areas. Examples of such resources can be found on calculator company websites (URLs are listed at the end of this chapter).

Digital photographs and movies also help bring into the mathematics classroom real-world situations that can be analysed with the aid of commercially available software or free software downloadable from the internet (see Hyde, 2004 and Pierce et al., 2005 for a list of sites). Data relevant to students' lives can be captured from pictures or movies to study a wide range of mathematics concepts, such as functions, ratio,

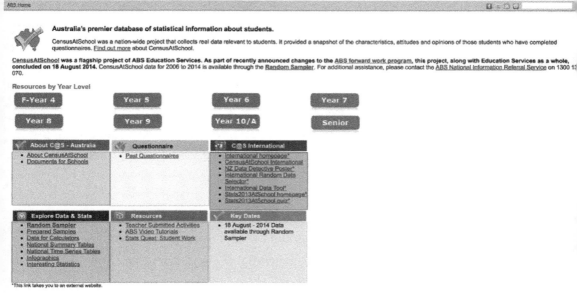

Source: Australian Bureau of Statistics (2014).

Figure 4.7 **CensusAtSchool archive**

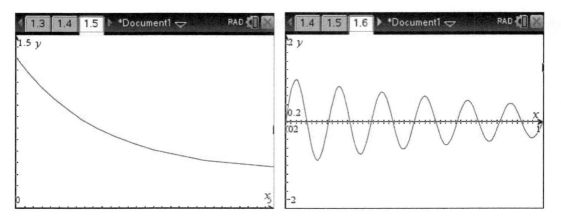

Figure 4.8a **Modelling cooling with an exponential function**

Figure 4.8b **Modelling periodic motion with a trigonometric function**

similarity and transformations, Pythagoras' Theorem, calculus and concepts of position, velocity and acceleration. Examples of such software include DigitiseImage (www.maths.sci.shu.ac.uk/digitiseimage/), LoggerPro (www.vernier.com/products/software/lp/) and Measurement in Motion (www.learninginmotion.com/products/measurement/index.html). Dynamic geometry packages like Cabri Geomètre (www.cabri.com) and Geometer's Sketchpad (www.keycurriculum.com) can also be used to input static and moving images in order to extract mathematical information that students can use to explore a situation.

Visualisation

A common characteristic of each of the above applications of technology is the opportunity for learners to visualise mathematical concepts. Students can observe changes in numbers, see patterns, and view images of geometric figures, relationships and data. Visualisation as a means of learning mathematics has gained more prominence through the use of technology, and visual reasoning has become more widely acknowledged as acceptable practice for mathematicians in the mathematical discovery process (Borba & Villarreal, 2005). Teachers may take it for granted that students are adept at visualisation and that visual images will help them to understand. However, not all students have highly developed visualisation skills, so teachers may need to direct students' attention to the important elements of the image. Since students may misinterpret an image and hold on to their misconceptions, teachers also need to scaffold students' use of visualisation and interpretation. Some examples of difficulties or misconceptions that may arise in these learning environments or the ways in which teachers may scaffold learning are provided in Chapters 8, 9, 10 and 11 in the contexts of algebra, geometry, measurement, and chance and data. Visualisation and the use of visual reasoning are discussed in Chapter 9.

> **REVIEW AND REFLECT :** Use a digital camera or smartphone to take photographs of a bridge in your local community or search the internet to obtain pictures of well-known bridges (e.g., Sydney Harbour Bridge, Story Bridge or Golden Gate Bridge).

> Print the images and use by-hand methods to find a function that models the main arch of the bridge.
>
> Download one of the free programs listed in the articles by Hyde (2004) and Pierce et al. (2005) and use this software to find a function that models the same bridge.
>
> Compare the functions you obtained by the two different methods, commenting on any similarities or differences. Discuss with fellow students any advantages and disadvantages of the by-hand and technology methods you used.

Finding and sharing mathematics

In previous sections, the internet has been represented as a library of resources for lesson preparation and classroom activities. However, teachers often have little time to search for ideas or information, or they may lack confidence in evaluating what they find on the internet. In attempting to provide guidance on how to select appropriate mathematics resources from the internet, Kissane (2009) provides five different categories of potential uses: (i) interactive opportunity, (ii) reading interesting materials, (iii) reference materials, (iv) communication, and (v) problem solving. In this paper, he provides examples of sites that exemplify each of these categories.

Examples of well-established sites that could be used to support student learning include:

- the NRICH site, which offers a range of resources for students and teachers including weekly challenges and problems: www.nrich.maths.org/frontpage
- the MacTutor History of Mathematics archive, which includes biographies of mathematicians as well as historical material: www-groups.dcs.st-and.ac.uk/~history
- the Math Forum site, which offers problems, puzzles, online mentoring and teacher discussion areas: www.mathforum.org
- scootle, which is an Australian Government Department of Education initiative that provides access to digital resources in support of the *Australian Curriculum* (employment system login required): www.scootle.edu.au/ec/p/home

- the University of Western Sydney list of apps for iPads, which is maintained by Catherine Attard: www.materdei.qld.edu.au/downloads/TechiPadAppsAnd Mathematics.pdf.

Since there is a plethora of resources now available on the internet, it is important to develop strategies that enable you to evaluate the quality of online materials. Handal et al. (2005) provide a practical approach to evaluating online mathematics resources. Some websites (e.g., the STEMCrAfT Project, www.utas.edu.au/education/research/research -groups/maths-education/stemcraft-project) also provide lists of teacher-reviewed STEM resources.

There are also a number of sites that offer students the opportunity to work with others (students and teachers) around the world. An example is *Ask Dr. Math* (www.mathforum. org), a question and answer service in the form of a searchable archive. It is also possible to post specific questions to the site if a query does not match items in the archive. This service is part of the Math Forum site supported by Drexel University that exists for the purpose of 'people learning mathematics together'. To a similar end, the NRICH (www. nrich.maths.org/frontpage) website encourages students to submit solutions to particular problems and publishes a selection of them with students' names.

The internet is also a networking tool that makes it possible for many people to collaborate and combine their efforts to solve mathematical problems. The Math Forum provides a list of such projects, some of which are ongoing and some of which have been completed, with data available for analysis (www.mathforum.org/workshops/sum96/data. collections/datalibrary/lesson.ideas.html). One such project is the Global Grocery List in which students share local grocery prices to build a growing table of data to be analysed for the purpose of drawing and testing conclusions.

Taking a completely different direction, Wolfram|Alpha (www.wolframalpha.com) provides answers to mathematics problems (and solutions to questions from other disciplines as well) by completing dynamic computations based on a massive collection of data, algorithms and methods rather than just searching the internet. Solutions to mathematics problems are presented in different representations or forms when available, as shown in Figure 4.9.

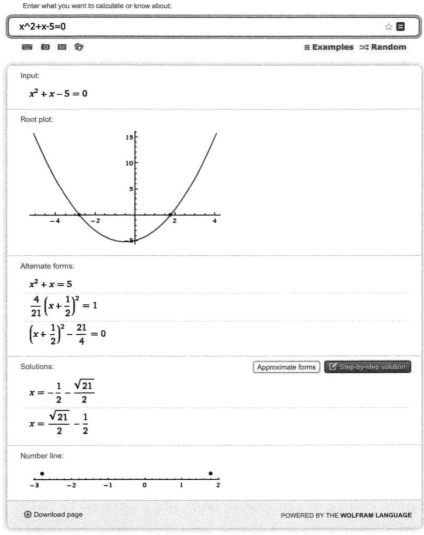

Source: © Wolfram Alpha.

Figure 4.9 **Solving a quadratic equation**

> **REVIEW AND REFLECT :** Do some research on approaches to evaluating the educational worth of internet sites (e.g., see Handal et al., 2005). Design an evaluation form and use it to evaluate some of the sites mentioned in this chapter.

Technology and curriculum

The introduction of new technologies into mathematics education inevitably raises questions about the kinds of knowledge and skills that are valued and worth keeping in the curriculum, and those that could perhaps be de-emphasised. However, this is not a new phenomenon. For example, manipulation of base ten logarithms, once taught as a labour-saving technique that enabled students to perform operations on large numbers, disappeared from the junior secondary curriculum when scientific calculators became widely available in schools in the mid-1970s.

> **REVIEW AND REFLECT :** Technology needs to be used intelligently with text resources, whether or not these resources have been designed with technology in mind.
>
> Many modern mathematics textbooks come with CDs and websites offering technology-based classroom activities. Evaluate some of these resources in the light of the potential benefits of learning mathematics with technology outlined above. To what extent do the CDs and websites add value to the textbook?
>
> Some schools may still be using older textbooks that do not explicitly support the integration of technology into mathematics education. Select a chapter from an older textbook and design appropriate technology-based activities to supplement or replace the existing text-based exercises and activities.

REVIEW AND REFLECT : Complete the following quiz (adapted from McKinlay, 2000) by indicating your preference along the continuum at the right of each statement.

Junior secondary mathematics students should be allowed to use digital technology (a graphics calculator, CAS calculator or computer software) to:

	Never	Always
Graph linear functions	←	→
Graph quadratic functions	←	→
Find the roots of quadratic equations	←	→
Perform linear regression	←	→
Perform quadratic regression	←	→
Calculate the mean of a set of raw data	←	→
Calculate the mean from a frequency table	←	→
Calculate the median of a set of raw data	←	→
Calculate the median from a frequency table	←	→
Construct a histogram	←	→
Write and run simple programs	←	→
Download programs from the internet	←	→
Expand and simplify algebraic expressions	←	→
Factorise algebraic expressions	←	→
Solve equations	←	→
Differentiate algebraic expressions	←	→
Integrate algebraic expressions	←	→
Solve differential equations	←	→
Manipulate and complete operations with matrices	←	→
Manipulate and complete operations with vectors	←	→
Draw geometric shapes	←	→
Verify geometric properties or relationships	←	→
Calculate with number including fractions	←	→
Measure angles and other attributes of geometric figures	←	→

Compare and discuss your responses with your group. How did individuals justify their opinions about student use of digital technology?

Recent discussions among Australian mathematics educators converge on the view that curricula must continue to focus on important concepts and skills of mathematics while emphasising mathematical communication and reasoning, and that good curriculum design should take advantage of all kinds of technology as tools for learning (Goos, 2012). While technology should not drive curriculum, it can certainly influence what mathematics is taught. There are three ways in which this might happen.

What can be omitted?

Some skills and procedures that can be performed using technology may require less emphasis, become optional or become redundant—especially if, in the past, they were the only methods available. For example, knowing how to manually complete the square is no longer the only way for pre-calculus students to find the turning point of a quadratic (see Figure 4.10a) and finding logarithmic solutions to exponential equations might not be considered essential when they can be obtained graphically (see Figure 4.10b).

Kissane (2001) has argued that, in a CAS environment, the traditional focus on teaching students how to manipulate expressions and solve equations algebraically should give way to a greater emphasis on helping students to express relationships algebraically, formulate equations and interpret the solutions. Some (e.g., Oates, 2009) have even gone as far as to say that content areas such as algebraic manipulation might become redundant, with the focus of teaching and learning on deeper and richer understanding of algebraic ideas.

What can be added?

In some cases, the curriculum can be extended to remove previous limitations—for example, having access to graphics calculators or computer software makes it possible to work with non-integer coefficients for quadratics, combinations of functions and large data sets. New approaches can also be introduced to tackle tasks that would not otherwise be accessible to students: curves can be fitted to real data, iterative procedures can provide scope for numerical analysis of problems not amenable to algebraic methods, and financial scenarios can be investigated using the graphics calculator's

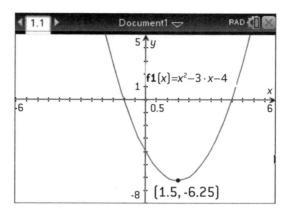

Figure 4.10a **Finding the turning point of a quadratic function**

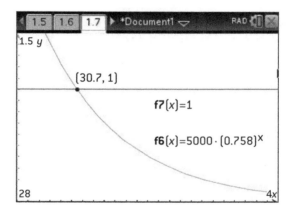

Figure 4.10b **Solving the equation 5000 × (0.758)^x = 1**

time-value-money module. Opportunities arise to investigate more complex application and modelling problems that demand consideration of assumptions, decisions about the appropriate degree of accuracy, and evaluation of the validity of models (e.g., Geiger, 2011).

How can the sequencing or treatment of topics be changed?

Access to graphics calculators makes it possible to use graphical approaches to build understanding before moving into analytical work, such as in solving systems of simultaneous equations. Before learning the standard algebraic solution techniques, students can first gain experience in drawing the graphs of two straight lines and finding their point of intersection, thus reinforcing the key concept that there is only one point whose coordinates satisfy both equations simultaneously. Similarly, graphical treatment of simple optimisation problems—such as finding the maximum area of a rectangle with fixed perimeter—makes these ideas accessible to junior secondary students without the need to invoke calculus concepts (see the box below). For senior secondary students, the availability of CAS facilities also makes it possible to explore models of real-life phenomena (including the use of differential equations) ahead of complex techniques of analysis that might otherwise be necessary (Oates, 2009). Such an approach also has the potential to provide students with a purpose for studying the mathematics 'behind' what can be facilitated with technology.

Finding the maximum area of a rectangle with fixed perimeter

I have 20 metres of wire with which to fence a rectangular garden. What are the dimensions of the largest area that can be enclosed?

A graphics calculator can be used to generate data on lengths and corresponding breadths and areas (Screen 1). The relationship between area and length is represented via a scatterplot (Screen 2) and the area function is plotted over these points (Screen 3). The maximum area is found by tracing along the curve or querying the calculator directly (Screen 4).

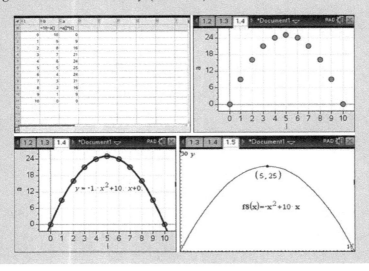

Although we have argued that the use of technology can have a positive effect on students' learning of mathematics, teachers also need to recognise that the features and functionalities built into new technologies can influence the curriculum in unanticipated and perhaps undesirable ways. To illustrate this point, consider the capacity of graphics calculators to fit regression model equations to data stored in lists and to calculate the corresponding R-squared values as a measure of goodness of fit (as in the cooling curve example shown in Figure 4.8a). Students who simply try out a range of regression models and use only the R-squared value to determine the most appropriate model are relying on the calculator's 'black box' algorithms rather than mathematical reasoning, and they risk choosing a model that does not make sense (e.g., in the case of a quadratic model that predicts a cup of hot water will boil after cooling). Without skilful teacher intervention, these easily accessible calculator functions can encourage students to take a purely empirical approach without learning how to justify their reasoning. Taking an even broader perspective on curriculum design, it is important for control of the curriculum to remain in the hands of educators rather than being unduly influenced by commercial interests or entities that produce educational resources. (Issues regarding curriculum content and curriculum decision making are explored further in Chapter 5.)

Technology and pedagogy

Many research studies have investigated the effects of technology usage on students' mathematical achievements and attitudes and their understanding of mathematical concepts, but less is known about how students actually use technology to learn mathematics in specific classroom contexts or about how the availability of technology has affected teaching approaches. In this section, we draw on our research in Australian secondary school mathematics classrooms to describe various modes of working with technology by using the metaphors of technology as *master, servant, partner* and *extension of self* (Geiger, 2006, 2009; Goos et al., 2003).

Teachers and students can see technology as a *master* if their knowledge and competence are limited to a narrow range of operations. In fact, students can become dependent on the technology if their lack of mathematical understanding prevents them from evaluating the accuracy of the output generated by the calculator or computer. As one student commented, 'sometimes you learn a technique using technology that you don't really understand, and then you don't grasp the concept'.

The way in which technology could prove the *master* for teachers became clear during observations of one of the research project classrooms. This teacher admitted very little expertise with using a graphics calculator, to the extent that he regularly called on a student 'expert' to demonstrate calculator procedures via the overhead projection panel. While the teacher lacked confidence in the use of technology, he nevertheless retained tight control of the lesson agenda through the medium of the student presenter, often providing the mathematical commentary and explanations accompanying the student's silent punching of the calculator keys. Due to syllabus and research project expectations, this teacher felt obliged to include technology-based learning activities in his lessons; however, his own lack of knowledge and experience in this area made him reluctant to allow students to use technology.

Technology is a *servant* if used by teachers or students only as a fast, reliable replacement for pen-and-paper calculations. For example, students in our research project commented that technology helped with large and repetitive calculations, allowed them to calculate more quickly and efficiently, reduced calculation errors, and was useful in checking answers. From the teacher's perspective, technology is a *servant* if it simply supports preferred teaching methods—for example, if the interactive whiteboard provides a medium for the teacher to demonstrate calculator operations to the class. We observed a more creative approach in one of the research project classrooms during a Year 11 lesson on matrix transformations. Students were supplied with the worksheet in Figure 4.11, and the teacher physically demonstrated the results of several matrix

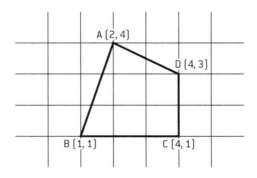

Figure 4.11 **Matrix transformation task**

transformations using transparent grid paper, plastic cut-out polygons and the over-head projector.

Students then investigated further with their own polygons and grid paper by record-ing the coordinates of the vertices before and after transformation, with the graphics calculator taking care of the matrix calculations so that conjectures on the geometric meaning of the transformations could be formulated and tested. Here, the technology became an intelligent servant that complemented the effective features of more conven-tional instruction.

Technology is a *partner* if it increases the power that students exercise over their learning by providing access to new kinds of tasks or new ways of approaching existing tasks. This may involve using technology to facilitate understanding or to explore different perspec-tives. Students participating in our research have commented that 'by displaying things in different ways, [technology] can help you to understand things more easily', and that 'technology may help you approach problems differently in the sense that you can visu-alise functions'.

Technology can also act as a partner by mediating mathematical discussions in the classroom—for example, when teachers and students use the interactive whiteboard to present and examine alternative mathematical conjectures. This is illustrated by the prac-tice we observed in one classroom of inviting students to compare and evaluate graphics calculator programs they had written to simplify routine calculations, such as finding the angle between two three-dimensional vectors $r_1 = \begin{pmatrix} a \\ b \\ c \end{pmatrix}$ and $r_2 = \begin{pmatrix} d \\ e \\ f \end{pmatrix}$ using the formula:

$$\vartheta = \cos^{-1}\left(\frac{r_1 \cdot r_2}{|r_1||r_2|}\right) = \cos^{-1}\left(\frac{ad+be+cf}{\sqrt{a^2+b^2+c^2}\sqrt{d^2+e^2+f^2}}\right)$$

The teacher provided only minimal instruction in basic programming techniques and expected individual students to consult more knowledgeable peers for assistance. Volunteers then demonstrated their programs via the overhead projection panel and examined the wide variation in command lines that peers had produced (see Figure 4.12 for examples).

This public inspection of student work also revealed programming errors that were challenged and subsequently corrected by other members of the class. When interviewed

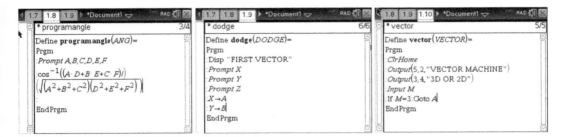

Figure 4.12 **Different student programs for finding angle between vectors**

after the lesson, students commented that programming helped them to develop a more thorough understanding of the underlying mathematical concepts, especially when there were opportunities to compare programs written by different people.

Technology becomes an extension of self when seamlessly incorporated into the user's pedagogical or mathematical repertoire, such as through the integration of a variety of technology resources into course planning and the everyday practices of the mathematics classroom. For students, this is a mind-expanding experience that accords them the freedom to explore at will. They explained this by saying that 'technology allows you to expand ideas and to do the work your own way', and 'it allows you to explore and go off in your own direction'. We observed such sophisticated use by teacher and students in a lesson involving use of iterative methods to find the approximate roots of a cubic equation. Students worked through a task in which they constructed spreadsheets to investigate whether or not the iteration process converged on a solution (as in Figure 4.13), with some also deciding to use function-plotting software to create an alternative, graphical representation of the problem.

This was a challenging task, and few students found all three roots of the equation. When one group of students did so, the teacher made a spur-of-the-moment decision to ask them to present their solution to the whole class via a laptop computer and data projector. With no time for rehearsal, the students shared the tasks of operating the computer keyboard, data projector remote control (which permitted scrolling and zooming independently of the computer) and laser pen, while coordinating their explanations and answering questions from their peers. Mathematical and communications technologies were smoothly incorporated into their unfolding argument, and were used to link

	A	B	C	D	E
1	x^3-8x-8=0 rearranged as x=x^3/8-1			x	f(x)
2				-1.5	-1.421875
3	x	f(x)		-1.421875	-1.35933065
4	-1.5	=(1/8)*((A4)^3)-1		-1.35933065	-1.31396797
5	=B4	=(1/8)*((A5)^3)-1		-1.31396797	-1.28357266
6	=B5			-1.28357266	-1.26434517
7				-1.26434517	-1.25264283
8				-1.25264283	-1.24569243
9				-1.24569243	-1.24162535
10				-1.24162535	-1.2392664
11				-1.2392664	-1.23790526
12				-1.23790526	-1.23712221

Figure 4.13 **Spreadsheet method for solving equation**

numerical and graphical representations of the equation-solving task, and to clarify and elaborate on points raised by fellow students and the teacher.

These examples show that introducing new mathematical and communication technologies into classrooms can change the ways that knowledge is produced. Implicit in these changes are a number of challenges for teachers, the most obvious of which involves becoming familiar with the technology itself. While this is important, some attention also needs to be given to the inherent mathematical and pedagogical challenges in technology-rich classrooms if the goal of a problem-solving and investigative learning environment is to be realised. For example, placing a digital tool in the hands of students gives them the power and freedom to explore mathematical territory that may be unfamiliar to the teacher and, for many teachers, this challenge to their mathematical expertise and authority is something to be avoided rather than embraced.

Other research reports that the introduction of technology into the mathematics classroom not only changes the nature of activities designed for learning but also the ways in which teachers and students interact (e.g., Heid, Thomas & Zbiek, 2013). In particular, technology-rich environments have the potential to promote more collaborative ways of working between students and between teachers and students (Beatty & Geiger, 2010;

Gadanidis & Geiger, 2010). For example, the ability to represent a function that models real-world data on a screen more readily facilitates interactions in which students and/or teachers seek and offer opinions or conjectures, provide suggestions for improvement, or share ideas about different directions to explore.

Other technology-enabled ways of working and collaborating on mathematics are emerging that will require careful thinking about managing students' learning. Increasingly, students and teachers are using elements of online learning, a format within which peer interaction requires promotion and management by the 'classroom' teacher (Mojica-Casey et al. 2014). For example, apps such as Padlet (www.padlet.com) allow students and teachers to share readings and other stimuli via a 'web wall' in order to begin collaborative discussions that draw upon class members' opinions, ideas and conjectures. Online systems are also being used by teachers to gain feedback about students' progress by real-time questioning and result aggregation (e.g., Socrative, www.socrative.com). Used effectively, such systems allow for the adaptive teaching that targets areas that students are finding difficult before moving on to new concepts and skills.

Schools are also trialling 'flipped classroom' approaches to teaching and learning, an alternative to traditional classroom instruction in which students typically prepare for class by working on resources before class (e.g., worksheets, investigations, online video material, teacher-generated video of how to solve problems) that have been placed in an online repository. *Flipping* the classroom has the potential to optimise teaching and learning time in class by permitting the teacher to assist with difficulties individuals or groups of students experience with preset work, rather than focus on whole-class presentation of examples and procedures. Thus, this approach can provide opportunities for teachers to adopt more student-centred teaching (Muir & Chick, 2014). For most teachers, however, adopting a flipped classroom approach requires a significant shift away from traditional teaching practices, and thus careful preparation is required (Muir & Geiger, 2015).

Perhaps the most significant challenge for teachers when adopting a technology-enabled or supported approach to teaching lies in sharing control of the technology with students and orchestrating the resulting classroom discussion.

REVIEW AND REFLECT : View these three videos about the way in which these mathematics teachers use digital technologies in their teaching of mathematics.

- *Dynamic geometry software*, AITSL illustration of practice, www.aitsl.edu. au/australian-professional-standards-for-teachers/illustrations-of-practice/ detail?id=IOP00175
- *GPS as a tool for teaching spatial awareness*, AITSL illustration of practice, www.aitsl.edu.au/australian-professional-standards-for-teachers/ illustrations-of-practice/detail?id=IOP00337
- *Maths on YouTube*, AITSL illustration of practice, www.aitsl.edu.au/ australian-professional-standards-for-teachers/illustrations-of-practice/ detail?id=IOP00173

How do each of these teachers use digital technologies for their teaching of mathematics? What are their beliefs about effective teaching of mathematics? How does their application of digital technologies reflect these beliefs?

Thinking about mathematics classes from your school experience, which of these approaches would you use? Why or why not?

Technology and assessment

The issue of using digital technologies in summative assessment that is assessment *of* learning is a contentious one. Opinions are divided on how freely available technology should be when students are expected to demonstrate their learning. The AAMT (2014) *Position Paper on Digital Learning in School Mathematics* adopts a balanced view on this matter in recommending:

- All students have ready access to appropriate technology and the associated virtual learning environments and communities that support and extend their mathematics learning experiences.
- Teachers at all levels are actively involved in exploring ways to take full advantage of the potential of technology for mathematics learning within the total curriculum. Professional learning should focus on new pedagogies for teaching mathematics.

- Students will be assessed in their understanding of mathematics both with, and without, digital technologies, and in their awareness of when to choose to use digital technologies.
- Education authorities make available to teachers, professional development opportunities to support the development of knowledge and skills necessary for the successful use of technologies in classrooms.`

Thus, it is expected that students have access to digital technologies that support their mathematics learning experiences and these technologies are also available to students during assessment when appropriate.

While trials of innovative practices involving digital technologies continue, for example, the use of computer-based interactive situations (Neumann et al., 2012), in practice, most recent developments in assessment policy have been concerned with access to and use of graphics calculators and CAS calculators or other devices that permit access to online resources via the internet in high-stakes assessment in the senior secondary years (Years 11 and 12).

In states with external examinations that assume access to graphics calculators, there has been a great deal of interest in the nature of assessment items—especially with respect to how these items may have changed since graphics calculators were introduced, the extent to which they require use of technology, and how technology might affect the solution method. Mueller and Forster (2000) analysed the Western Australia Tertiary Entrance calculus examination before and after government-mandated implementation of graphics calculators and found that the percentage of algorithmic, procedurally oriented questions declined but there was an increase in the number of applications questions in which technology could take care of complex calculations. In Victoria, Flynn and colleagues (Flynn & Asp, 2002; Flynn & McCrae, 2001) have investigated the impact of permitting CAS calculators on the kind of questions asked in the Victorian Certificate of Education examinations. Queensland has a long history of school-based assessment and teacher ownership of curriculum, and thus there is greater diversity of practice in technology use.

More recently, there has been research into the influence of using CAS on students by-hand skills. In Victoria, where CAS is an expected technology in some courses, it was found that there were positive influences of CAS on student achievement with little evidence for the diminishing of by-hand skills (Evans et al., 2008; Leigh-Lancaster, Les & Evans, 2010).

Researchers have also been interested in investigating differences between levels of calculator use in assessment. Kissane, Bradley & Kemp (1994) distinguished between three choices that could be made regarding calculator use in formal assessment: required (assumes all students have access); allowed (some students will not have access); or disallowed. They later developed a typology of student use of graphics calculators in examinations and within courses, shown in Table 4.1, which can be used to design examinations that capitalise on the capabilities of this technology. This typology is also applicable to more powerful handheld technologies that have CAS as an available facility. Other examples of schemes for classifying examination questions are reported in Flynn and McCrae (2001).

Table 4.1 A typology of expected use of graphics calculators in assessment

Calculator use is expected

1. Students are explicitly advised or even told to use graphics calculators.
2. Alternatives to graphics calculators are very inefficient.
3. Graphics calculators are used as scientific calculators only.

Calculator is used by some students, but not by others

4. Use and non-use of graphics calculators are both suitable.

Calculator use is not expected

5. Exact answers are required.
6. Symbolic answers are required.
7. Written explanations of reasoning are required.
8. Task involves extracting the mathematics from a situation or representing a situation mathematically.
9. Graphics calculator use is inefficient.
10. Task requires that a representation of a graphics calculator screen will be interpreted.

> **REVIEW AND REFLECT :** Find out about policies regarding use of technology for assessment in the senior secondary mathematics subjects taught in your state or territory. For which subjects is technology use required, allowed or disallowed? For what kind of assessment tasks is technology use required, allowed or disallowed? What kind of technology is covered by these policies?
>
> Use the typology provided in Table 4.1 to analyse a Year 11 or Year 12 mathematics examination paper. Comment on the balance you find between the three types of technology use expected.
>
> Investigate recent research on the impact of CAS technology on mathematics assessment. Compare the schemes for classifying examination questions reported by Flynn and McCrae (2001) with the typology provided in Table 4.1 (which was devised for graphics calculators without CAS capabilities). Use Flynn and Asp's (2002) scheme to analyse an examination paper that permits the use of CAS.
>
> Find out about schools' policies on the use of technology for assessment in junior secondary mathematics. How is students' work in mathematics expected to contribute to their learning about technology? How do expectations about technology use in mathematics compare to those in other subjects? Conduct an inquiry at your school and compare your findings with those of your pre-service colleagues.

The requirement for students to use graphics calculators—including those with CAS capabilities—in formal summative assessment tasks has given rise to many other issues beyond that of level of use discussed above. One issue concerns our expectations as to what students should record to 'show their working' with a calculator (Ball & Stacey, 2003). A list of keystrokes is usually not very helpful, so what might be a reasonable alternative? This is really a question about the forms of mathematical representation and communication we value and want our students to demonstrate. One approach is to require students to show enough work so that the reasoning processes can be followed throughout the solution (United States College Board, 2016). In practice, this means that students must show the mathematical set-up (e.g., the equation to be solved or graphed or the derivative to be evaluated) and steps that lead to the solution, in addition to results produced by the calculator. Whatever approach is adopted, it must take account of the range of techniques used to solve problems within CAS-active classrooms where students have been reported to have very personal preferences for using different combinations of technology and pen-and-paper techniques (Cameron & Ball, 2015). See Chapter 6 for further discussion of assessment.

Challenges in creating technology-rich mathematics learning environments

Thus far, we have painted a very positive picture of technology use in mathematics class-rooms in terms of benefits for student learning and innovative teaching approaches. Yet, there are many challenges in creating technology-rich mathematics learning environments. Research in this area has identified a range of factors that influence whether and how mathematics teachers use technology: their skill and previous experience in using technology; time and opportunities to learn (pre-service education, professional development); access to hardware and software in the school; availability of appropriate teaching materials; technical support; curriculum and assessment requirements; institutional culture; knowledge of how to integrate technology into mathematics teaching; and beliefs about mathematics and how it is learned (e.g., Attard & Orlando, 2014; Bate, Day & Macnish, 2013; Goos & Bennison, 2008; Pierce & Stacey, 2013; Zuber & Anderson, 2013). We can classify these factors as being related to the teacher's *knowledge and beliefs*, the *professional context* or *sources of assistance*, as shown in Table 4.2. Let us consider each of these categories in turn, and

Table 4.2 Factors influencing technology use in mathematics education

Knowledge and beliefs	Skill/experience in working with technology
	Pedagogical knowledge (technology integration)
	General pedagogical beliefs
Professional context	Access to hardware, software, teaching materials
	Technical support
	Curriculum and assessment requirements
	Students (perceived abilities, motivation, behaviour)
Sources of assistance	Pre-service education (university program)
	Practicum and beginning teaching experience
	Professional development

identify implications for you as a mathematics teacher. (We will return to these categories in Chapter 17 when we examine professional learning and development.)

Many studies have demonstrated close links between teachers' *knowledge and beliefs* about learning, their pedagogical practices and their orientation towards technology. Teachers with a transmission view of mathematics learning tend to display teacher-centred practices and to use technology mainly for calculation or demonstration, while those with constructivist beliefs take a more learner-centred approach and use technology for concept development and mathematical exploration. You might like to revisit your responses to the quiz in the section on technology and curriculum (above), as these reflect your own pedagogical knowledge and beliefs.

Factors related to *sources of assistance* are very relevant to your experience as a pre-service teacher, but because technology changes so rapidly, you will need to continually update your knowledge throughout your teaching career. Professional development in technology in mathematics education is offered by mathematics teacher associations and through journals, websites and conferences (see Chapter 17 for more information).

Often, it seems that the teaching *context* plays an overriding role in supporting or hindering teachers' efforts to create technology-rich learning environments. We have already discussed the impact of curriculum and assessment policies on teachers and students' use of handheld technologies. By mandating, permitting or prohibiting handheld technologies in different mathematics subjects, these policies can also influence which students get to use technology. For example, it is common to find higher use among senior secondary classes taking advanced mathematics subjects because this is required by the relevant syllabuses. Conversely, younger students and those enrolled in lower-status mathematics subjects that do not lead to tertiary study are often disadvantaged because access to technology is not given a priority by the school for these classes.

From a practical point of view, gaining access to technology resources for the classes you teach may be a significant challenge for you as a mathematics teacher. In some schools, students will have ready access to technology such as laptop computers or iPads. However, in others, classrooms may not be equipped with computers or data projectors, and access to computer labs or class sets of laptops or handheld technologies may be limited. There may also be restricted access to the internet. In these circumstances, you need to be realistic about what is possible and concentrate on what you can do with the resources you have.

Alejandre (2005) has identified some common situations faced by mathematics teachers who want to incorporate technology into their classroom practice, and she makes the following suggestions on how to make the best of them.

If you have *one computer (desktop or laptop) in your classroom, without internet access,* you can use the computer:

- As a reference station—ask a student to check some facts related to the task that the class is completing by using CD-ROMs.
- To accompany and enhance a mathematical investigation lesson—organise a task in which students have to collect data (e.g., to investigate the relationship between their height and their arm span) and then enter it into a spreadsheet that can be used to summarise and analyse trends across the whole class.
- As one of several stations or task centres within a classroom 'menu'—arrange the classroom so students work in groups of three or four on each activity.

If you have *one computer (desktop or laptop) in your classroom, with internet access,* you can use the computer in much the same way as described above, but its reference capacities are now expanded because you can give students access to the type of mathematical websites we discussed earlier in the chapter.

If you have *one computer (desktop or laptop) in your classroom and a data projector* or some other kind of display, you can use the computer:

- To introduce or reinforce concepts by displaying and manipulating dynamic images and objects, such as drawings, diagrams or graphs, that you might otherwise have presented via an overhead projector or the whiteboard.
- To introduce work the students will have to do when they go to the computer laboratory.

If you have *a cluster of four to six computers in your classroom,* you can use the computers:

- As a reference area—if students are working on a task in small groups, one student from each group could go to the computers to search for information to help them complete the task (as there are likely to be more than four to six groups, this strategy could also accommodate pairs of students at each computer).

- As stations or task centres offering different technology-based activities through which groups of students rotate.

If you have no computer in your classroom and have to *book a computer laboratory*, be sure to have a back-up plan in case unforeseen problems occur. You may find it easier to use graphics calculators, which can perform many of the functions of mathematical software.

REVIEW AND REFLECT : Work with a partner to develop a corresponding list of suggestions for teachers who have limited or uncertain access to digital technologies. What strategies can you suggest in the following circumstances?

- You have to book a class set of graphics calculators/iPads/laptops for each lesson in which you plan to use them, and there are several class sets in the school.
- There are only one or two class sets of graphics calculators/iPads/laptops, and they are usually reserved for the senior classes.
- Access to digital technologies is very difficult, but you have your own calculator/iPad/laptop and a display (data projector).
- There are a couple of class sets of old-model graphics calculators.

Conclusion

Early recommendations about preparation for teaching with technology assumed that teachers needed only general technological literacy. We now recognise that knowing how to use computers and other forms of technology is not the same as knowing how to teach effectively with technology, since PCK is required to integrate technology into the curriculum in specific subject domains, such as mathematics. PCK, which enables teachers to create mathematical representations that connect students with the subject matter, is at the heart of teaching effectively with technology, and is embodied in such metaphors for technology as *master, servant, partner* and *extension of self* that we introduced in this chapter. The opportunities that teachers provide for technology-enriched student learning are also affected by ways in which they draw on their own knowledge and beliefs about the role of technology in mathematics education, and by how they interpret aspects of their teaching contexts that support or hinder their use of technology.

Recommended reading

Ball, L. & Stacey, K. (2003). What should students record when solving problems with CAS? In J. Fey, A. Cuoco, C. Kieran, L. McMullin & R. Zblek (Eds), *Computer algebra systems in secondary school mathematics education* (pp. 289–303). Reston, VA: NCTM.

Goos, M., Galbraith, P., Renshaw, P. & Geiger, V. (2003). Perspectives on technology mediated learning in secondary school mathematics classrooms. *Journal of Mathematical Behavior*, 22, 73–89.

Handal, B., Handal, P. & Herrington, T. (2005). Evaluating online mathematics resources: A practical approach for teachers. In M. Coupland, J. Anderson & T. Spencer (Eds), *Making mathematics vital* (Proceedings of the 20th biennial conference of the Australian Association of Mathematics Teachers, pp. 153–165). Adelaide: AAMT.

Hoyles, C. & Lagrange, J.-B. (2010). Introduction. In C. Hoyles & J.-B. Lagrange (Eds), *Mathematics education and technology—Rethinking the terrain* (pp. 1–11). New York: Springer.

Kissane, B. (2009). What does the Internet offer for students? In C. Hurst, M. Kemp, B. Kissane, L. Sparrow & T. Spencer (Eds) *Mathematics: It's mine* (Proceedings of the 22nd biennial conference of the Australian Association of Mathematics Teachers, pp. 135–144). Adelaide: AAMT.

Resources and websites

Casio: www.casio.edu.shriro.com.au

Texas Instruments: https://education.ti.com/en-GB/aus-nz/home

WolframAlpha: wolframalpha.com

CHAPTER 5
Mathematics curriculum models

From the 1870s when public education systems were established in the Australian colonies until the early 21st century, the official curriculum was a state-based curriculum rather than a national curriculum prescribed by a central authority. Under the Australian constitution, education is the responsibility of state and territory governments rather than the federal government. However, the recent emergence of a national, rather than a state-based, education policy environment has paved the way for development of a national curriculum for Australian schools in the primary and secondary years. Nevertheless, states and territories still have some say in how and when they will implement the *Australian Curriculum* up to Year 10. In addition, in the senior secondary years, state-based curriculum authorities are responsible for the structure and organisation of their senior secondary courses and for determining how they will integrate the *Australian Curriculum* content and achievement standards into their courses. Due to the rapidly changing curriculum landscape in Australia, it is beyond the scope of this chapter to explore the content and structure of secondary school mathematics curricula in the *Australian Curriculum* as well as its implementation in each state and territory; nor is this the place to discuss general theories of curriculum development and change. Instead, we focus on some of the 'big ideas' behind curriculum decision making in the context of Australian mathematics education. The first part of the chapter lays the groundwork by introducing some fundamental curriculum concepts. Next, we consider how decisions are made about what mathematics should be taught in secondary schools, and we provide a brief historical overview of mathematics curriculum development in Australia.

Finally, we describe different models for organising the secondary school mathematics curriculum.

Curriculum concepts

Curriculum can be defined in many ways, with some definitions referring only to educational intentions and others to the reality of what actually happens in schools. For example, some people would say that a curriculum is a plan for learning (the intention), while others would see curriculum as the set of all educational experiences offered to students by teachers and schools (the reality). Stenhouse (1975) brought these two ideas together in his definition of curriculum as 'an attempt to communicate the essential principles and features of an educational proposal in such a form that it is open to critical scrutiny and capable of effective translation into practice' (p. 4). *Curriculum* can also be represented in different ways according to the perspectives of the various participants in curricular activities (Cai & Howson, 2013):

1. At the system level, the intended curriculum represents the vision laid out by the curriculum designers in written curriculum documents and materials.
2. At the classroom level, the implemented curriculum represents teachers' interpretation of the formal written documents and the way they enact this in the classroom.
3. At the student level, the attained curriculum represents the learning experiences as perceived by students as well as what students actually learn.

Stein, Remillard and Smith (2007) use slightly different terminology to insert another level between the system-level curriculum document and its classroom implementation. In doing so, they distinguish between the formal *written* curriculum and the *intended* curriculum that refers to the teacher's plans for instruction. Many factors mediate the transformation of the written and intended curricula into the enacted and attained curricula. These include:

- teachers' mathematical knowledge and their beliefs about mathematics and how it is learned
- teachers' professional identities that develop from a sense of roles and relationships with others

- teacher perceptions of students and their abilities and motivations
- organisational and policy contexts that affect the time and support available for planning and interaction with colleagues.

A curriculum also has a number of components that address the purpose, content, organisation and assessment of student learning. Van den Akker (2003) provides the following list of essential curriculum components and the questions they address:

- Rationale: what educational purposes and principles underpin the curriculum?
- Aims and objectives: towards which specific learning goals are students working?
- Content: what are students learning and how is this sequenced?
- Learning activities: how are students learning?
- Teacher role: how is the teacher facilitating learning?
- Materials and resources: with what are students learning?
- Grouping: how are students allocated to various learning pathways and how are they organised for learning within the classroom?
- Location: what are the social and physical characteristics of the learning environment?
- Time: how much time is available for specific topics and learning tasks?
- Assessment: how shall we know how far learning has progressed?

The relevance of these individual components will vary according to whether curriculum planning and implementation take place at the level of the education system, the school or the classroom. For example, system-level curriculum documents usually pay most attention to the rationale, aims and objectives, and content, and often provide suggestions for content sequencing and time allocation. At the classroom level, the individual teacher is typically most concerned with learning activities, teacher role, and materials and resources. All ten components need to be addressed coherently and systematically in school-level curriculum planning and implementation. It is also important to maintain a close alignment and balance between all components, even though the process of curriculum change may emphasise only some specific components at a particular time (e.g., through the introduction of new learning goals, new content or new assessment approaches).

What mathematics should we teach, and why?

Education is concerned with selecting and making available to the next generation those aspects of culture—knowledge, skills, beliefs, values and customs—thought by our society to be most worthwhile. However, people can vary in their opinions as to what constitutes worthwhile curriculum content.

REVIEW AND REFLECT : What mathematics do you think that all students should know and be able to use after ten years of schooling? Make a list of what you consider the mathematical topics, concepts, skills and ways of thinking that are critical for students to succeed in further education, training, employment and adult life beyond Year 10. Give reasons to support your selection.

In pairs or small groups, compare your list with other pre-service teachers. Comment on reasons for any similarities and differences between your lists.

Compare your list with the content structure of the *Australian Curriculum: Mathematics* or the version of the national curriculum implemented in your own state or territory.

Democratic access to powerful mathematical ideas

The task above challenges teachers to justify their selection of 'essential' mathematics. Curriculum choices need to be founded on an understanding of why mathematics is important, especially in these times of rapid social, economic and scientific change. The *National Statement on Mathematics for Australian Schools* (Curriculum Corporation, 1991) gives four reasons why everyone—not only the select few—needs to understand and be able to use mathematics. Firstly, mathematics is used *in daily living*, whether this involves managing our home lives (measuring quantities for cooking, gardening, sewing, carrying out maintenance or repairs to buildings or vehicles), our personal finances (spending, saving, budgeting, taking out loans, planning for retirement), or our leisure activities (travelling, reading maps, playing games). Secondly, mathematics is necessary for intelligent *participation in civic life*. Often, this requires interpreting data in order to make informed decisions about economic, social, political, health or environmental issues. Thirdly, mathematics is used *at work*. Although a high level of mathematics is needed in certain professions (e.g., engineering, science, information technology, economics), a foundation of mathematical knowledge underpins a very wide range of careers in industry, trades, communication, design, planning and agriculture. Finally, mathematics is *part of our cultural heritage*—it is one of humankind's greatest intellectual and cultural achievements and deserves to be part of a liberal education for all.

Another way of looking at the question of what mathematics to teach, and why, comes from those who argue that it is a fundamental human right for all students to have democratic access to powerful mathematical ideas (Malloy, 2002). In the US, this ideal was exemplified by publication of the NCTM's (1989) *Curriculum and Evaluation Standards for School Mathematics*, and by the early 1990s, many countries around the world had developed strong national programs that emphasised the breadth and connectedness of mathematical content and the processes of mathematical thinking. The components of powerful mathematical ideas emphasised by these programs are reflected in the second version of the NCTM's curriculum guidelines, the *Principles and Standards for School Mathematics* (NCTM, 2000). This document outlines curriculum standards—statements of what mathematics teaching should enable students to know and do—for mathematical content in the areas of number, algebra, geometry, measurement, and data analysis and probability, and for the mathematical processes of problem solving, reasoning and proof, connections, communication and representation. But what is meant by 'democratic access' to these ideas? Malloy (2002) proposes that the literature on democratic education identifies four distinguishing characteristics that provide a rationale for democratic access to the curriculum. Firstly, a *problem-solving curriculum* should develop students' ability to draw on their mathematical knowledge to solve problems of personal and social relevance. Secondly, *inclusivity and rights* should be promoted by presenting mathematics from multiple perspectives that affirm the worth of individuals and groups from diverse backgrounds. Thirdly, there should be *equal participation in decisions that affect students' lives*, so that students use the classroom as a forum for public discussion of their own and others' ideas. Fourthly, students should experience *equal encouragement for success* through access to materials that develop critical habits of mind and engage them actively in learning mathematics. Many of these ideas are discussed in more detail in Part IV of this book (Chapters 13, 14 and 15), in which we examine issues of equity and diversity in teaching mathematics to all students.

Who makes curriculum decisions?

Van den Akker (2003) proposes that decisions about what to include in the mathematics curriculum (and what to exclude from it) may be influenced by three major orientations or their respective proponents (see Figure 5.1). The first of these orientations is represented by mathematics as an academic discipline that has its own cultural heritage, thus curriculum choices from this orientation are based on claims about the structure of the discipline and what counts as essential mathematical knowledge. For example, we might argue that an understanding of functions is essential because this is fundamental to the study of mathematical relationships and representations. The second orientation takes account of societal claims about relevant problems and issues, and curriculum choices from this orientation may be influenced, for example, by the needs of employers in the commercial, technical, financial and industrial sectors for workers who can apply mathematics to practical tasks and problems. A third orientation takes the learners' perspective in emphasising curriculum content and learning experiences that are personally meaningful, challenging and intrinsically motivating. From this orientation, curriculum choices might reflect a desire to help learners become confident and critical users of mathematics in their everyday lives. The rationale of a mathematics curriculum may reflect all of these orientations. However, in practice, curriculum decisions often involve compromises to accommodate the interests of their various proponents.

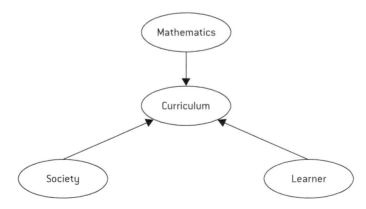

Figure 5.1 **Sources of influence on mathematics curriculum content**

There are many potential stakeholders involved in developing mathematics curricula in Australia. The following list is adapted from Harris and Marsh (2005, p. 19):

- Ministers for Education (state and territory, federal)
- federal agencies such as the Department of Education
- Department of Education senior officers (state and territory)
- Catholic Education senior officers (state and territory)
- Independent School senior officers (state and territory)
- ACARA
- state and territory curriculum and assessment authorities
- teachers' unions (state and territory, federal)
- school councils
- principals
- teachers
- parent organisations
- students
- university academics (mathematicians, mathematics educators)
- employers
- business organisations
- mathematics teacher professional associations (such as the AAMT) and mathematics education research organisations (the Mathematics Education Research Group of Australasia)
- textbook writers and publishers
- Education Services Australia
- ACER
- media
- educational consultants
- lobby groups.

The level of influence and activity of these potential stakeholders varies over time and in different contexts. Harris and Marsh (2005) argue that their roles can be understood by examining a high control model of curriculum decision making, which, although developed several decades ago (Rogers & Shoemaker, 1971), is consistent with the current emphasis on top-down accountability in education systems. This is known as the authority–innovation–decision-making model, in which stakeholders are divided into a superordinate group that makes the major decisions in initiating and directing the curriculum development process, and a subordinate group that implements the decisions made by the higher status group. The functions of these two groups are represented by Figure 5.2.

While the authority–innovation–decision-making model is useful in alerting us to the typically hierarchical nature of curriculum decision-making, it nevertheless underestimates the agency of teachers by overlooking the significant power that teachers have in implementing (or rejecting) change, and in participating in the change process through

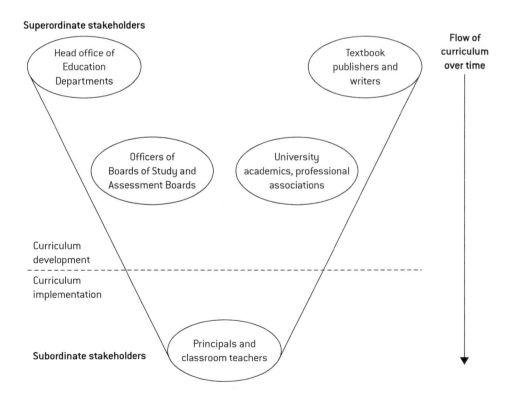

Figure 5.2 **Superordinate and subordinate functions in curriculum development and implementation**

their membership of professional associations, teacher unions, school clusters, and subject departments. Harris and Marsh (2005) suggest that the most important aspect of agency is that of translation: teachers exert significant agency in *translating* the intended curriculum into the implemented curriculum. Stein, Remillard and Smith (2007) go even further to claim that teachers have a role in *transforming* the curriculum into a form they believe is workable in their classrooms. The authority–innovation–decision-making model is also limited in that it recognises only the system contexts of curriculum (education departments and curriculum and assessment authorities, universities, textbook publishers, schools) but not the broader social and cultural context of curriculum change—represented by parents, students and employers—and the roles of the media and lobby groups in publicising and diffusing concerns about curriculum content and standards.

REVIEW AND REFLECT : Find out how mathematics curriculum decisions are made in your state or territory. What is the role of the local curriculum and assessment authority in relation to the national authority (ACARA)? What is its relationship to the state/territory education department and the Catholic and Independent schools sectors? What formal roles do teachers, parents, teacher unions, business organisations and university academics play in mathematics curriculum development?

Conduct a media survey over a semester to identify news items about mathematics education. What views are expressed about mathematics curricula? By whom are they expressed? To what extent are these views critical or supportive of current curricula?

Historical overview of mathematics curriculum development in Australia

School mathematics in Australia has been shaped by a range of forces since European settlement in 1788. Clements, Grimison and Ellerton (1989) have argued that colonialism was the major influence on mathematics education over the first 200 years, in that both the colonial power and those who were being colonised accepted without question that mathematics curricula and teaching methods should replicate British models. In England, the three main classes of society had access to different types of education: elite public schools for children of the aristocracy; grammar schools and public schools for children of the middle and professional classes; and charity schools for children of lower social class families. This pattern strongly influenced the development of school and university mathematics education in Australia.

In early colonial times in Australia (1788 to the mid-1850s), almost all children were of convict or lower social class origins and primary schools adopted curricula very similar to those used in charity schools in England. Children were taught reading, writing and arithmetic, with boys additionally being offered training in agriculture and the mechanical arts and girls receiving instruction in needlework. The scope of arithmetic, mirroring that taught in English schools, was limited to the four operations, vulgar fractions, an understanding of proportion, and simple applications of these skills to practical tasks. As the non-convict population grew, there was a corresponding increase in demand for education suited to children of the wealthy and the middle class. Schools established in response to this demand adopted curricula almost identical to the public and grammar schools in England, which emphasised classics (Greek and Latin) and mathematics (formal algebra and Euclidean geometry) in order to prepare the sons of the upper and middle classes for entry to English universities.

In later colonial times (mid-1850s to Federation), the newly established universities in the Australian colonies exerted a significant influence on school mathematics curricula through their control of matriculation and public examinations. Again, the colonial influence was evident: for example, when the universities of Sydney and Melbourne were founded in 1851 and 1853 respectively, they adopted academic programs comparable to those offered by Oxford and Cambridge universities in order to claim equivalent status with these elite English institutions, and their entrance examinations were almost identical to those set in England in requiring candidates to pass in Greek, Latin, Euclidean geometry, algebra, arithmetic and English. Preparing students for these examinations became the overriding goal of secondary schools and, because the examination questions were narrow in scope and rewarded rote learning, the mathematics teaching in secondary schools was oriented towards reproduction of learned facts and procedures. However, it is important to note that during this period, fewer than 2 per cent of all Australian children attended secondary school, so most of the population had no access to mathematics education beyond the rudiments of arithmetic.

Despite the impetus for social and political reform brought on by Federation in 1901, mathematics education in Australia in the first half of the twentieth century continued to mimic English curriculum models and the mathematics taught in schools remained largely unaffected by developments in mathematics itself. This stagnation is reflected in the use, until at least the 1950s, of secondary school arithmetic, algebra and geometry textbooks published in England at the beginning of the twentieth century. Mathematics teaching was still formal and rigid, emphasising memorisation and drilling in routine skills in preparation for written examinations. Although the new Australian constitution had given responsibility for education to the states, mathematics curricula in schools across Australia still depended on English ideas and hence were remarkably similar.

The 'New Mathematics' movement of the 1960s brought an end to curriculum inertia (Pitman, 1989). The Soviet launch of *Sputnik* 1 in 1957 and the reaction to this perceived military threat by non-Communist Western nations, and especially the USA, provided the political context for major curriculum reform, resulting in the allocation of vast amounts of funding to curriculum development projects in the physical sciences and mathematics in order to prepare better scientists and engineers and thus re-establish military superiority over the Communist bloc countries. The intellectual foundations for the 'New Mathematics' movement lay in the work of the Bourbakists in France, who redefined mathematics in terms of abstract notions of sets, functions, axioms and formal logic. Translation of this view into school mathematics curricula was supported by university mathematicians and mediated by mathematics educators who saw connections between formal mathematical structures and prevailing cognitive developmental theories of learning. Secondary mathematics curricula in Australia, as in many other countries, changed to reflect a new view of mathematical knowledge as being based on axiomatic structures of formal logic. However, in practice, neither students nor teachers were able to cope with such abstract ideas and, by the mid-1970s, evidence was also emerging that the 'New Mathematics' curriculum not only failed to prepare students adequately for tertiary mathematics, but also left children without fundamental arithmetical skills.

Mathematics curriculum initiatives in Australia in the 1980s were influenced by:

- concerns that young people were unable to apply their mathematical knowledge and skills in real-world situations

- advances in technology (e.g., computers and calculators) that made it possible to reduce the emphasis on pen-and-paper computation
- economic changes that led to the demise of old industries and creation of new ones that needed new types of workers
- calls for development of a core mathematics curriculum accessible to all children (Pitman, 1989).

The formalism of 'New Mathematics' was replaced by a focus on mathematics as problem solving and the desire to achieve a better balance between equity and quality in mathematics education. These goals drove many international, national and state-based mathematics curriculum development projects in the 1980s. Perhaps the best known Australian initiative was the Mathematics Curriculum and Teaching Program (MCTP), which identified, documented and disseminated examples of good practice among mathematics teachers to encourage them to model their own practice on 'what worked' successfully in other classrooms (Lovitt & Clarke, 1992).

In this brief historical overview, it is possible to trace the effect of academic mathematics and mathematicians, changing social, political and economic conditions, and concerns for equity and relevance in the curriculum—the three broad orientations claimed by van den Akker (2003) to be the major influences on curriculum decision making. However, since the late 1980s, curriculum development in Australia has been a highly politicised issue because of unresolved tensions between the states' and territories' constitutional autonomy in education matters and the Commonwealth's desire for a national curriculum to achieve greater consistency between state-based education systems. (See Marsh, 1994, and Reid, 2005 for analyses of the politics of national curriculum collaboration.)

The Commonwealth began funding school education in 1963, and from this time until 1988, exerted only indirect influence on school curricula via funding for curriculum projects and resource development. More direct Commonwealth intervention began in 1988 when John Dawkins, Commonwealth Minister for Employment, Education and Training, issued a policy statement arguing for a single national curriculum framework that could be adapted by the states and territories to inform local syllabus development. The following year, the Australian Education Council, comprising Commonwealth and state Ministers of Education, announced the Hobart Declaration that endorsed agreed

national goals for schooling and launched the process of national collaborative curriculum development. By 1991, the curriculum had been organised into eight learning
areas (later known as key learning areas), and teams of writers were commissioned to
produce *Statements and Profiles* for each learning area. The *Profiles* described what students
were expected to know and do, and thus marked a shift from the conventional curriculum focus on content to be taught, or 'inputs', towards the desired 'outcomes' for
student learning. Within each *Profile*, the outcomes were arranged into content and process strands in a sequence that depicted progressive growth of student understanding. In
mathematics, the content strands were number, space, measurement, algebra, chance
and data, and the process strand was labelled 'Working Mathematically' (Australian
Education Council, 1994). By the time the national curriculum framework was submit-
ted to the Australian Education Council meeting of June 1993, the political complex-
ions of the state governments had changed and this, together with intensive lobbying
by mathematics academics and others in the mathematics education community who
were critical of the new curriculum, led ministers to vote to defer acceptance of the
Statements and *Profiles*. Ultimately, the national curriculum documents were referred
back to the states and territories for determination of the extent, nature and timing of
any implementation.

The initial failed attempt at a national curriculum nevertheless provided a common
framework that was adopted or adapted during the 1990s by all states and territories to
produce outcomes-based curricula in the key learning areas. This approach was endorsed
by the Adelaide Declaration on National Goals for Schooling in the 21st century, which
maintained the spirit of federalist collaboration.

In 2003, the prospect of national curriculum collaboration emerged once again when
Brendan Nelson, Commonwealth Minister for Education, Science and Training, issued
a strong call for removal of inconsistencies between state and territory curricula, school
starting ages, and Year 12 assessment procedures. At that time, the feeling among experts
in curriculum theory was that any new attempt at national curriculum collaboration
would fail unless it recognised the political realities of Australia's federal system, articulated a clear rationale and conceptual base that did not rely solely on claims about consistency, economic efficiencies or national identity, and engaged the professional community
to build a genuine constituency of support (Reid, 2005).

However, in 2008, the path towards a national curriculum was firmly set when the Commonwealth and state Ministers for Education issued the *Melbourne Declaration* (MCEETYA, 2008). The *Melbourne Declaration* gave a commitment to promoting world-class curriculum and assessment, requiring implementation of a national curriculum that was also to be specified at state, territory and local levels. In the same year, ACARA was established as an independent statutory authority to manage the development of the national curriculum and associated assessment and reporting structures. In 2015, the Education Council endorsed the *Australian Curriculum* in eight learning areas up to Year 10. Senior secondary subjects in English, mathematics, science, history and geography have also been endorsed by education ministers for integration into courses offered by state- and territory-based curriculum authorities. The shift from a state-based to a federalist approach to curriculum development gives the national government a much greater role while allowing some flexibility for states and territories in local implementation (Stephens, 2014).

How can the mathematics curriculum be organised?

The introduction of the *Australian Curriculum* is having a significant impact on standardising curriculum organisation to Year 10 in all Australian states and territories. However, in the senior secondary years, a range of mathematics subjects is typically offered to cater for the differing post-school pathways that students may follow. This remains the case for the senior secondary mathematics subjects that are part of the *Australian Curriculum*.

REVIEW AND REFLECT : Visit the ACARA website and your local curriculum authority website and investigate the organisation of the mathematics curriculum in the junior and senior secondary years.

Identify the extent to which the ten curriculum components listed by van den Akker (2003) are represented in these curriculum documents. Are there any differences in emphasis between the compulsory and post-compulsory years of schooling?

What mathematics subjects are offered in Years 11 and 12? What is the rationale for each of these subjects and who are the intended clientele? How does the mathematical content differ in these subjects? What are the consequences of differentiated content? (E.g., is it possible for students to switch

from one subject to another? How do subjects prepare students for different post-school destinations?) How much flexibility are teachers allowed in
selecting content?

The *Australian Curriculum: Mathematics* has attracted much critical attention from the
mathematics education research community in Australia (see Atweh, Goos, Jorgensen &
Siemon, 2012), especially in relation to how it is organised. Atweh and Goos (2011) argue
that the *Australian Curriculum: Mathematics* has a content-based organisation, reflecting a view
of mathematics as a fixed, universal and objective body of knowledge. This approach envi-
sions the role of mathematics education as enabling students to acquire knowledge and
skills in preparation for work. However, this view has been challenged by researchers who
call for school mathematics to develop students' capacities for problem solving and critical thinking—not only to enable students to understand how the world works, but also to
develop the agency to transform aspects of their lives as informed and active citizens. The
sections that follow illustrate three different ways of organising the mathematics curriculum that foreground mathematical thinking processes, numeracy across the curriculum
as an interdisciplinary approach to dealing with authentic problems, and critical mathematics for social justice.

Process-driven curriculum models

In a previous section, we traced the historical development of the mathematics curriculum in Australia, noting how the emphasis has shifted over time from mathematical content and skills to mathematical structures and, by the 1980s, to mathematical processes.
The call by the NCTM in the USA for problem solving to be the guiding principle for
mathematics education in the 1980s (NCTM, 1980) became the catalyst for process-driven
curriculum development in Australia, leading to problem solving and applications achieving new prominence in mathematics curricula in almost every part of Australia (Stacey &
Groves, 1984).

During the 1980s, several curriculum development projects across Australia focused
on problem solving, applications or modelling. In South Australia, Gaffney and Treilibs
(1982) had developed mathematical modelling curriculum materials focusing on real-
life problems, while in Western Australia, the Curriculum Branch of the Education

Department (1984) had developed materials for an Applying Mathematics course. Several projects were also underway in Victoria. The Reality in Mathematics Education (RIME) Teacher Development Project was attempting to improve the quality and relevance of mathematics teaching in Years 7–10 through active investigation of mathematical and real-world situations (Lowe, 1984). In 1985, the MCTP (Lovitt, Clarke & Stephens, 1986) began to develop and trial sample applications lessons for Years 9 and 10 as well as the primary classroom, and in early 1986, the Australian Academy of Science decided to transform its *Mathematics at Work* series (Treilibs, 1980–81) into two volumes (Lowe, 1988, 1991) with an emphasis on applications and mathematical modelling for the senior secondary level. In Queensland, Clatworthy and Galbraith (1987) began a mathematical modelling course at a senior secondary college, where the modelling component was conducted in parallel with conventional topics taught and tested in traditional format.

By the early 1990s, senior secondary mathematics curricula in Australia were under-going major revision in almost every state and territory to make mathematics more accessible to all students and increase the emphasis on problem solving, investigation and modelling (Stephens, 1990). Attempts to introduce formal school-based or external assessment of these mathematical processes, whether through examinations, centrally set common tasks, problems selected from a common bank, or projects and extended investi-gations developed at the school level, met with varying degrees of success. Concerns over teacher and student workload and the authenticity of student work completed under non-supervised conditions, combined with the extent and pace of change, threatened the integrity and sustainability of these process-driven curriculum models in some parts of Australia (Stillman, 2001, 2007).

> **REVIEW AND REFLECT :** How are the process aspects of mathematics conceptu-alised by senior secondary mathematics curricula developed by ACARA and your local jurisdiction? To what extent are problem solving, applications, modelling and investigations represented in these documents? What advice is provided about how to teach these processes?
>
> What kinds of tasks are used to assess these processes? Under what condi-tions do students complete these assessment tasks?

> Discuss your findings with peers in the light of the concerns about curriculum sustainability raised by Stillman (2001, 2007).

The *Australian Curriculum: Mathematics* for Foundation–Year 10 identifies four 'proficiency' strands that aim to describe the thinking and doing of mathematics: understanding, fluency, problem solving and reasoning. The proficiency strands are described in short overview statements but are not elaborated further, for example, by year level, in the same way as the content strands. This content-based organisation can present challenges for teachers in developing students' mathematical thinking while exploring mathematical content.

> **REVIEW AND REFLECT :** Do a library search to obtain some of the process-driven curriculum resources developed in the 1980s (e.g., RIME, MCTP, *Mathematics at Work*); or find out if local schools have these resources. Select one and use the ideas in Figure 5.3 to evaluate its relevance to the current junior and senior secondary mathematics curricula in your state or territory.
>
> Compare the relative emphases on mathematical processes in the junior and senior secondary mathematics curricula in your state or territory. How similar or different is the conceptualisation of 'processes' in these curricula? Discuss with your peers any implications for continuity in curriculum planning across the secondary school years.

Numeracy across the curriculum

The *Melbourne Declaration* (MCEETYA, 2008) introduced the idea of 'general capabilities that underpin flexible and analytical thinking, a capacity to work with others and an ability to move across subject disciplines to develop new expertise' (p. 13). The *Australian Curriculum* identifies seven general capabilities that are to be developed and used by students across all learning areas: literacy, numeracy, ICT capability, critical and creative thinking, personal and social capability, ethical understanding and intercultural understanding. However, the *Australian Curriculum* pays little systematic attention to using these general capabilities as

Mathematics—proficiency strands

UNDERSTANDING

1. **Making connections**
   - How does this fit with what you already know?
   - How is this topic/problem like things you have studied before?
   - How is the work we have done today (on …) related to work we did last week/last term (on …)?

2. **Applying familiar knowledge in new contexts**
   - The new context may be mathematical, another school subject, or in the real world.

Mathematical context	Other school subjects	Real world
Using fractions and percentages in probability	Recording and analysing data in science experiments	Using trigonometry to find the height of a building

3. **Translating between representations**
   - Using a fraction wall, number line, pieces of a rectangle, sectors of a circle to represent fractions
   - Using a table of values, graph, equation to represent a function
   - Translating a word problem into an equation or diagram

4. **Explaining mathematical thinking**
   - Explaining 'why' is closely linked to reasoning.

FLUENCY

1. **Knowing—that**
   - Recalling facts, definitions, concepts

2. **Knowing—how**
   - Choosing appropriate procedures
   - Calculating, manipulating and solving accurately and efficiently (using written, mental, and technology-based strategies)
   - Estimating sensibly

REASONING

1. **Making and investigating conjectures**
   - What is the pattern?
   - What will happen next?
   - What would happen if …?

   AC examples:
   Investigate number sequences to find patterns
   Investigate the properties of odd and even numbers
   Investigate combinations of translations, reflections and rotations
   Investigate the angle sum of triangles

2. **Developing and evaluating mathematical arguments**
   Does this always work? Sometimes? Never? Convince us?
   - Does multiplication make things bigger? How do you know?
   - What kind of number do you get when you add two odd numbers? Why do you think that?
   - What would happen to the area of a square if we doubled the side lengths? How could you explain your thinking?
   - If two rectangles have the same perimeter, will they have the same area? Does this always work? Sometimes? Never? Why?

PROBLEM SOLVING

1. **Problem solving**
   A task is a problem if the person attempting it does not know the solution method in advance.

   Three curricular alternatives to problem solving:
   - Teaching for problem solving (teaching mathematical content for later use in solving mathematical problems);
   - Teaching about problem solving (teaching heuristic strategies to improve generic ability to solve problems);
   - Teaching through problem solving (teaching standard mathematical content by presenting non-routine problems involving this content). (Stacey, 2005)

   Table 2.4 Teacher scaffolding questions during problem-solving

Getting started	While students are working	After students are finished
What are the important ideas here?	Tell me what you are doing.	Have you answered the problem?
Can you rephrase the problem in your own words?	Why did you think of that?	Have you considered all the cases?
What is this asking us to find?	Why are you doing this?	Have you checked your solution?
What information is given?	What are you going to do with the result once you have it?	Does it look reasonable?
What conditions apply?	Why do you think that that stage is reasonable?	Is there another solution?
Anyone want to guess the answer?	Why is that idea better than that one?	Could you explain your answer to the class?
Anyone seen a problem like this before?	You've been trying that idea for five minutes. Are you getting anywhere with it?	Is there another way to solve the problem?
What strategy could we use to get started?	Do you really understand what the problem is about?	Could you generalise the problem?
Which one of these ideas should we pursue?	Can you justify that step?	Can you extend the problem to cover different situations?
	Are you convinced that bit is correct?	Can you make up another similar problem?
	Can you find a counter-example?	

   (Holton & Clarke, 2006)

   During questioning ask:
   How do you know?
   Why do you think that?
   How could you justify that?

2. **Investigation**
   Mathematical investigations are 'contextualised problem solving tasks through which students can speculate, test ideas and argue with others to defend their solutions' (Diezmann, Watters, & English, 2001, p. 170).

   AC Examples:
   - Investigating 'best buys'
   - Investigating issues involving numerical data collected from primary and secondary sources
   - Investigating reports of surveys in the media
     (Note that most uses of the term 'investigate' in the AC — *Mathematics* refer to building mathematical reasoning.)

3. **Modelling**
   The mathematical modelling process is driven by the desire to obtain a mathematically productive outcome for a problem with life-like or genuinely real-world motivation.
   It involves:
   - Identifying a real-world problem
   - Making simplifying assumptions
   - Formulating the mathematical problem, identifying important variables
   - Solving the mathematical problem
   - Interpreting the solution
   - Verifying the model
   - Reporting, explaining, predicting

Figure 5.3 **Elaborations of the four proficiency strands in the current *Australian Curriculum: Mathematics***

a tool for curriculum design, possibly as a consequence of its content-based organisation that maintains strong boundaries between school subjects. Nevertheless, recent research and development activity by Goos, Geiger and Dole (2014) on embedding numeracy across the school curriculum in Australia has shown that this approach can be implemented by teachers at all levels of schooling.

Goos et al. (2014) investigated the effectiveness of a teacher professional learning program aimed at enhancing numeracy teaching across a range of school subjects, including history, science, English, health and physical education, and studies of society and environment. This program was based on a multifaceted model of numeracy that represents a synthesis of research related to effective numeracy practice (see Figure 5.4). The model was designed to be accessible to teachers to support their curriculum planning and reflection. It incorporates the dimensions of *mathematical knowledge, dispositions, tools* and *contexts*, embedded in a *critical orientation* to using mathematics.

A numerate person requires *mathematical knowledge*. In a numeracy context, mathematical knowledge includes not only fluency with accessing concepts and skills, but also problem-solving strategies and the ability to make sensible estimations (Zevenbergen, 2004).

A numerate person has *positive dispositions*—a willingness and confidence to engage with tasks and apply their mathematical knowledge flexibly and adaptively. Affective issues have long been held to play a central role in mathematics learning and teaching (Leder & Forgasz, 2006), and the importance of developing positive attitudes towards mathematics is emphasised in national curriculum documents (e.g., National Curriculum Board, 2009).

Being numerate involves using *tools*. In school and workplace contexts, tools may be representational (symbol systems, graphs, maps, diagrams, drawings, tables), physical (models, measuring instruments), and digital (computers, software, calculators, internet) (Noss, Hoyles & Pozzi, 2000; Zevenbergen, 2004).

As numeracy is about using mathematics to act in and on the world, people need to be numerate in a range of *contexts* (Steen, 2001). These may be real-life contexts (personal, citizenship-related, work-related) or curriculum contexts.

The model is grounded in *a critical orientation* to numeracy since numerate people not only know and use efficient methods, they also evaluate the reasonableness of the results obtained and are aware of appropriate and inappropriate uses of mathematical thinking to analyse situations and draw conclusions. They also need to recognise how mathematical information and practices can be used to persuade, manipulate, disadvantage or shape opinions about social or political issues (Jablonka, 2003).

This model has been used to identify the numeracy demands of non-mathematics subjects in the *Australian Curriculum*, investigate teachers' understanding of numeracy, and analyse teachers' capacity to recognise and take advantage of numeracy opportunities in the subjects they teach (Goos, Dole & Geiger, 2012a; Goos, Geiger & Dole, 2011, 2014). You can read teachers' own accounts of their efforts to embed numeracy across the curriculum in a special issue of the *Australian Mathematics Teacher* (Cooper, Dole, Geiger & Goos,

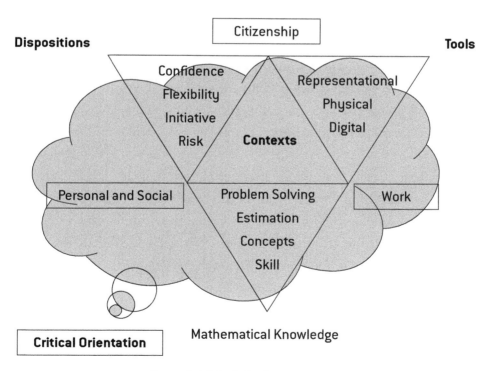

Figure 5.4 **Model of numeracy**

2012; Gibbs, Goos, Geiger & Dole, 2012; Goos, Dole & Geiger, 2012b; Peters, Geiger, Goos & Dole, 2012; Willis, Geiger, Goos & Dole, 2012).

REVIEW AND REFLECT : Visit the *Australian Curriculum* website and find the definition of numeracy as a general capability to be embedded in all learning areas. Compare this definition with the numeracy model shown in Figure 5.4, and discuss any similarities and differences between them.

Where is numeracy in the Australian Curriculum: History?

The rationale for the *Australian Curriculum: History* promotes a *critical orientation* to inquiry, stating that 'history is a disciplined process of inquiry into the past' (ACARA, 2012a, p. 3). This is reinforced by the general capabilities section, which claims that 'critical thinking is essential to the historical inquiry process' (p. 10). *Context* is an organising device for teaching historical concepts, with curriculum contexts becoming progressively broader through the years from Foundation to Year 10. The curriculum supports discipline-specific *dispositions* by encouraging students to develop empathy for others and explore the perspectives, beliefs and values of different societies and cultures. Key aspects of *mathematical knowledge* supporting historical inquiry include data analysis and chronology. Maps and timelines are the most common representational *tools* referred to in the history curriculum.

A teacher devised a history task for her Year 6 class to explore the development of Australia as a nation during the twentieth century. She prepared a worksheet with questions that required students to access census data from the Australian Bureau of Statistics website. Before setting the class to work, she overheard one student ask a peer 'are Aboriginal people counted in the census?' The teacher recognised this as an opportunity for developing a critical orientation, and instead of using her worksheet, she asked students to write down the questions that *they* wanted to ask. One student wondered, 'how many children are there in Australia?' A question like this opens up many possibilities for exploring the age distribution of the population and how this has changed over time. Figure 5.5 provides the most recent census data. Which age groups should be counted as 'children'? Why has the age distribution changed so much between 1960 and 2010? What

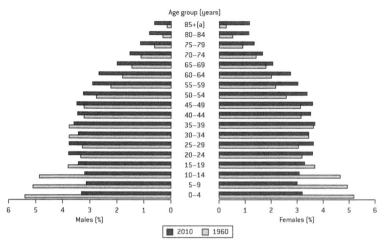

Age group (years)

(a) The 85+ age group includes all ages 85 years and over and is not directly comparable with the other age groups.

Figure 5.5 **Age distribution of the Australian population, 1960 and 2010**

Source: Australian Historical Population Statistics (3105.0.65.001); Australian Demographic Statistics (3101.0).

implications would this change have for the provision of social services and infrastructure, such as hospitals and schools?

> **REVIEW AND REFLECT :** With a partner, read the example on numeracy in the history curriculum shown above and explain how the numeracy model has been used to analyse the numeracy opportunities that the teacher created for her students.
>
> Read the short articles in *The Australian Mathematics Teacher,* volume 68, issue 1, written by teachers who have used the numeracy model presented in this chapter to embed numeracy across the curriculum. What challenges did they face in implementing this approach to curriculum design? What supports enabled these teachers to succeed in overcoming these challenges?
>
> Review the video vignettes on the Queensland College of Teachers ClassMovies website at www.classmoviestv.com/qctuq. Use the discussion questions

> provided with each vignette to analyse the approaches to curriculum design used by these teachers.
>
> Drawing on the resources listed above, work in pairs to select a topic from an *Australian Curriculum F–10* other than mathematics and design a task, lesson or learning sequence that enhances numeracy in that curriculum area.

Social justice curriculum models

Teaching mathematics for social justice involves achieving equity not only *in*, but also *through*, mathematics learning. (Social and cultural issues in mathematics education are discussed in more detail in Chapter 14.) This involves teachers using mathematics as a tool to help students interpret and address injustice and inequities in society. There are few examples of teaching mathematics for social justice in the literature, especially at the school level, but those that have been reported provide useful insights into what such a curriculum might be able to achieve. (See Gutstein & Peterson, 2006 for an excellent collection of essays and examples of how to weave social justice issues throughout the mathematics curriculum.)

Frankenstein (1997, 2001) has taught mathematics for social justice at the college level for many years in the USA. Her students are mainly working-class adults who failed to achieve success in secondary school mathematics. She defines her curriculum goals in terms of developing students' ability to pose mathematical questions in order to deepen their appreciation of social issues and to challenge people's perceptions of those issues. Gutstein (2003) took a similar approach to teaching mathematics to Years 7 and 8 Latino (Mexican-American) students in an urban, working-class community in the US. Gutstein had three goals in teaching for social justice: he wanted to develop students' socio-political consciousness to help them understand the forces that shaped their world, a sense of agency in believing that they could make a difference in the world, and positive social and cultural identities by affirming their language, culture and history. Like Frankenstein, Gutstein also had specifically mathematical objectives that supported the larger goal of teaching for social justice: he wanted his students to be able to 'read the world' using mathematics, to develop mathematical power, and to become more positively disposed towards mathematics. To achieve his curriculum goals and objectives, Gutstein devised a series of real-world projects, each lasting from a couple of days to a couple of weeks, in which students used mathematics to analyse social issues and understand power relations.

Mathematics for social justice

One of the real-world issues investigated by Frankenstein's (2001) students involves challenging official interpretations of the unemployment rate. We have adapted this investigation by providing recent data from the Australian Bureau of Statistics in Table 5.1. The Australian Bureau of Statistics uses the following definitions in determining the unemployment rate:

- Employed: persons aged 15 years and over who worked for any length of time during the reference week for pay, profit, commission, payment in kind or without pay in a family business, or who had a job but were not at work.
- Unemployed: persons aged 15 years and over who were not employed during the reference week, but who had actively looked for work and were available to start work.
- Labour force: all persons aged 15 years and over who, during the reference week, were employed or unemployed.
- Marginally attached to the labour force: people who wanted to work and were either actively looking for work but not available to start work in the reference week, or were available to start work within four weeks but were not actively looking for work.
- Discouraged jobseekers: people who are marginally attached to the labour force, wanted to work and who were available to start work within four weeks but whose main reason for not taking active steps to find work was that they believed they would not be able to find a job because of reasons of age, language or ethnicity, schooling, training, skills or experience, there being no

jobs in their locality or line of work, or they considered that there were no jobs at all available.

- Unemployment rate: the number of unemployed expressed as a proportion of the labour force.
- Participation rate: the labour force expressed as a percentage of the civilian population.

Table 5.1 Civilian population aged 15–69 years, labour force status—September 2010

8.38 LABOUR FORCE STATUS(a)—September 2010

	Males'000	Females'000	Persons'000
Civilian population aged 15 years and over	8 865.5	9 022.2	17 887.7
Persons in the labour force	6 546.7	5 460.3	12 007.0
Employed	6 220.9	5 176.9	11 397.7
Unemployed	325.8	283.4	609.2
Persons not in the labour force	2 318.8	3 561.9	5 880.7
With marginal attachment to the labour force	349.4	576.4	925.9
Wanted to work and were actively looking for work	32.9	41.9	74.9
Were available to start work within four weeks	23.1	30.2	53.2
Were not available to start work within four weeks	9.9	11.8	21.7
Wanted to work but were not actively looking for work and were available to start work within four weeks	316.5	534.5	851.0
Discouraged jobseekers	39.9	62.3	102.1
Other	276.6	472.3	748.9

8.38 LABOUR FORCE STATUS(a)—September 2010			
Without marginal attachment to the labour force	1 960.3	2 985.5	4 954.8
Wanted to work but were not actively looking for work and were not available to start work within four weeks	126.6	239.9	366.5
Did not want to work	1 637.5	2 562.0	4 199.5
Permanently unable to work	205.3	183.5	388.8

(a) Civilian population aged 15 years and over

Source: Australian Bureau of Statistics, Persons Not in the Labour Force, Australia, September 2010 (Yearbook Australia 2012, 6220.0). www.abs.gov.au

---

REVIEW AND REFLECT : 1. Calculate the participation rate and unemployment rate using Table 5.1 and the official definitions provided above.

2. In your opinion, which groups listed in Table 4.1 should be considered unemployed? Why? Which should be considered part of the labour force? Why?

3. Given your selections from Question 2, recalculate the participation and unemployment rates and compare your answers with those obtained in Question Explain how changes in the numerator and denominator affected the answers.

Conclusion

In this chapter, we have looked at mathematics curriculum development and curriculum models from an international and national perspective and identified some of the historical, social and political forces that shape curriculum change. Curriculum is not simply a 'product', created by 'experts' and handed over to teachers for classroom implementation; instead, curriculum decision making is a dynamic process in which teachers can play a significant role within and beyond the classroom.

Recommended reading

Atweh, B., Goos, M., Jorgensen, R. & Siemon, D. (Eds). (2012). *Engaging the Australian curriculum: Mathematics—Perspectives from the field*. Mathematics Education Research Group of Australasia. Available from http://www.merga.net.au/node/223

Goos, M., Dole, S. & Geiger, V. (2012b). Numeracy across the curriculum. *Australian Mathematics Teacher*, 68(1), 3–7.

Gutstein, E. & Peterson, B. (Eds). (2006). *Rethinking mathematics: Teaching social justice by the numbers*. Milwaukee, WI: Re-thinking Schools Ltd.

Stacey, K. (2005). The place of problem solving in contemporary mathematics curriculum documents. *Journal of Mathematical Behavior*, 24, 341–350.

CHAPTER 6
Assessing mathematics learning

Assessment involves students reflecting on their learning and teachers collecting and interpreting evidence of students' learning, using this evidence to make judgements about the quality of learning to inform planning and teaching, and communicating these judgements to different audiences: the learners, their parents, other teachers, school leadership, governments and other education institutions including tertiary institutions. 'Assessment in mathematics has political, ideological and cultural aspects, as well as educational aspects' (Galbraith, 1995, p. 274). Assessment is political, as it brings the issue of power into the teaching and learning environment. The political nature of assessment has become more evident since the introduction of the *National Assessment Program—Literacy and Numeracy (NAPLAN)* in 2008 for students in Years 3, 5, 7 and 9 (Lowrie, Greenlees & Logan, 2012). In Australia's publicly funded education system, schools are held accountable to the community and governments through the assessment of their students. Lowrie et al. (2012) report on studies that show that teachers' assessment practices are unduly influenced by the *NAPLAN* model of assessment. Governments use assessment data from *NAPLAN* and other international assessment programs such as the TIMSS (see Thomson, Hillman & Wernet, 2012) and the *OECD Program of International Student Assessment* (PISA) (see Thomson et al., 2013a) to inform education and curriculum policy and to direct resources to programs and schools, sometimes according to ideology so that school communities in need are denied sufficient resources to make a difference—such as in the debate over the 'Gonski model' of funding schools.

In 2015, ACARA published senior secondary curricula for some subjects, including mathematics. However, state governments currently control the post-compulsory schooling curriculum, and assessment policy and debate continues about the nature of assessment and implications for mathematics learning and equity (Stephens, Clarke & Wallbridge, 1994; Teese, 2013). According to Stephens et al. (1994), changing assessment requirements and methods at Year 12 level impact content and assessment at all levels of secondary schooling. Therefore, controlling assessment exercises power over mathematics curriculum and teaching more broadly. Teese (2013) argued that changes in high-stakes curriculum and assessment at the end of secondary schooling typically benefit the most socially advantaged students and thus maintain structural inequities. In this chapter, we discuss the purpose of assessment, the different types of assessment, and their relevance to assessing mathematics learning. Recording, analysing and reporting assessment of student learning is also discussed. We invite you to inquire into the political, ideological and cultural dimensions of assessment in mathematics as well.

Purposes of assessment

According to Clarke (1996a), the fundamental purposes of assessment are 'to model, to monitor and to inform' (p. 328). A regime of assessment employed by an educational system is thus seen as valid and reliable depending on the extent to which it: (a) provides 'an effective model of valued performance in mathematics and an effective model of educational practice' (p. 328); (b) monitors 'these valued performances' by providing all students with sufficient 'opportunities to display their capabilities in forms that can be documented' (p. 328); and (c) informs the actions of stakeholders, such as students, teachers, parents/carers, employers, subsequent educational providers (e.g., TAFE or universities), systems and governments.

Assessment acts as a model for the discipline of mathematics because 'irrespective of the purposes we might have for assessment, it is through our assessment that we communicate most clearly to students which activities and learning outcomes we value. It is important, therefore, that our assessment be comprehensive, and give recognition to all valued learning experiences' (Clarke, 1988, p. 1). Assessment is part of the learning process. This is recognised in the statements about the purpose of assessment from various education departments around Australia; for example, 'the primary purpose of assessment is to improve student learning' (Department of Education, Queensland, 2016) and 'assessment should be an integral part of teaching and learning' (Western Australian Board of Studies, 2014). Hattie and Timperley (2009) highlight the significance of feedback that occurs through assessment for improved student learning. Through interaction with other students and the teacher, students reflect on their work and use metacognitive skills to self-correct and direct their problem solving. Evidence of students' learning, thinking and knowledge provides feedback for teachers and assists them to focus on what matters, and the skills and concepts that need their attention in the moment and as they plan for future learning. It allows them to build on students' mathematical and cultural knowledge, and skills and target specific learning needs. Various forms of assessment data are needed if teachers are to best meet students' needs.

The departments of education across Australia also document a number of principles of assessment to achieve these goals; for example, 'assessment should be fair' and 'assessment should be designed to meet their specific purposes' (Western Australian Board of Studies, 2014); 'assessment should be ongoing and based on a continuum of learning/development' and 'assessment should build positive attitudes and self-confidence' (Victorian Curriculum and Assessment Authority, 2014); and 'assessment for improved performance involves feedback and reflection' (Department of Education and Training, Victoria, 2013b).

Generally, the states in Australia identify three types of assessment as part of the learning process: *assessment for learning* (AfL, formative assessment), *assessment as learning* (student self-assessment), and *assessment of learning* (summative assessment). These types of assessment are defined in Figure 6.1.

Formative assessment has the purposes of advancing students' learning and informing teachers' instructional decisions (Callingham, 2008; Even, 2005; Klenowski, 2009; Stiggins et al., 2014). It may occur at the beginning and throughout a teaching cycle using informal methods such as classroom questioning and observation as well as various formal methods such as pre-tests, 'warm up' tasks at the beginning of a lesson, assignments, problem-solving tasks and investigations (Suurmatamm, Koch & Arden, 2010; A. Watson, 2006b). A framework involving a continuum of learning is used to analyse responses for formal methods of formative assessment and so reflects constructivist or socio-cultural theories of learning (Shephard, 2001). Student self-assessment (*assessment as learning*) supports students'

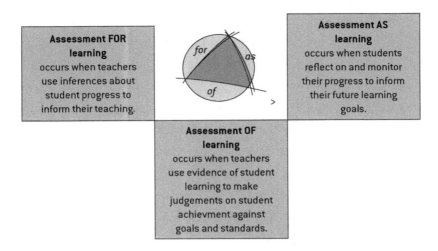

Source: Department of Education and Training, Victoria (2013a).

Figure 6.1 **Purposes of assessment**

metacognition and enables them to construct new meaning (Earl, 2012). Teachers also use student self-assessment to inform planning and teaching.

Performance assessment occurs as students work on various tasks. In performance assessment, 'the teacher observes and makes a judgement about the student's demonstration of a skill or competency in creating a product, constructing a response, or making a presentation' (McMillan, 2004, p. 198). Performance assessment tasks are also used in various senior secondary mathematics units across Australia (Clarke & Stephens, 1996; Queensland Curriculum and Assessment Authority, 2015). Typically, socio-cultural theories of learning underpin performance assessment, and criterion-referenced frameworks regarding problem solving, modelling, reasoning and communicating typically inform the design and analysis of this form of assessment.

Summative assessment is used to indicate the achievement status or level of performance of a student. This usually occurs on completion of a unit or course of study. Teachers normally use standards-referenced models, applying the standards documents in the curriculum for F–10 students. In senior secondary school, both standards-referenced and norm-referenced methods are used for summative assessment. Written tests or examinations in a limited timeframe tend to follow behaviourist theories of learning.

National Assessment Program (NAPLAN)

All Australian students in Years 3, 5, 7 and 9 complete the four *NAPLAN* tests, usually in the second week of May of each year. One of these is the *NAPLAN* numeracy test, which uses standards of the *Australian Curriculum: Mathematics* to design multiple-choice items. Results are returned to schools towards the end of Term 3. The purpose of this assessment program is to 'drive improvements in student outcomes and provide increased accountability for the community' (ACARA, 2013b). Schools are able to evaluate their performance by comparing their performance with other schools, socio-cultural 'like' schools in their state and across the nation. Schools use these data gathered over time for different cohorts to evaluate their performance and provide evidence for designing and evaluating their school improvement goals. However, *NAPLAN* results do not enable schools to monitor the progress of the same cohort of students as they progress through secondary school. Parents are able to compare their child's performance with others in their school, and all other students in the same year level and thereby hold schools accountable. *NAPLAN* allows comparison across Australia, and thus it is a system of accountability for meeting priorities set by state governments and the federal government.

However, as noted earlier, teachers and mathematics educators are concerned that the performativity culture of the education system is pressuring teachers to return to traditional teaching and assessment methods rather than connect assessment to student-centred teaching (Harris & Brown, 2008; Klenowski, 2011). Moreover, the model of testing used for *NAPLAN* focuses on a narrow range of mathematical experiences that ask students to recall content and use procedures, at the expense of developing understanding, fluency in flexible thinking, problem solving and reasoning (Dimarco, 2009; Williams, 2012). Klenowski (2009) argued that 'mere performance on a test does not necessarily mean that learning has occurred. Learners can be taught how to score well on tests without much underlying learning' (p. 263). Examples of items from *NAPLAN* and international tests are discussed later in this chapter.

> **REVIEW AND REFLECT :** Research the results of *NAPLAN* testing of students in a secondary school with which you are familiar. Interview the mathematics coordinator, curriculum coordinator and/or mathematics teachers.

- Who is provided with results of these tests and how does this happen?
- How do the school, mathematics faculty and mathematics teachers use these results to inform the school and classroom mathematics program?
- What changes to curriculum have been made to the school program in an attempt to improve the school's performance?
- What changes have been made to the teaching of mathematics to improve students' performance?

Evidence-based judgement models

Evidence of student learning can be interpreted in different ways. For example, is student knowledge measured in some way or is it inferred from observations of performance? In objective tests, measures are derived from counting the number of correct items or aggregating marks allocated for various parts of items that are correct. In other forms of assessment, such as the report of a mathematical investigation, numbers (e.g., 20 out of 25) may be assigned according to a global judgement about the quality of the report as a whole or by aggregating marks allocated for successful completion of various predetermined requirements of the assessment piece that can involve local judgements of quality. The two most common frameworks used for interpreting the results of the assessment are norm referencing and criterion referencing.

Norm-referenced judgement models

Norm referencing compares student performance to that of other students on the same or similar tasks. Students are ranked from lowest to highest performance, and grades are then awarded based on some predetermined distribution. Grades thus have relative rather than absolute meaning, making it difficult to detect changes in achievement patterns over time. Sadler (1987) also notes that this system of grading 'may in fact be used as a means of distributing "merit" in such a way as to artificially create (and maintain) a shortage of high grades' (p. 192).

Criterion-referenced judgement models

Conversely, criterion referencing judges scores or performances in relation to a set of absolutes—an external, predetermined set of criteria. If a student meets all criteria for a particular grade, then that grade must be awarded no matter how many other students have also met the criteria. However, as Sadler (1996) points out, this often does not occur in practice due to confusion between criteria and standards. Sadler (1996) clarifies the distinction between the terms in an assessment context: 'by CRITERION is meant a property or characteristic by which the quality of something may be judged. By a STANDARD is meant a definite level of achievement aspired to or attained' (p. 2).

Standards-referenced judgement models

There are four basic methods of specifying and publicly declaring standards: by using numerical cut-offs, tacit knowledge, exemplars or verbal descriptors. Numerical cut-offs have the appeal of being simple and appearing to signify sharp boundaries. On a criterion designated 'mathematical techniques', for example, the boundaries for standards could be set at 85 per cent, 70 per cent, 50 per cent and 25 per cent. However, this is only another form of norm-referenced assessment, as the model does not define the knowledge to be evidenced. Tacit (or mental) standards can be used as the basis of post-assessment moderation by consensus of teachers or by a district moderator. Such standards are not articulated, but reside in the heads of the assessors. In contrast, standards-referenced assessment can be strengthened by using a combination of exemplars of student work and verbal descriptions. 'Exemplars are key examples chosen so as to be typical of designated levels of quality or competence' (Sadler, 1996, p. 200). These can be annotated to show the context in which the example was produced, the qualities that the example displays to show how it meets a particular level of quality, and what additional characteristic(s) it would need to be designated to be at the next level. Exemplars are provided by various authorities using standards-referenced assessment (e.g., *Assessment for Learning in a Standards-referenced Framework—Mathematics* available from www.shop.bos.nsw.edu.au). Verbal descriptors of standards are statements of the properties that characterise the designated level of quality. The example in Figure 6.2 is for Stage 4 of the New South Wales Mathematics Years 7–10 Syllabus. Here, descriptors of standards are given to distinguish three levels of performance on the task.

Competency-based assessment is 'a special case of standards based assessment where the proficiency continuum has been reduced to a simple dichotomy, "competent" versus "not competent", with only one "standard"' (Maxwell, 1997, p. 72). Competencies are used in vocational education.

Activity: diagonals of a quadrilateral

If the diagonals of a quadrilateral bisect each other, what type of quadrilateral could it be?

Give reasons for your answer and illustrate by drawing diagrams.

Is there more than one type of quadrilateral in which the diagonals bisect each other?

What conclusions can you make?

Criteria for assessing learning

Students will be assessed on their ability to:

- Demonstrate knowledge and understanding of the nature of different quadrilaterals.
- Draw a valid conclusion about the diagonals of quadrilaterals.
- Communicate mathematical ideas.

Guidelines for marking

Range	Students in this range
High	Identify and accurately draw different types of special quadrilaterals: trapezium, parallelogram, rectangle, rhombus, square, kite. Make a valid conclusion about the bisection of diagonals for the different types of quadrilaterals.
Satisfactory	Identify and accurately draw at least two special quadrilaterals. Identify some quadrilaterals for which the diagonals bisect each other.
Progressing	Identify and draw one quadrilateral for which the diagonals bisect each other.

Source: Board of Studies New South Wales (2002), Mathematics Years 7–10 Syllabus.

Figure 6.2 **Verbal descriptors of standards exemplar, Mathematics Stage 4**

REVIEW AND REFLECT : Gather and analyse assessment materials used by mathematics teachers in your secondary school. Collect materials used for different year levels (junior and senior mathematics) and different types of assessment tasks.

What theories of mathematics learning are reflected in the design of the mathematics tasks used? What judgement models does the mathematics faculty use?

Choose one assessment task. Make recommendations for improving the task for measuring student achievement using the assessment guidelines in your state.

Developmental-based assessment

Developmental-based assessment was developed and trialled in secondary mathematics classrooms by Pegg and his team (Pegg & Panizzon, 2004, 2007–08) in rural New South Wales. The impetus for teachers using developmental-based assessment techniques 'arose directly from changing assessment practices required to satisfy the requirements of new syllabus documents introduced into New South Wales for mathematics and science in 2000' (Pegg et al., 2003, p. 4). This form of assessment relies on teachers interpreting students' responses within a framework of cognitive developmental growth, namely, Biggs and Collis's (1991) Structure of the Observed Learning Outcome (SOLO) model, but incorporating more recent developments in SOLO (Pegg, 2003). The structure of a response represents the learning cycle (unistructural, multistructural to relational) within a mode (see Table 6.1).

'Essentially SOLO is concerned with specifying "how well" (qualitative) something is learned rather than "how much" (quantitative)' (Pegg & Panizzon, 2007–08, p. 67). Teachers 'place learners along a developmental continuum' (Pegg et al., 2003, p. 4) and use the framework as a source of 'advice … on possible pathways for future teaching endeavours' (p. 4).

A number of tests and individual interview scripts have been designed for primary students using a learning continuum or development framework to support teachers'

Table 6.1 SOLO taxonomy

Response level	Meaning
Unistructural	Responses use single elements of the task, often with contradictions between them.
Multistructural	Responses use multiple elements of the task.
Relational	Responses create connections among elements of the task to form an integrated whole.

analysis of results. Some have been designed for students in the middle years. For example, the Department of Education and Training, Victoria, provides development-based tools with online analysis support: *Fractions and Decimals Online Interview* (2014a), *Assessment for Common Misunderstandings* (2014b) and *Scaffolding Numeracy in the Middle Years—Multiplicative Thinking* (2013b).

Collecting and interpreting evidence of student learning

Secondary mathematics teachers use a wide range of assessment methods, though it would be fair to say that tests and exams predominate (Suurmatamm et al. 2010; Watt, 2005). Watt (2005) reported that secondary mathematics teachers in NSW were initially reluctant to adopt alternate means of assessment, believing them to be too subjective and thus unreliable. More recently, Sullivan et al., (2013) found that secondary mathematics teachers across Australia are more likely to use their own assessment records than externally set assessment results when planning their teaching. The shift to *assessment for learning* and *assessment as learning* has led to the use of a wide range of assessment practices and tools with the potential to better meet the learning preferences of students of diverse socio-cultural contexts and gender (Klenowski, 2009; Leder et al., 1999; Watt, 2005). Different forms of assessment are also useful for the different purposes of assessment, the different stages of learning a new topic, higher-order processes and assessing mathematics in context. Given mathematics inquiry, problem solving and modelling involves the use of digital technologies, technologies should be incorporated in assessment (Callingham, 2011; D. Leigh-Lancaster, 2010). Assessment tasks in themselves should be worthwhile learning activities for students. A representative collection suggested in a variety of mathematics curriculum documents from various Australian states includes those shown in Table 6.2.

Advice regarding the use of these forms of assessments, particularly when they are contributing to exit scores or exit levels in post-compulsory courses of study, varies in terms of whether or not the assessment must be done under fully supervised conditions in class or in unsupervised conditions in the students' own time. Advice about the extent to which authorship needs to be verified by teachers also varies.

In the remainder of this section, we describe a selection of methods of assessment including informal assessment (classroom questions, student self-assessment), alternative

Table 6.2 Assessment tasks

Rich tasks (the New Basics)	Reports (written, oral, multimedia or combinations
Rich assessment tasks	of these) on:
Concept maps	• an experiment or survey
Student self-assessment	• a mathematical investigation
Exit tickets	• a field activity
Teachers' questions during lessons	• a project
Open-ended questions	Practical tasks such as measurement activities
Debates	Extended analysis tasks
Journals	Extended modelling and problem-solving tasks
Topic tests	Portfolios
• subject examinations	• diagnostic tests

assessment (open-ended questions, rich assessment tasks, performance assessment) and traditional assessment instruments (tests).

Teacher classroom questions

'Assessing the "quality" of learning … is better situated in the classroom, where teachers make judgements on a day-to-day basis about what their students know and can do' (Callingham, 2008, p. 18). Teacher questions that serve the purpose of determining what students understand and can do are part of the normal interactions in the classroom. They scaffold students' learning and focus their attention on what matters when problem solving (see the lesson scenarios in Chapters 2 and 3). During independent or group work, the teacher circulates, monitoring progress and assisting students individually or in small groups. The questions used should require students to explain what they are doing and how they found their solution so the teacher can gauge their level of understanding and the sophistication of the solution process that they are using. What have you found? How did you get your solution? Why do you think that method will work? These kinds of questions should also be used during classroom presentation and discussion of solutions to non-routine problems. In this situation, the questions should also focus on comparing strategies and solutions, and students justifying their approach as well as explaining it. Have we found all the possible solutions? How do you know? Are everyone's solutions the same or different? Why/why not? *Assessment for learning: Aspects of strategic questioning* (Education

Services Australia, n.d.) has practical advice and examples of strategic questioning. Digital tools such as *Socrative* and *Padlet* can be used by the teacher to gather, record and present students' responses to questions or solutions to problems digitally.

When interacting with students who are experiencing difficulty with a task, the temptation is to explain the process again. A better approach is to find out what they understand by asking them to restate the problem or explain the process they are using. A. Watson (2006b) observed two experienced teachers of mathematics who used questioning for formative assessment. Neither teacher used recall nor closed questions; rather, both teachers focused their probing questions on inquiry or problem-solving procedures, affect (attitude) and metacognition. Both teachers supported their students to develop strategies to enable them to work independently on the tasks, but neither of them used high-order questions to scaffold understanding of the mathematics. A. Watson (2006b) advised that teachers needed to be mindful of the mathematics learning objectives and the common misconceptions when posing both closed and open questions in dialogue with students. The SOLO taxonomy has proven invaluable in helping teachers realise how limiting many of the questions they were asking students were in providing insight into the degree of mathematical understanding held by students (Pegg & Panizzon, 2004).

Self-assessment tasks

'How would he know? He never asks me.' This response was given by a student to one of the authors when conducting a research project asking for comment on what the student's mathematics teacher knew about the student's mathematics work. Her response highlights the importance of teachers talking to students in order to form a productive relationship to facilitate their learning. This student's response also suggests that a teacher's judgement about a student's strengths, weaknesses and needs may be quite different from that of the student. Many mathematics teachers understand the role of students' awareness of their own strengths and weaknesses for providing motivation and direction for their learning and have included student self-assessment methods in their practice (McDonough & Sullivan, 2008). Self-assessment, or assessment *as* learning, also contributes to their learning. This 'occurs when students personally monitor what they are learning and use this feedback from their monitoring to make adjustments, adaptations, and even major changes in what they understand' (Earl, 2012, p. 28).

The IMPACT procedure (Clarke, 1988; see Chapter 17) is one approach that teachers can use to gather general information about students' perceptions of their learning and progress. Other approaches include self-assessment questionnaires, for example, KWL charts (see Figure 6.3), graphic organisers or checklists about particular mathematics topics (see the 'Just right' tasks self-assessment tool, Figure 13.2). Written reflections or journal entries are another student self-assessment strategy. These are sometimes called 'exit tickets'. Examples of prompts for journal entries, or exit ticket tasks, include:

- What is one thing you have learned today? What helped you to find that out?
- What is one question you still want to ask? How can you help yourself to find out more about this?
- Create a problem based on today's lesson that could be used in a test. Record the working and solution for this problem.
- Explain how [the mathematics of today's lesson] could be used in the real world.

Open-ended questions

- The volume of a rectangular prism is 32 cubic units. What could be the dimensions of this prism?
- The average age of three people is 18 years. If one person is 12 years old, what could be the age of the other two people?
- The gradient of a straight line is −3. Sketch four different straight-line graphs with a gradient of −3.
- Find ten fractions between $\frac{1}{6}$ and $\frac{1}{3}$.

	KWL Chart: Topic	
Know about	Want to learn about	Have learned about

Figure 6.3 **KWL chart**

Open-ended questions such as these are very useful as catalysts for learning activities (Sullivan & Clarke, 1991), and as items in assessment instruments. There are multiple solutions and methods for solving them, so they are particularly useful in all classrooms, as students have different levels of skill and understanding of concepts. They can be used to find out what students know about a concept when beginning a topic, investigating a concept, applying a skill, or ascertaining students' understanding or skills at the end of a teaching sequence. Since there are multiple solutions and methods for finding solutions to these kinds of questions, a student's solution and method of solution may be located on the continuum of standards used in state curriculum and assessment documents (where these apply).

Rich assessment tasks

Plummer (1999), writing within the context of quality assessment for the revised School Certificate for New South Wales secondary schools, considers assessment tasks to be rich 'if they provide assessment information across a range of course outcomes within one task, optimising students' expression of their learning' (p. 15) and thus reducing the need for additional assessment. According to Plummer (1999), rich assessment tasks:

- Explicitly describe the expectations of the task to the learner.
- Engage the learner.
- Connect naturally with what has been taught.
- Provide opportunities for students to make a start.
- Are learning activities.
- Provide opportunities for students to demonstrate subject knowledge, skills and understandings.
- Focus on the critical areas of learning within a subject.
- Assist teachers to determine the specific help which students may require in content areas. (p. 15)

Importantly with rich assessment tasks, individual students access these tasks at their own knowledge and skill level and pursue the task in widely differing ways (Gough, 2006).

Performance assessment

These tasks can be highly structured, teacher-directed assignments or semi-structured or more open tasks in which students are required to define the goals of the task, the method of investigation and means of reporting. Cognitive research and practical experience suggest that in worthwhile performance-based tasks:

- Students perform, create, construct, produce, or do something.
- Deep understanding and/or reasoning skills are needed and assessed.
- Sustained work is involved—often days and weeks.
- Students are asked to explain, justify and defend.
- Performance is directly observable.
- Engaging ideas of importance and substance are involved.
- Trained professional judgements are required for scoring.
- Multiple criteria and standards are pre-specified and public.
- There is no single 'correct' answer.
- If authentic (i.e., the task is similar to that encountered in real-life contexts), the performance is grounded in real-world contexts and constraints. (McMillan, 2004, p. 199)

Examples of performance tasks

- Carparks: redesign the school carpark to improve its efficiency, capacity and safety.
- Soft drink cans: design an aluminium soft drink can that holds an appropriate volume and is attractive, easy to handle and store, and economical to manufacture.
- Excursion: plan and estimate costs for an excursion for your class to a local place of interest.
- Game: design a board game, including the playing board and a clear set of rules.
- Vacation: plan and develop a budget for a vacation trip for a group of four people to a location within Australia or to another country.

Source: Adapted from Clarke (1997, p. 33).

Tests

Tests and examinations are individual tasks normally undertaken in a defined period under supervised conditions with any access to information and resources (e.g., 'open book') clearly defined. They can be oral, written, digital or online. Tests can also be taken in unsupervised conditions (e.g., take-home exams and online websites). Items or questions in tests should follow the content and activities of mathematics lessons, and students should have access to the resources that have been used during learning. Some online test instruments (e.g., Victorian Curriculum and Assessment Authority, 2015) set items according to responses to prior items. If students have been using technology such as a graphics or CAS calculator, then assessment conditions should allow access to these tools so that their understanding and skills can be assessed comprehensively (D. Leigh-Lancaster, 2010). The use of digital technologies for assessment *of* learning (formal summative assessment) is discussed more fully in Chapter 4.

A variety of question and item types are used in mathematics tests, including multiple-choice, short-answer, extended multi-step analysis and problem-solving questions. Questions can be closed or more open, with students required to do some analysis or problem solving. Examples of questions can be found in past examination papers, state, national and international tests and research articles, as well as in teacher resources (e.g., Beesey et al., 2001). Advice on designing different types of test questions abounds in most classroom assessment texts (e.g., McMillan, 2004). When selecting questions and items for the test, it is important to be aware of the common errors or misconceptions that students may display. As these can be masked by the phrasing of questions or use of particular examples, teachers should include more than one question about a concept to gather more comprehensive information. Deizman and Lowrie (2009) identified that poorly designed graphical images also contribute to student errors on test items. Before being administered, the test as a whole needs to be evaluated and validated (see Figure 6.8).

Commercial textbook electronic resources often provide item banks and templates for constructing tests. These can be very useful, but it is important to check that the structure, language and graphics of these questions are consistent with the teaching and learning activities used in lessons.

As well as ensuring that the test is consistent with the content of teaching, it is also necessary to select items and structure these items to gauge the diversity of understanding, levels of mathematical thinking and performance. This involves more than including simple and difficult questions or questions relating to different standards in the curriculum. Items testing higher-order thinking and relational thinking must be included, as proposed in the SOLO taxonomy (Pegg, 2003). Kastberg (2003) illustrates how Bloom's Taxonomy can be used for designing mathematics tests, and includes algebra examples for each of the levels of thinking in this taxonomy (knowledge, comprehension, application, analysis, synthesis and evaluation).

Involving students in designing the test is a very useful and successful assessment strategy. Students are asked to construct questions for the test (see the example of self-assessment and exit tickets above), with the teacher either providing guidelines or leaving the requirements open ended. Students' questions provide insight into the depth and breadth of their knowledge and level of thinking, as well as the skills and understanding revealed in the answers they provide for their question. Using their questions in the test is also affirming for students.

Test items

Four test items concerning decimals and percentages from *NAPLAN*, PISA, TIMSS and Department of Education and Training, Victoria, are provided in Figures 6.4–6.7. Two of these are multiple-choice items, and two are short-answer questions. The *NAPLAN* test uses both multiple-choice and short-answer items. The items cover each of the strands in the mathematics curriculum and at different minimum year level standards, so that Year 9 test items include items from Year 5 to advanced Year 10. The PISA study for 15-year-old students includes items for four content strands: uncertainty and data, quantity, space and shape, and change and relationships. The items for each content strand are also designed to concern three types of mathematical thinking: interpreting, employing (using) or formulating (modelling). Likewise, the TIMSS items also concern content and mathematical thinking. The content strands are number, algebra, geometry, and data and chance; the mathematical thinking strands are knowing, applying and reasoning. Figure 6.7 is an interview item, that is, the teacher conducts individual interviews with students using these items, thus providing an opportunity for the teacher to ask further questions to elicit the students' thinking when finding the solution.

29	A ticket costs $75,
	A fee of 10% is added to the price.
	Which calculation will give the new price?

75 + 10   75 + 0.1   75 × 0.1   75 × 1.1
  ◯   ◯   ◯   ◯

Source: ACARA (n.d.), NAPLAN Year 9 Numeracy Non-calculator Example Test.

Figure 6.4 **NAPLAN Year 9 multiple-choice item**

Which car?

Chris has just received her car driving licence and wants to buy her first car.
This table below shows the details of four cars she finds at a local car dealer.

Model:	Alpha	Bolte	Castel	Dezal
Year	2003	2000	2001	1999
Advertised price (zeds)	1800	4450	4250	3990
Distance travelled (kilometres)	105 000	115 000	128 000	109 000
Engine capacity (litres)	1.79	1.796	1.82	1.783

Source: Thomson et al. (2013a, pp. 302–303).

Figure 6.5 **PISA (15-yr-old) Quantity and Employ test item**

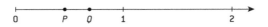

P and *Q* represents two fractions on the number line above.
P x *Q* = *N*.
Which of these shows the location of *N* on the number line?

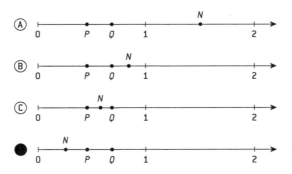

Source: Thomson et al. (2012, p. 141).

Figure 6.6 **TIMSS Year 8 Number and Reasoning
multiple-choice test item**

18. Reserve bank and Chocolate Milk

Give the student some blank paper and a pen.

Not long ago, the Reserve Bank announced that interest rates were going up one quarter of one percent.

a) How would you write one quarter of one percent with numbers?
b) Could you write it a different way, that is still equal to one quarter of one percent?
c) Show the student the chocolate milk drink and percentage card.
  Pen and paper can be provided if this is helpful.

Source: Department of Education and Training, Victoria (2014a).

Figure 6.7 **Fraction and Decimal interview item**

REVIEW AND REFLECT: Compare these test items (Figures 6.4–6.7):

What could you find out about your students' knowledge of decimals and/or percentage when using each of these items?

Identify the mathematical learning outcomes assessed by each of these items.

Which mathematical proficiency (understanding, fluency, problem solving or reasoning) do these items assess?

What are the limitations, or weaknesses, in each of these items?

In what situations, and for what purposes, would you use multiple-choice, short-answer or extended-response items in a mathematics test?

***

REVIEW AND REFLECT: Consider the different forms of assessment listed and discussed above.

Write a list of advantages and disadvantages for each form of assessment.

Identify the most appropriate purpose for which this form of assessment could be used.

Which of these forms of assessment are used in your secondary school and why?

To what extent and how do the conditions and policy of the school and state assessment authority limit a teacher's choice of assessment?

Preparing assessment tasks for students

The Australian Curriculum, Assessment and Certification Authorities ACACA, (1996) drew up guidelines for both quality and equity in assessment. These guidelines are reproduced in the box below and reflect the current assessment principles documented by the various state authorities around Australia. They should be borne in mind when making decisions about the type of assessment task to use, designing and constructing tasks or tests, and selecting items and questions for assessment purposes.

Guidelines for assessment quality and equity

1. An assessment item should assess what the item writer intends it to assess and only what on face value it purports to assess.

2. Students should not, unless there is a specific and justifiable reason for doing so, have to decode some hidden instructions or clues on how to answer an assessment item.

3. Specialist language or jargon in an assessment item should only be used to aid clarity and accuracy and if that specialist language is an integral part of the teaching and learning in that subject.

4. The reproduction of gender, socioeconomic, ethnic or other cultural stereotypes should only be used in assessment items after careful consideration as to its necessity.

5. To allow students to demonstrate their command of what the item is supposed to assess, the item should be presented clearly through an appropriate choice of layout, cues, visual design, format and choice of words, and state its requirements explicitly and directly.

6. The use of background material and requirement of assumed knowledge in an assessment item should only be used when the item writer can reasonably presume all students have ready access to these.

7. Assessment criteria should be explicit, clear, unambiguous and declared in advance.

8. The criteria should allow students to identify appropriate ways to demonstrate command of the required knowledge and skills.

9. The criteria should also allow the marker to recognise, where appropriate, different ways in which a student may demonstrate command of the required knowledge and skills.

Source: (ACARA, 1996).

> **REVIEW AND REFLECT :** Discuss the above guidelines and the equity issues concerning assessment. Why are these important?
>
> Using the guidelines, write a critical reflection on an assessment task that you have collected from your secondary school or designed and used in your teaching.

Once an assessment task has been found, modified or designed, it is important to ensure that student responses provide useful evidence of their learning. Typical quality assurance steps to ensure this happens are summarised in the flow chart in Figure 6.8. (See Bush & Greer, 1999 for more information on preparing assessment tasks.)

Strengthening the consistency of teacher judgements

Teachers have always had complete autonomy when it comes to using assessment for formative purposes. With the advent of a national mathematics curriculum for Years 7–10, and the demands on teacher accountability, teachers, depending on the state system, are under more pressure to use the year level standards when analysing student achievement for summative purposes to ensure consistency of teacher judgements. Some states (e.g., Queensland) require teachers to use standards-based criterion referencing when analysing performance and summative student assessment.

Consistency of teacher judgements is very important for high-stakes assessment that is instrumental in determining which students win sought-after places at high-status tertiary institutions. Since the 1980s, school-based assessment has risen in prominence as either a replacement for, or an adjunct to, external public examinations at the end of secondary schooling. School-based assessment relies on teachers' ability to make assessment judgements that are consistent across students and tasks, and consistent with the judgements made by other teachers within and outside the school. Morgan and Watson (2002) note that interpretive assessment judgements are 'influenced by the resources individual teachers bring to the assessment task' (p. 103), and that this may lead to inequity when different interpretations are made of students' achievements. However, the difficulties of being able to make a sound judgement based on equitable practices should not be a reason for doing nothing, as a range of strategies is available for strengthening the consistency of teacher judgements. Two such strategies—moderation and the use of rubrics—are discussed below.

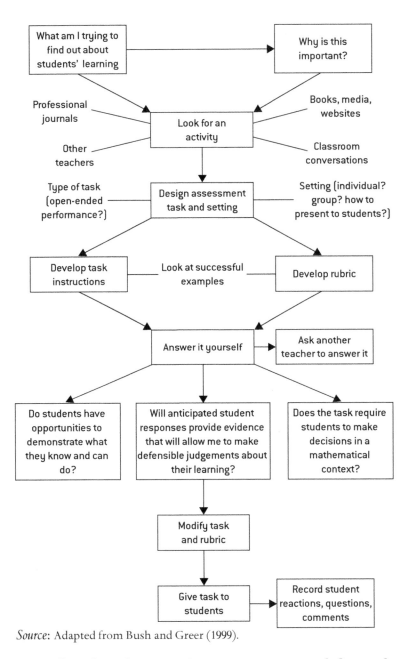

Source: Adapted from Bush and Greer (1999).

Figure 6.8 **Flowchart for preparing an assessment task for students**

Moderation

In some states and territories (e.g., Queensland and the ACT), teachers' voices are given high status and 'teachers' qualitative judgements' are considered an important component of teachers' professionalism underpinning school-based assessment systems (Sadler, 1987, p. 193). In Queensland, for example, 'teachers, informed by the syllabus principles of exit assessment and using evidence collected over time across a range of techniques and contexts, are best placed to make judgements about students' levels of achievement' (Queensland Curriculum and Assessment Authority, 2015). An externally moderated entirely school-based assessment system is used. The external experts are practising teachers from other schools and tertiary educators.

Rubrics

Assessment rubrics are tools for rating the quality of student performance that identify the anticipated evidence that will be used for making judgements. They are used for all types of assessment. Publishing these rubrics for students along with the assessment task makes the expectations of the task explicit for students, and encourages students to be self-directed and reflective in their learning.

Generic, or holistic, rubrics can apply to a broad spectrum of tasks. Holistic rubrics commonly use general descriptors for levels of performance, such as exemplary, excellent, good, satisfactory and not satisfactory. The example in Table 6.3 illustrates how teachers might use assessment judgements to inform their teaching.

Table 6.3 Everyday rubric grading

E	Excellent example	Meets or exceeds expectations. Complete, clear communication. Clear understanding. Any error is trivial.
M	Meets expectations	Understanding is evident. Needs some revision or expansion, but written comments are enough. No additional teaching needed.
R	Needs revision	Partial understanding is evident, but significant gap(s) remain. Needs more work/teaching/communication.
F	Fragmentary	Clearly misunderstands. Insubstantial attempt made.

Source: Stutzman and Race (2004, p. 36).

Task-specific rubrics apply to a specific task. The rubric in Table 6.4 and the following box was used in the Middle Years Numeracy Research Project (Siemon, personal communication, 2002) for an algebraic problem-solving task involving the use of inverse relationships, simple ratios, division and application in a familiar context.

Medicine doses

Occasionally, medical staff need to calculate the child dose of a particular medicine, using the stated dose for adults. The rule used is as follows:

$$\text{Child dose} = \text{Adult dose} \times \frac{Age}{Age + 12}$$

A. If the adult dose for a particular medicine is 15 mL, what would be the appropriate dose for a six-year-old child?
B. A nurse uses the formula to work out the dose for an eight-year-old boy. She correctly calculates it as 6 mL. What was the adult dose?
C. At what age would the adult dose be the same as the child dose? Explain your reasoning.

Table 6.4 Rubric for medicine doses

	Descriptor	Score
A	No response or incorrect	0
	Information from formula used but incorrect or incomplete calculation	1
	Correct (5 mL), appropriate use of formula or recognition of proportion	2
B	Incorrect or no response	0
	Fraction component identified but incomplete, e.g., recorded as 6 divided by 6 + 12	1
	Fraction correct $\left(\frac{6}{18}\right)$ but not interpreted appropriate to context	2
	Fraction given as $\frac{1}{3}$	3
C	No response or incorrect	0
	Information from formula used but incorrect or incomplete calculation	1
	Correct (20 mL), appropriate use of formula	2

Source: Beesey et al. (2001); Siemon (personal communication, 2002).

Multiple analytic rubrics describe levels of performance for individual aspects of performance that could stand alone. These rubrics include scoring systems for test or exam tasks, as well as for performance and problem-solving tasks. An example of an analytic rubric is provided in Table 6.5. It was developed by the Mathematical Association of Victoria (2006) for its *Maths Talent Quest*, which is an open-ended mathematical performance task involving students in mathematical inquiry, investigation and problem solving. Students are able to present their entry in a variety of formats. The rubric used for competition judging has descriptors and scores for communication, understanding, originality and presentation. Table 6.5 shows the descriptors for only the communication category.

REVIEW AND REFLECT : Comment on the strengths and weaknesses of the sample rubrics for making judgements about students' performance.

Collect other examples of assessment rubrics from schools, the internet or other references. Select a rubric and explain why it would be useful for assessment in your mathematics teaching.

A. Watson (2006a, p. 153) argues that assessment should focus on what mathematicians do when inquiring and constructing meaning. She proposes the following verbs for some of these actions:

exemplifying	specialising	completing
deleting	correcting	comparing
sorting	organising	changing
reversing	varying	generalising
conjecturing	explaining	justifying
verifying	convincing	refuting

Use a selection of these or other verbs that describe what mathematicians do to design an assessment rubric for a non-routine problem-solving task.

Source: Board of Studies New South Wales (2002), Mathematics Years 7–10 Syllabus.

Table 6.5 Rubric for communication category judging and scoring criteria for *Maths Talent Quest*

Sub-category	High	Score	Medium	Score	Low	Score	Not evident	Score
A **Approach to the investigation**	The approach to the investigation is explicit with aims and goals, a thorough plan for solving and conclusions clearly stated.	4	The approach to the investigation is often clear with aims, goals, a plan for solving and with conclusions stated.	3	The approach to the investigation and aims and goals are stated.	2	There are no aims and goals given and the approach to and planning for the investigation is unclear. Conclusions were unclear or not stated.	0
B **Use of mathematical terminology**	Precise and appropriate mathematical terminology and notation is used to support mathematical thinking and communicate ideas.	4	Mathematical terminology and notation in the solution is used to share ideas.	3	Some mathematical terminology and notation to communicate is used.	2	No mathematical terminology and notation is used or every day, familiar language to communicate ideas is used.	0
C **Clear and detailed explanations**	Explanations with clear and effective detail of how and why solutions were made are given.	4	Explanations with detail are mostly given about how and why most solutions were made.	3	Explanations are incomplete or ineffective as to why the solution makes sense.	2	Does not explain the solution or the explanation cannot be understood or related to the investigation.	0
D **Use of diary or learning journal**	The diary or learning journal is an explicit, reflective account of the processing and inquiry in the investigation.	4	The diary or learning journal gives some explanation of the mathematical thinking and learning involved.	3	The diary or learning journal offers very little explanation of the mathematical thinking and learning involved.	2	There is no diary or learning journal.	0

Source: MAV (2006).

Recording, profiling and reporting

Recording

Keeping records of students' progress and achievement is a teaching responsibility and an important part of the process of providing feedback to students and informing parents and other teachers. These data are also critical for evaluating one's own performance as a teacher of mathematics and helping make decisions about the focus of lessons and suitable learning activities for students.

Recording of student performance can take many forms. For example, teachers may wish to allocate marks in relation to the criteria for assessing learning in the *Diagonals of a quadrilateral* activity (see Figure 6.2). This could be recorded as in Figure 6.9. Alternatively, the teacher may use specific learning objectives (or intentions) to document students' achievement of these objectives, as shown in Figure 6.10.

Spreadsheets and electronic records provided online or commercially can be especially useful for recording achievement data. When selecting or designing these forms of records, it is important to ensure that the format is flexible enough for the diversity of assessment activities used and that it will produce summative assessment data appropriate for the mode of reporting required by the school.

Reporting

Reporting provides feedback for stakeholders and teachers should be mindful of the importance of feedback for students' mathematical dispositions and their mathematical thinking and behaviours (Hattie & Timperley, 2009). We need to know the audience (student, teacher, parent, school leaders) and take the audience into account in both how and what we report. Modes of reporting include:

- written reports (either paper or electronic)
- interviews with parents or carers and/or students
- culminating performances or presentations
- student folios.

REVIEW AND REFLECT : Consult your local curriculum and assessment authority website and determine the regulations with regard to assessment and reporting. Find out about one secondary school's reporting policy, process and requirements. Share your findings with your pre-service colleagues and discuss the merits of different formats and approaches.

Parents and carers of students in Years 7 and 9 participating in state and territory numeracy assessments receive individual reports that show where the results place a student's achievement in relation to the national numeracy benchmarks and in comparison to other students in the year level. State and territory results from the international mathematics assessment programs such as TIMSS and PISA can be found in specific international

Name: Maria Activity: Diagonals of a Quadrilateral	Class: 8 F
Criteria	**Performance**
Demonstrates knowledge and understanding of the nature of different quadrilaterals	4/5
Draws a valid conclusion about the diagonals of quadrilaterals	2/2
Communicates mathematical ideas	3/3

Figure 6.9 **Recording feedback and assessment using scores**

Party Planet TEACHER TRACKING DOCUMENT

Student name	Can identify tenths, hundredths and thousandths	Can divide a rectangle into halves	Can divide a shape into thirds	Can divide a shape into fifths	Can identify the numerator and the denominator	Can order and compare fractions	Can identify equivalent fractions	Can place percentages, fractions and decimals on the same number line	Can add & subtract fractions with the same denominator	Can add & subtract fractions with a different denominator	Can multiply fractions	Can divide fractions	Comments

Source: Parsons and Reilly (2012, p. 6).

Figure 6.10 **Teacher tracking document**

and national reports, some of which are available from websites such as www.acer.edu.au and www.pisa.oecd.org. Summaries of the results are provided in the annual *National Report on Schooling in Australia*, which is available on the MCEETYA website (www.mceetya.edu.au).

REVIEW AND REFLECT : Read the following quotation from a Queensland teacher: Like, it's like marks, I mean most maths teachers in Knowledge and Procedures want to use marks. Why? What's a mark? What does a half a mark or a mark here or there mean? And yet they insist on using marks to grade kids and the only real reason is because it is easy. The same as exams, what do exams tell you? The only time, the only reason exams are any good are for teachers for marking. They are not good for kids, they are not good for developing understanding, so what's the point? (Extract from CCiSM project interview, 2005)

- Discuss this teacher's beliefs about assessment.
- Write a statement explaining your beliefs about mathematics assessment and the practice that you will use when teaching mathematics.

Conclusion

In this chapter, we have discussed the purpose and methods of assessing students' mathematical learning. It is clear that monitoring student progress and discussing understanding and achievement are critical for students' learning, and that mathematics teachers use many different methods to gather information about students' understanding and skills. This information should inform teachers' decisions in the moment and for further teaching and learning activities for the current focus or future topics. Given the advantages and disadvantages of the various assessment and reporting methods discussed in this chapter, it is important for teachers to be able to justify the practices that they use.

Recommended reading

Callingham, R. (2011). Assessing statistical understanding in middle schools: Emerging issues in a technology-rich environment. *Technological Innovations in Statistics Education, 5*(1), 1–12.

Clarke, D. J. (1997). *Constructive assessment in mathematics: Practical steps for classroom teachers.* Berkeley, CA: Key Curriculum Press.

Earl, L. M. (2012). *Assessment as learning: Using classroom assessment to maximize student learning.* Thousand Oaks, CA: Corwin Press.

Education Services Australia. (n.d.). *Assessment for learning.* Retrieved from www.assessmentforlearning.edu.au/professional_learning/

Plummer, F. (1999). Rich assessment tasks: Exploring quality assessment for the School Certificate. *SCAN, 18*(1), 14–19.

Stiggins, R., Chappuis, J., Chappuis, S. & Arter, J. (2014). Assessment *for* learning and *of* learning. In *Classroom assessment for student learning* (Chapter 2, pp. 29–46). Upper Saddle River, NJ: Pearson.

PART 3

Teaching and learning mathematical content

CHAPTER 7
Teaching and learning number

When students begin secondary school, they have already had many years of school and life experiences with number. Most students strongly associate mathematics with numbers and readily appreciate the application of number to real problems. In secondary school, as in the primary years, number concepts and skills are needed for all other mathematics domains. By the end of primary schooling, students have learned the underpinning concepts of whole numbers and fractions, and the skills to operate with numbers and think mathematically to solve problems. However, for most students, their understanding of these concepts and facility with operations is unlikely to be secure. At the beginning of secondary school, there is a diverse range of competencies and strategies used by students, and the mathematical skills and achievement of far too many students actually decline in their first year of secondary school (Siemon et al., 2001; Vale et al., 2013). As their secondary mathematics teacher, you need to begin by finding out what your students know and how they reason and solve problems for each topic in number. With this knowledge, you can then pay attention to consolidating their understanding and build on their knowledge to develop more efficient strategies, improve skills, operate with the full range of numbers in our number system and use effective strategies for more complex problem-solving situations.

In this chapter, we will assist you to 'unpack' some of the key number concepts and describe teaching approaches that both connect with students' primary experiences of learning and support mathematical understanding, fluency and flexibility with procedures, problem solving and reasoning. We will focus on concepts, skills and reasoning that are important for mathematical literacy and success in senior secondary mathematics at the expense of discussing particular topics in senior secondary mathematics, such as complex numbers or matrices. The key topics include place value with decimals, fractions, mental computation, multiplicative thinking (including operations with fractions and decimals), proportional thinking (including ratio and percentage), exponential thinking, irrational numbers and integers. Of course, students taking vocational mathematics subjects in senior secondary mathematics will revisit many of the topics discussed in this chapter to solve problems for particular contexts. Since it is important for secondary teachers to understand the development of concepts and mathematical thinking over the long term of a child's mathematical education, we will first present an overview of students' number learning in primary school.

Number in the primary years

Primary teachers use a range of materials and learning aids to assist children's learning of number (Booker, Bond, Sparrow & Swan, 2004). Concrete materials enable students to visualise concepts in multiple ways. The materials are important for connecting mathematical symbols and language with concepts. Initially, young children count by rote, and then they use one-to-one correspondence and touch objects to count by ones and count all when adding. In the early years, primary teachers then build more efficient strategies for addition using the part–part–whole concept of number, counting on from largest number, doubles, near-doubles and building to ten to establish addition and subtraction facts (Department of Education and Training, New South Wales, 1998; Wright, 1998). You may observe that some secondary students who are not proficient still use counting on by ones (counting on fingers or with tally marks) for addition.

Children begin to understand fractions and decimals in context and model these using materials in primary school, normally from Year 2 or 3. Linear, region and area models are used and they continue to need to use concrete materials and visual representations in early secondary school to assist them to understand equivalence of fractions and decimals, to compare and order fractions and decimals, and to understand operations with fractions and decimals. For example, Figure 7.1 shows a fraction wall made of fraction strips (a linear

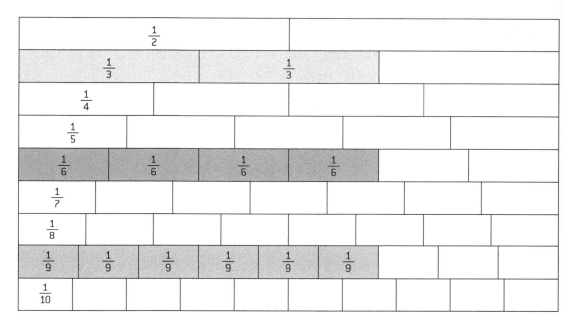

Figure 7.1 **A fraction wall showing** $\frac{2}{3} = \frac{4}{6} = \frac{6}{9}$

model), and children can use this wall to find equivalent fractions and play a game to practise finding equivalent fractions (Pearn, 2007; Way, n.d.).

In the upper primary school, children extend their knowledge of the 'Base 10' number system to include large numbers and decimal fractions. Through a range of application tasks, they develop a sense of very large and very small numbers and learn to compare them, though these concepts may not be consolidated for many students commencing secondary school. Children commonly explore the concept of a million through problems such as: How long would it take one person to hand out one million pamphlets if you walked door to door in Sydney?

Developing number sense and skills with all types of numbers continues during secondary schooling, and it is important to continue the process of developing and connecting the language, symbols and physical representations of whole numbers, fractions and decimals using the concrete materials and visual representations commenced in primary school and applying this approach to new concepts such as integers.

Number in the secondary years

In the secondary years, students consolidate and extend their number sense for solving problems and making judgements (McIntosh, Reys & Reys, 1997). Students with number sense are able to:

- Calculate mentally
- Use approximate values to estimate
- Understand the connections and relationships between numbers and use them when calculating
- Have a sense of the size of a number in relation to other numbers
- Switch between equivalent representations of numbers
- Assess the reasonableness of a solution when solving problems

Students in secondary school also learn about other types, or sets, of numbers:

- integers (positive and negative numbers)
- ratios
- rational numbers (numbers that can be expressed as a vulgar fraction)
- irrational numbers (numbers that cannot be expressed as a vulgar fraction)
- real numbers (the set of rational numbers, irrational numbers and zero)
- (for some students) complex numbers.

Recent Australian research and results of international studies of mathematics achievement shed light on common misconceptions and the concepts and types of number problems that cause the most difficulty for secondary students. A large study of students in the middle years of school—that is, Years 5–8—in Victoria found that many students had difficulty with:

- explaining and justifying mathematical thinking
- reading, manipulating and using common fractions, decimals, ratio, proportion and formulae
- thinking multiplicatively
- generalising simple patterns
- interpreting results in context (Siemon et al., 2001)

Further, there was a considerable range of achievement levels for students in a single-year level, up to seven years of schooling. Other researchers have observed weaknesses in number knowledge and skills, including mental computation (Callingham & McIntosh, 2002), comparing decimals (Moloney & Stacey, 1997; Steinle, Stacey & Chambers, 2002), decimal density and negative decimals (Widjaja, Stacey & Steinle, 2008, 2011) and ratio and proportion (Dole, Cooper, Baturo & Conpolia, 1997; Livy & Vale, 2011). Performance by 15-year-old Australian students on problems concerning quantity (number) for the 2012 PISA was not as strong as performance in algebra and chance and data items (Thomson et al., 2013a).

Siemon et al. (2001) recommended that teachers in the middle years focus the speaking and listening among students and teachers in mathematics classrooms on building meaning and making connections between ideas. This is because asking students to represent concepts in multiple ways, to explain their reasoning and to justify their reasoning, contributes to concept and skill development. More than this, effective teachers orchestrate whole-class discussion by sequencing presentation and discussion of strategies and reasoning beginning with the least efficient and building to the most efficient (Stein et al., 2008). In this way, each student's solution is valued and students can more easily discern differences in solution strategies and identify a more efficient strategy than their own that they could use in future. We will now discuss the teaching and learning of foundation number topics in secondary school. You will need to use some particular approaches for students who are well behind their peers. These are discussed in Chapter 13.

Place value

Understanding the value of digits according to their place in a number is the basis of counting and comparing all numbers. Children in the primary years learn that the values are structured on multiples of 10 (Base 10 number system), that zero is a placeholder, and that the decimal point is a marker that separates the whole elements from fractional elements of the number. In other cultures, a comma is sometimes used instead of a full stop. With the aid of a place value chart (see Figure 7.2), saying aloud the place value when reading decimal numbers will help students who have difficulty comparing decimals. For example, the number 31.265 should be read as thirty-one and two tenths, six hundredths and five thousandths, or thirty-one and two hundred and sixty-five thousandths, and not as

Tens	Ones •	Tenths	Hundredths	Thousandths
3	1 •	2	6	5

Figure 7.2 **Place value chart**

thirty-one point two six five. This approach will also help students to interpret large numbers that are common in the media—for example, $3.6 billion—and be aware that the scientific notation is 3.6×10^9. Number expanders are useful for renaming, and learning is enhanced when students make their own number expander for decimals (Booker et al., 2004). Moloney and Stacey (1997) found that only 75 per cent of Year 10 students could compare decimals with 80 per cent accuracy; higher proportions of students make errors in the earlier secondary grades. Students also have difficulty relating their everyday use of decimals to school mathematics (Irwin, 2001).

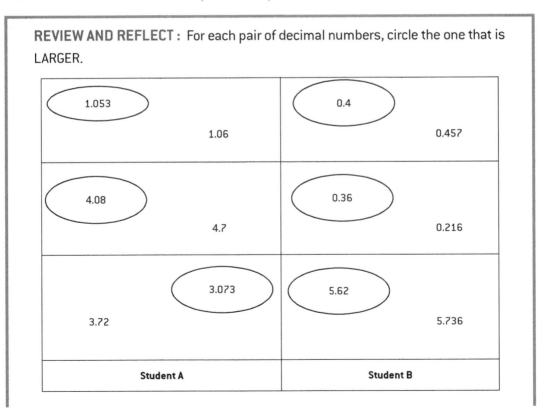

REVIEW AND REFLECT : For each pair of decimal numbers, circle the one that is LARGER.

Student A	Student B
(1.053) 1.06	(0.4) 0.457
(4.08) 4.7	(0.36) 0.216
3.72 (3.073)	(5.62) 5.736

The above figure shows responses from two students in research conducted by Steinle et al. (2002) for three items.

Analyse these two students' responses. How is their thinking similar and/or different? Comment on their level of skill.

Write a pair of decimal numbers that Student A would be likely to compare incorrectly, but Student B would compare correctly.

Write another pair of decimals Student B would be likely to compare incorrectly while Student A would compare correctly.

How could you show these students that their thinking is incorrect?

Misconceptions arise for a number of reasons. Steinle et al. (2002) observed two main persistent misconceptions among a number of others. One is the 'longer is larger' fallacy. Students who think longer is larger are using 'whole-number thinking' (also prevalent as the source of misconceptions with fractions). The 'shorter is larger' fallacy often occurs after students are introduced to negative integers, and these students have difficulty placing decimals smaller than one on a number line labelled with integers (see Figure 7.3). These students seem to think that decimal numbers are below zero or negative (Irwin, 2001; Widjaja et al., 2008). Other students have difficulty identifying and interpreting the decimal numbers between decimals (Widjaja et al., 2011). Steinle and Stacey

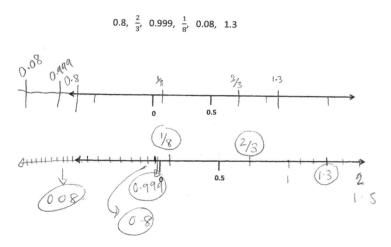

Figure 7.3 **Two examples of student decimal misconceptions**

have designed the Online Fraction and Decimal Interview (Department of Education and Early Childhood Development, 2013) so that teachers could use this formative assessment tool to identify strengths and weaknesses in students' understanding. Linear attribute blocks that show students the value of digits in decimal numbers, together with greater familiarity with number lines, enhances understanding of the order and density of decimal numbers including negative decimals (Steinle et al., 2002). Various online interactive learning objects are useful for number line tasks including exploring numbers 'in between'. Open-ended investigations, games and application problems related to familiar contexts for students improve students' knowledge of decimals (Irwin, 2001).

REVIEW AND REFLECT : Develop a set of materials and tasks that would be suitable to use with the students whose misconceptions are recorded in the examples above.

Find out about and make some linear attribute blocks (Steinle et al., 2002).

Make a number expander for decimals with four decimal places (Department of Education and Training, Victoria www.education.vic.gov.au/school/teachers/teachingresources/discipline/maths/continuum/Pages/numexpand.aspx).

Search for games and other resources including online interactive learning objects for teaching decimal place value.

Fractions

While students have been learning about fractions since the early years of schooling, conceptual understandings and operations with fractions are difficult for many students in secondary school (Brown & Quinlan, 2006; Callingham & McIntosh, 2002; Pearn & Stephens, 2004; Siemon et al., 2001). A focus on rules without understanding of fractions and the persistence of whole-number thinking are sources of many misconceptions and errors in students' work (Gould, 2005; Pearn & Stephens, 2004).

Many students understand a fraction to be part of a whole rather than seeing a fraction as also meaning a part of a collection and the operation of division (Gould, 2005). Students at risk in secondary schools do not accurately illustrate equal parts when drawing models of proper fractions, especially when trying to use circles, and many students have difficulty modelling improper fractions using a number line (Gould, Outhred & Mitchelmore, 2006; Pearn & Stephens, 2004).

REVIEW AND REFLECT : The two tasks below require students to place fractions on a number line, but the second is more difficult. Propose an explanation for this difference.

1. Put a cross where you think $\frac{8}{5}$ would be on this number line.

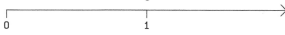

2. Put a cross where you think 1 would be on this number line.

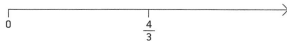

Students also need to understand that $\frac{5}{6}$ means $5 \div 6$—that is, 5 divided by 6 or 5 shared between 6. Consider the problem and three typical solutions shown in Figure 7.4. While many students will place cuts in the pizzas to show that each family would get $\frac{5}{6}$ of a pizza, either by slicing each pizza into six pieces or by slicing $\frac{1}{6}$ off each pizza, one or two students in your class will probably slice three pizzas in half and two into thirds to give a result of $\frac{1}{2}+\frac{1}{3}$. These students use an Ancient Egyptian notion of fractions with numerators only represented by 1 unconsciously in this real-life situation. Capable students could be challenged to design similar problems and show that a fraction can be represented by a sum of other fractions with numerators of 1. (Note that the Ancient Egyptians used only one fraction with a numerator other than 1 and that was $\frac{2}{3}$.)

Making sense of fractions also means that students are able to compare fractions and find equivalent representations of fractions. Fractions modelled using paper folding to create area models, such as fraction strips and rectangles, and fraction walls made of multiple fraction strips, can be used for open-ended investigations such as 'find ten fractions that are equivalent to $\frac{2}{3}$' or 'find ten fractions between $\frac{1}{4}$ and $\frac{3}{4}$' (see Figure 7.1). Formalising these investigations through discussion of the findings enables students to appreciate that equivalent fractions are formed when multiplying or dividing by one $\left(\frac{5}{6}\times\frac{3}{3}=\frac{15}{18}\ and\ \frac{21}{35}\div\frac{7}{7}=\frac{3}{5}\right)$.

Six families want to share five rectangular pizzas. How would you cut the pizzas so that each family has an equal amount?

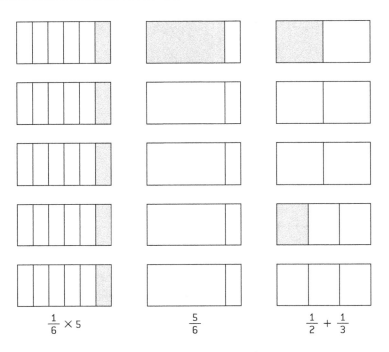

$$\frac{1}{6} \times 5 \qquad\qquad \frac{5}{6} \qquad\qquad \frac{1}{2} + \frac{1}{3}$$

Figure 7.4 **Three solutions for 5 ÷ 6**

Some students use whole-number thinking when comparing fractions with unlike denominators. For example, a student may argue that $\frac{2}{3}$ is larger than $\frac{7}{10}$ because there is only one difference between the numerator and the denominator in this fraction (Pearn & Stephens, 2004). These students perceive fractions as two whole numbers. To develop their fraction sense, students need to be able to visualise fractions that are near 0 or 1 or some other common fraction, such as $\frac{1}{2}$. Having a mental image of fractions with the same numerator becoming smaller and smaller as the denominator increases will assist students who are whole-number thinkers. Students can then check their reasoning by finding equivalent fractions.

For these reasons, it is not wise to approach the teaching of comparing and adding fractions using the cross-multiplication algorithm $\left(\frac{a}{b} + \frac{c}{d} = \frac{ad+bc}{bd}\right)$ before finding out about students' conceptual understanding of fractions. In the *Australian Curriculum: Mathematics* (ACARA, 2012b) students in primary school are expected to add and subtract fractions with the same or related denominators. The learning sequence for adding and subtracting fractions should proceed by first modelling using materials and number lines, then estimating, and deriving algorithms with a well-designed investigation (Brown & Quinn, 2006; Way, n.d.).

REVIEW AND REFLECT : The following two problems encourage the use of visual images and estimation:

Picture these two sets of fractions in your mind.

$$\frac{4}{10} \quad \frac{4}{47} \quad \frac{4}{8} \qquad\qquad \frac{99}{100} \quad \frac{6}{7} \quad \frac{15}{16}$$

Order the fractions in each set from smallest to largest. Explain the visual images, or number sense that you used to order these fractions.

Use four of these digits: 1, 3, 4, 5, 6 or 7 (only once), to create a sum of two fractions that is less than but as close as possible to 1. (Try writing these digits on small square pieces of paper so that you can try digits in different places using estimation to find a solution.)

$$\frac{\square}{\square} + \frac{\square}{\square} < 1$$

Document your thinking processes. In a group, share and compare your strategies and discuss why you chose particular ways to solve these problems. Discuss the mental pictures of fractions, the fraction concepts and the skills that you used.

Use these problems with a group of secondary students, assess the students' understanding of fractions and report your findings.

Prepare a set of learning activities to address any misconceptions that you observe. Include multiple representations of fractions and games for practice (see Oliver, 2005 for an example).

Mental computation

Mental computation is the ability to use number facts and number sense to solve operations without the aid of computational tools, or pen and paper. It is more than automatic recall, though knowledge of the basic number facts for addition, subtraction, multiplication and division is needed for efficient thinking strategies. Mental computation has received a lot of attention in mathematics curriculum documents since the early 1990s. The emphasis resulted from research that showed that adults used mental strategies and computational tools such as calculators more often than pen-and-paper algorithms to estimate and calculate with accuracy (e.g., Northcote & McIntosh, 1999). Teaching materials have been developed for the primary and middle years, but secondary textbook writers have paid scant attention to the continuing development of flexible, efficient mental computation strategies with whole numbers, fractions, decimals and percentages.

Some students seem to develop very flexible ways of thinking with number in spite of an absence of structured learning opportunities in the classroom (Downton, 2009). Callingham and McIntosh (2002), in their study of students from Year 3 to Year 10, found that students' mental computation competence 'drops sharply between Years 6 and 7' (p. 159). Forty per cent of Year 7 students were not yet competent with table facts and inverses, adding and subtracting two-digit numbers, multiplying two-digit numbers by a single-digit number, or halving even two-digit numbers. Year 7 students also made errors adding and subtracting decimals to one decimal place, adding halves and quarters beyond one, subtracting familiar unit fractions from one, and finding a half, a quarter or 25 per cent of two- or three-digit numbers. McIntosh (2002) further explains that many students' mental computation errors with whole numbers were procedural, whereas the errors made with decimals, fractions and percentages were because students did not understand the concept.

Students who use efficient mental strategies demonstrate understanding of the field laws (illustrated in Figure 7.5) and the order of operations to partition numbers and/or reorganise the sequence of thinking, and so lighten the cognitive load (Wright & Downton, n.d.).

Commutative	Associative	Distributive
$a + b = b + a$	$(a + b) + c = a + (b + c)$	$(a + b) \times c = (a \times c) + (b \times c)$
$a \times b = b \times a$	$(a \times b) \times c = a \times (b \times c)$	$(a \div b) \times c = (a \div c) + (b \div c)$

Figure 7.5 **Field laws**

REVIEW AND REFLECT :

1. The scores for a football match are shown on the television screen as:

    Ess   *57*

    N Melb   *76*

    How far behind is Essendon?

2. There are 24 lollies in a packet of MintChocs. If I have 8 packets of MintChocs, how many lollies are there altogether?

   Solve these two problems mentally and then document your thinking process for each. In a group, share and compare your strategies and discuss why you chose particular ways to solve the problem. Use a number line to model the subtraction strategies.

***

Consider the following question and student responses (Downton, 2009):

Sam read 72 books during the readathon, which was 4 times as many books as Jack. How many books did Jack read?

Bindy: 6 x 12 is 72 and 6 x 6 is 36 and half of 72 is 36 and because it's 4 times I need to halve 36 and that's 18, so that means 4 x 18 is 72 so Sam would read 18 books.

Mark: 12 x 6 is 72, but that's 6 times as many and it has to be something times 4 is 72. 12 x 4 is 48 and double that would be 24 x 4, which is 96. I still think the doubling strategy could work. 8 x 4 is 32, 16 x 4 would be 64, 20 x 4 is 80 but that's too much so it has to be between 12 x 4 and 20 x 4. 20 x 4 is 2 x 4 too much so I need to take that away so it's 18 x 4, so Sam read 18 books. (p. 167)

> Discuss these two students' responses and the field laws that they used for their mental strategy (see Figure 7.5). How would you record their reasoning on a whiteboard so that other students could see and make sense of each strategy?

Not surprisingly, mental computation with fractions, decimals and percentages is more cognitively demanding than for whole numbers. Callingham and Watson (2004) found that students first use mental strategies with simple fractions, decimals with the same number of places and common percentages, and take some years to develop strategies for finding percentages of a number, such as 90 per cent or 15 per cent. Unfortunately, many secondary mathematics textbooks just don't give students opportunities to do mental computations and develop these skills. Sadly, the secrets of the field laws are often kept from students with the least developed number skills when teachers persist with requiring these students to spend more time on pen-and-paper algorithms rather than on developing number sense and mental strategies. However, there are lots of opportunities to include practice in mental computation and it is vital to be aware of the importance of teaching and practising mental strategies and to take these opportunities when they arise—for example, calculating the missing angles in a triangle using mental computation with the whole class as an oral activity rather than asking the students to write equations for the exercise. Online quiz and test programs often also encourage the use of mental strategies.

Lessons in mental computation based on students sharing strategies with the whole group validate everyone's thinking as effective, demonstrate more efficient strategies, give ownership of ideas to students, recognise that there is more than one way to work something out, and illustrate that different ways are used for different numbers.

> **REVIEW AND REFLECT** : Find out what mental strategies secondary students use. Use the sample problems above. Record their thinking strategies and compare with the findings of colleagues.
>
> Plan a mental computation lesson for whole numbers, fractions or decimals that is appropriate for the students you interviewed (see Wright & Downton, n.d. for teaching approaches).

Multiplicative thinking

Automatic recall of multiplication facts is not secure for 40 per cent of Year 7 students (Callingham & McIntosh, 2002). Further, thinking multiplicatively is more than remembering multiplication facts. We have seen that our number system, including decimals, is a multiplicative structure and that division underpins the meaning of fractions. Understanding multiplication, its relationship to other operations, and the field laws that govern the operation are foundation concepts for many other concepts and skills in secondary mathematics: proportional reasoning, ratio, exponential reasoning, transforming

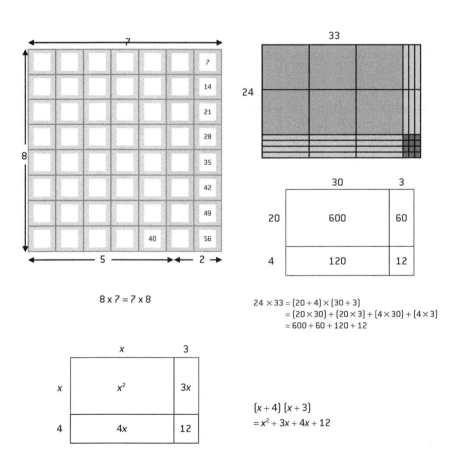

Figure 7.6 **Arrays and area diagrams illustrating the distributive law for multiplication**

algebraic expressions, and functions that define proportional relationships in data. Weaknesses in secondary students' multiplicative thinking lead to poor understanding of ratio and proportion: 'the essence of proportional reasoning lies in understanding the multiplicative structure of proportional situations ... for example, 4 in relation to 8 as multiplying by 2 rather than adding 4' (Shield & Dole, 2002, p. 609).

Arrays and area diagrams are commonly used in primary classrooms to help children learn multiplication facts (Booker et al., 2004; Reys et al., 2012; Young-Loveridge, 2005). They illustrate the commutative, associative and distributive laws for multiplication (see Figure 7.6), and help students to make sense of the multiplication of fractions and decimals. Arrays of algebra blocks and area diagrams can be used to model the distributive law and factorisation of quadratic expressions in algebra, so illustration using whole numbers facilitates algebra learning (Leong et al., 2010; Leigh-Lancaster & Leigh-Lancaster, 2002).

Multiplication of fractions and decimals

When using calculators to investigate the multiplication and division of decimal numbers, students are usually surprised to find that multiplication by a number less than one results in a smaller number and that division results in a larger number. Students need multiple approaches and representations, such as paper folding, to make linear and rectangular area models for multiplying fractions and decimals so that they understand where the algorithms and rules come from. For example, to make sixths using a strip of paper, you

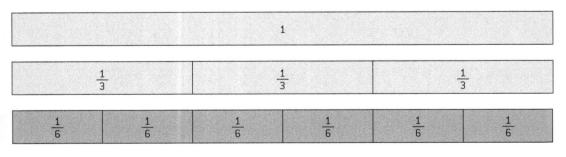

From the first folds, $1 \div 3 = \frac{1}{3}$

From the second fold, $\frac{1}{2}$ of $\frac{1}{3} = \frac{1}{6}$, that is, $\frac{1}{2} \times \frac{1}{3} = \frac{1}{6}$. Or $\frac{1}{3} \div 2 = \frac{1}{6}$.

Figure 7.7 **Paper folding into sixths**

need to fold it into thirds and then fold it in half again. Students arrive at this process intuitively (see Figure 7.7).

Folding paper in this way shows that multiplying fractions makes a smaller fraction and that the numerators are multiplied and the denominators are multiplied. It also illustrates that multiplication by a fraction is the same as division by the reciprocal of the fraction, in this case a whole number $\frac{1}{3} \div 2 = \frac{1}{3} \times \frac{1}{2}$.

Figure 7.8 illustrates an array for multiplying decimals and shows that tenths multiplied by tenths equal hundredths. In addition to deriving the procedure for multiplying decimals, you should also teach students to estimate and check solutions using problems such as:

The decimal point is not working on the calculator, where does it go in these solutions?

- $534.6 \times 0.545 = 2\,9\,1\,3\,5\,7$
- $49.05 \times 6.044 = 2\,9\,6\,4\,5\,8\,2$

Division by fractions and decimals

Division is usually introduced to primary children as 'shared between', the partition meaning of division. However, they also need to understand division as 'how many groups

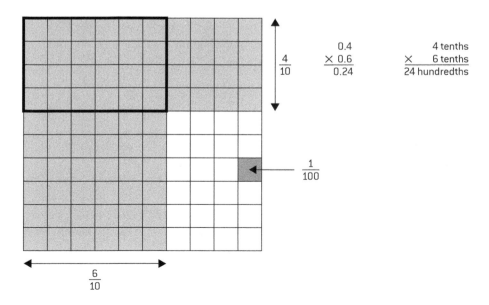

Figure 7.8 **0.4 × 0.6 = 0.24**

of', the quotition or measurement meaning of division. Otherwise, students have difficulty thinking about dividing by fractions, for example, understanding what $9 \div \frac{1}{4}$ means (Gould, 2005). The fraction strips discussed above can be used to illustrate the division algorithm for fractions. Having made the strips, we can ask, 'How many thirds in a sixth?' This is the same as three 'lots' of sixths—that is, $\frac{1}{6} \times 3 = \frac{1}{2}$. So dividing by a fraction is the same as multiplying by its reciprocal.

When students solve fraction division problems placed in a context, they make sense of the problem and generate a strategy for solving it (Gould, 2005). Context suggests particular interpretations, and hence strategies and reasoning for solving problems (Sinicrope, Mick & Kolb, 2002). Capable students may discover the strategy for finding a whole given a part (i.e., finding the unit rate) or finding the missing factor using an area model for division that enables division by fractions to make sense (Flores, 2002; Pagni, 1998; Sinicrope et al., 2002). Example problems are shown in Table 7.1. The sharing or unit rate uses the notion of proportion.

REVIEW AND REFLECT : Search for materials and resources, including online interactive resources, to develop students' understanding of multiplication and division by fractions.

Search for activities to use in practising the estimation of solutions.

Search for contexts and applications of division by fractions or decimals and design investigation or problem-solving tasks for students.

Table 7.1 Division of fraction problem types

	Quotition or measurement	Sharing or unit rate	Missing factor
Context	If I need $\frac{3}{4}$ cup of flour to make a cake, and I have $2\frac{1}{2}$ cups of flour, how many cakes can I make for the fete?	If it takes me $\frac{3}{4}$ hour to paint $2\frac{1}{2}$ doors, how many doors can I paint in one hour?	If I have $\frac{3}{4}$ square metre of material and the length of one side is $\frac{2}{5}$ metre, what is the length of the other side?
Guiding question or expression	How many groups of $\frac{3}{4}$ in $2\frac{1}{2}$?	$\frac{3}{4} : 2\frac{1}{2} = 1 : \square$	$\frac{2}{5} \times \square = \frac{3}{4}$

Proportional thinking

Proportional thinking involves making sense of quantitative relationships and comparing quantities that have a multiplicative relationship (Cai & Sun, 2002; Shield & Dole, 2008). Ratio and proportion are used in many real situations, and they underpin and connect many mathematical ideas in secondary school mathematics (Shield & Dole, 2008). Learning about ratio, proportion and percentages provides lots of opportunities for engaging students in authentic tasks and problem solving. Rate problems are a familiar context for proportional reasoning that demonstrate that multiplicative thinking underpins proportion. Rate problems, such as those involving speed or the price of products, are one type of multiplication problem and one type of proportion problem (Reys et al., 2012). Many measurement and geometry concepts rely on proportional understanding, such as scales and scale factors, relationships between attributes of shapes (e.g., pi, π and the Golden Ratio, φ), speed and trigonometry. Understanding proportional relationships as directly or inversely proportional is necessary for developing function sense (Cai & Sun, 2002; Shield & Dole, 2008). The concepts of rate, direct proportion and inverse proportion, gradient and slope, and non-linear relationships all rely on understanding the multiplicative structure of proportion.

Students make errors when working on ratio and proportion problems when they don't distinguish between proportional and non-proportional relations and apply additive or difference strategies when they should be using multiplication and vice versa, or when they apply taught algorithms inaccurately (Hilton et al., 2013; Kastberg et al., 2012). Although problems may share similar structural features, careful interpretation is needed to discern the difference between a multiplicative relationship and an additive relationship, as in the following examples:

1. In the shop, four packs of pencils cost $8. The teacher wants to buy a pack for every pupil. She needs 24 packs. How much must she pay?
2. Today Bert becomes two years old and Leslie becomes six years old. When Bert is 12 years old, how old will Leslie be?

Using qualitative reasoning to explain solutions or estimates is also prevalent among incorrect responses by secondary students for proportion problems involving comparing (Hilton et al., 2013; Kastberg et al., 2012). See the problem in the box below.

End-of-term activities

This table shows the end-of-term activities voted by Year 5 and Year 6 students.

Year level	Students who chose the beach	Students who chose the movies	Total students
Year 5	8	14	22
Year 6	7	6	13

Going to the beach is a relatively more popular choice with Year 6 students than Year 5 students. True or False? Because (choose the best reason):

- More students in Year 5 chose the beach.
- Only 6 students in Year 6 chose not to go to the beach.
- Fewer students in Year 6 chose the beach but there are fewer students in the class.
- More than half of the Year 6 students chose the beach and less than half of the Year 5 students chose the beach.

Source: Hilton et al. (2013, p. 541).

There are four general types of proportional reasoning problems: rates or direct proportions (e.g., Question 1 above); comparing two parts of a whole (e.g., *End-of-term activities*); comparing rates or densities (e.g., comparing speeds, kilometres per hour); and scaling problems. Teachers need to emphasise the multiplicative structure of ratio and proportion, as some textbooks neglect this concept when defining and giving examples of ratio and proportion (Shield & Dole, 2013). Textbooks tend to be dominated by missing number ratio problems and the application of the 'cross-multiplication' algorithm and neglect comparison problems. Students need to experience proportion in a range of contexts and multiple representations. Care does need to be taken when selecting problems, as students from different socio-cultural backgrounds will bring different assumptions to particular real contexts when solving these problems (Peled & Suzan, 2011; see Chapter 14 for examples).

A number of strategies can be used to solve ratio and proportion problems. These include finding the unit rate, using equivalent fractions, and using a scale factor (or size

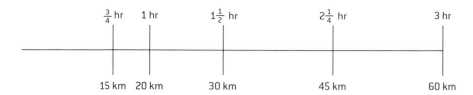

Figure 7.9 **A double number line**

time	$1\frac{1}{2}$ hr	$\frac{1}{2}$ hr	1 hr	3 min	
distance	30 km	10 km		1 km	45 km

Figure 7.10 **A ratio table**

changing). Shield and Dole (2008) present a range of solution strategies and Kastberg et al. (2012) discuss the diverse strategies used by students when solving a comparing density problem. Visual models that support proportional reasoning include the double number line (or proportional number line) and ratio tables (Gravemeijer, van Galen & Keijzer, 2005; Shield & Dole, 2008). Ratio tables are pairs of rows in a multiplication table. These models, or tools, enable students to see situations when proportional reasoning is and is not justified and they provide students with the opportunity of reinventing for themselves the cross-multiply and divide algorithm generally used for proportion problems. These two different models are illustrated in Figures 7.9 and 7.10, respectively, for the following problem:

Monica and Kim were riding from Echuca to Heathcote in the Great Victorian Bike Ride. After one and a half hours, they passed a sign post that showed that they had ridden 30 kilometres and that they still had 45 kilometres to ride. Monica said, 'we're doing well.' Why did she say this? How long would it take them to reach Heathcote?

REVIEW AND REFLECT : How do these models help to solve the problem above? Discuss with your peers.

Ratios that are equivalent are proportional. Application tasks and projects using scales—such as scale drawing, model building and using maps and plans—provide opportunities for students to develop this concept. Spreadsheets are very useful for investigating proportional relationships. In a spreadsheet, the concept of proportion can be presented dynamically, as a sequence of constant ratios obtained by applying the same rule to numerous pairs of numbers or quantities that have been gathered from investigations of real or mathematical contexts (Friedlander & Arcavi, 2005). Two examples, one about π and the other trigonometry, are included in Chapter 10. Other exemplary tasks that involve students actively collecting data to investigate ratio to find out whether they are proportional or not include 'Planets' (*Maths300*, n.d.), 'Triple jump ratio' (Clarke, 1996b), the 'Murdered mammoth mystery' (Goos, 2002) and the 'Baby and the heatwave' (see Chapter 15). Experiencing directly proportional relationships in which the rate is constant is important for understanding linear functions. However, students find it difficult to transfer their understanding of rate from familiar contexts such as speed and linear functions to other contexts (Herbert & Pierce, 2011). Exploring contexts and problems in which the rate is variable is a valuable experience for pre-calculus learning. These are discussed further in Chapter 12.

Percentage

Per cent is a particular type of ratio, and it is normally introduced in primary school as a special fraction, in which models are used to show meaning and equivalence and apply procedures for switching between fractions, decimals and percentage. While we encounter percentages such as $8\frac{1}{4}$ per cent interest and 0.05 blood alcohol content, converting these percentages to fractions or decimals is difficult for students because many students think that percentages are less than one and can only be hundredths. You should approach the teaching of percentage by modelling percentages with visual materials including estimation tasks and using application problems drawn from contexts that are familiar to and engaging for students. Students who have established concepts and skills with equivalent fractions are already familiar with processes that can be used to solve many percentage problems. Students should be encouraged to recall common conversions from percentages to fractions or decimals automatically.

Calculating the percentage of an amount is typically presented in textbooks as a procedure to learn and memorise rather than one to make sense of, and to make connections with fractions and proportional thinking. For simple calculations of per cent as shown

Written algorithm (typical textbook example)	Mental computation	Calculator operation
20% of 35 $$= \frac{20}{\cancel{100}_{20}} \times \frac{\cancel{35}^{7}}{1}$$ $$= \frac{\cancel{20}^{1}}{\cancel{20}_{1}} \times \frac{7}{1}$$ $$= \frac{7}{1}$$ $$= ?$$	20% of 35 $$= \frac{1}{5} \text{ of } 35$$ $$= ?$$	20% of 35 $\boxed{20}\ \boxed{\times}\ \boxed{35}\ \boxed{\%}$ $= ?$

Figure 7.11 **Procedures for finding the percentage of amounts**

in Figure 7.11, students should use mental computation and explore the function of the percentage key on a calculator. Capable students should be encouraged to develop mental strategies for calculating percentages such as 11 per cent and 15 per cent.

Ratio, proportion and percentage are important in senior secondary vocational mathematics, such as business mathematics subjects. Online interactive learning objects and commercial ready reckoners are relevant and useful resources for students' investigation of applications in the workplace.

Exponential thinking

The operation of repeating multiplication is known as raising a number by a power, or exponentiation—for example, $2 \times 2 \times 2 \times 2 \times 2$ is expressed as 2^5. The inverse operation is finding the root. Clearly, multiplicative thinking is critical for developing an intuitive understanding of exponentiation (Munoz & Mullet, 1998). Using and understanding exponentiation, and its inverse, are important for solving a range of problems in measurement and geometry, and series of numbers generated by repeated multiplication have many real applications for describing relationships. Students first encounter these ideas when exploring the multiplication facts represented by square arrays, and when finding the volume of cubes. These experiences provide students with concrete representations of square and cubic numbers (e.g., $5^2 = 25$ and $3^3 = 27$), but expressions in which the exponent is larger than three are abstractions. Solving problems such as the one below based on exponential series provides opportunities for students to investigate powers greater than three. Such problems are typically used for generating patterns in early algebra activities.

Aunty Sue has offered to give an allowance to her niece/nephew for the next five years. They can choose one of the following scenarios:

- $500 in the first year and then an extra $100 in each subsequent year
- $100 in the first year and then half as much more in each subsequent year
- $20 in the first year and then double the amount each subsequent year.

Which offer should they choose? What if the offer was extended to ten years?

Using spreadsheets or graphics calculators to generate tables of values to solve these types of problems provides a type of graphic organiser to support students' understanding (Ives & Hoy, 2003), shows the difference between additive, multiplicative and exponential operations, and avoids or challenges misconceptions such as $3^5 = 3 + 3 + 3 + 3 + 3$ or 3×5. Graphing exponential series (i.e., functions) provides a visual representation so that the students can see the relative magnitude of these numbers (Munoz & Mullet, 1998).

Also challenging for students are the ideas of negative and zero exponents. Ives and Hoy (2003) recommend using a graphic organiser in the form of a table so that students investigate the pattern of numbers to discover that $a^0 = 1$ and that $a^{-2} = 1/a^2$ (see Figure 7.12). Once again, a graph of the exponential function provides a visual representation to reinforce this learning. (Teaching and learning function concepts are discussed further in Chapter 8.)

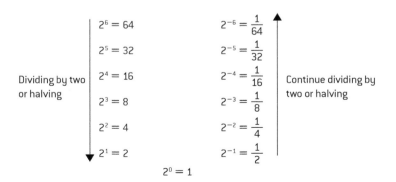

Figure 7.12 **Graphic organiser for index numbers**

REVIEW AND REFLECT : Discuss ways of generating the results with students for the graphic organiser shown above.

What does it mean to find the square root of a number?

Why does $\sqrt{x} = x^{\frac{1}{2}}$?

Evaluate $\sqrt{32}$ without using a calculator. You may need to do some research to find methods for evaluating square roots.

Investigate applications of $\sqrt{2}$, for example, paper sizes (see Scott, 2011).

Operating with numbers and algebraic symbols expressed in exponential form is a focus for number learning that underpins abstract thinking and reasoning in algebra. Errors in algebraic reasoning in senior secondary mathematics are often attributed to errors or misconceptions when operating with expressions in exponential form (Barnes, 2007; Gough, 2001). Some common errors include:

$3^2 + 5^2 = 8^2$ $2^3 \times 2^4 = 4^7$ $3^2.3^3 = 95$ $a^2 \times b^5 = ab^7$

$3^{-2} = 9$ $4x^{-2} = \frac{1}{4}x^2$ $(3a)^4 = 3a^4$ $(a^2)^5 = a^7$

A procedural approach that emphasises memorising index laws contributes to these errors (Barnes, 2007). Using expansion to evaluate and simplify such expressions improves student performance and leads to discovery and ownership of these laws by students.

Integers

On the surface, this topic provides secondary teachers with an opportunity to teach students something new about numbers. However, many students have already encountered negative numbers; even very young children can describe 'underground numbers', and most curricula require primary students to be able to model negative numbers using number lines and explore various contexts in order to develop a concept of negative number. Therefore, it is important to commence with finding out about your learners, even when teaching new topics. What do they know about integers? Where do they encounter them in real situations?

Operations with integers are typically introduced with a focus on rules and procedures to adopt with little opportunity for students to derive strategies for making sense of these numbers and operating with them (Beswick, 2011b). Learning can be enhanced

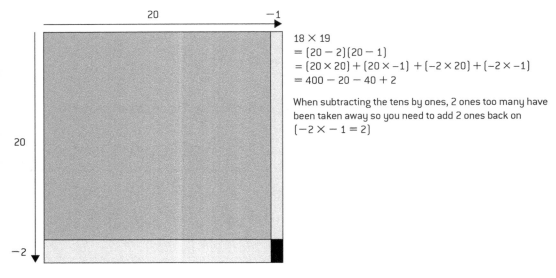

Figure 7.13 **18 × 19 illustrating multiplying negative integers**

by carefully selecting examples so that students can generate these rules for themselves (Runnesson, Kulberg & Maunula, 2011). Number lines and materials such as algebra blocks (Algebra Experience Materials, AEM) are used to model integers and the addition and subtraction algorithms (Leigh-Lancaster & Leigh-Lancaster, 2002). Selecting examples that focus on keeping something unchanged while changing one element at a time will assist students to use these tools to develop strategies for adding and subtracting integers. For example, Beswick (2011b) describes a task using a number line for which students had to derive equations with a difference of 7. For this investigation, the difference remained unchanged and students generated multiple equations including both positive and negative numbers with a difference of 7. Walking forwards and backwards along number lines will assist students to realise that subtraction of negative numbers is the same as adding positive numbers, which teachers use as metaphor with number lines (Beswick, 2011b; see 'Walk the plank', *Maths300*, n.d.).

Multiplication and division by negative integers is more complex, and senior students continue to make errors when applying the distributive law in algebra (Gough, 2001). Clausen-May (2005) recommends using arrays to show the multiplication of negative numbers. The area diagram in Figure 7.13 shows a mental computation strategy using the idea of compensation when applying the distributive law and illustrates the rule for multiplying negative numbers.

> **REVIEW AND REFLECT :** Find two secondary mathematics textbooks. Compare the ways in which operations with integers are presented in these textbooks.
>
> How do they define integers? What models or representations are used? What activities are included? What approach is used for learning?
>
> Compare the examples provided. What are the learning objectives for these examples?
>
> Which do you think are likely to be the most effective textbook examples, activities and exercises? Why?

Irrational number

Surds are likely to be students' introduction to irrational numbers, even though they will have been using pi (π) in measurement contexts or have explored digit patterns in decimal numbers to find those that never repeat. They will encounter surds when applying Pythagoras' Theorem and evaluating square roots to find the length of the hypotenuse of a right-angled triangle. However, because calculators give decimal answers to a given number of places, students may not appreciate that there is anything special about irrational numbers. Irrational numbers cannot be measured as quantities accurately, cannot be expressed as a fraction, and have an infinite non-repeating digit pattern when expressed as a decimal. CAS calculators use rational and irrational number notation (see Figure 7.14),

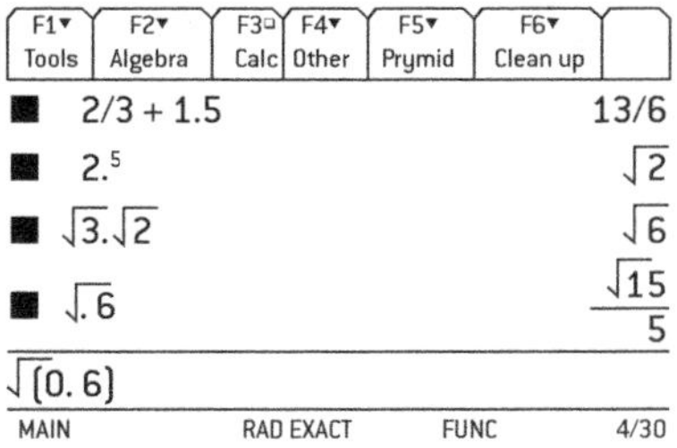

Source: Ball and Stacey (2005).

Figure 7.14 **Arithmetic calculations on CAS calculator**

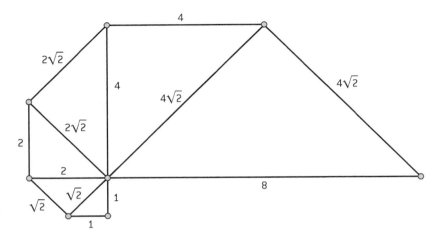

Figure 7.15 **Right-angled isosceles triangle construction for a logarithmic spiral**

so these tools may be used to explore irrational numbers and to generate algorithms for operating with irrational numbers.

Arnold (2001) argues that students have difficulty with irrational numbers because they are defined by what they are *not*. He proposes that we should focus students' attention on what irrational numbers *are* by exploring visual images of the most common and beautiful irrationals, π, $\sqrt{2}$ and the Golden Ratio, $\phi\left(\left[1+\sqrt{5}\right]/2\right)$. Using compass and ruler, or with the aid of geometric software tools or CAS, students can explore contexts such as spirals and Golden Rectangles to make more sense of irrational numbers (Stacey & Price, 2005). A simple geometric construction of a right-angled isosceles triangle (1, 1, 2) illustrates the irrational number and is the basis for constructing a logarithmic spiral (see Figure 7.15). Simplifying the surds for each hypotenuse in the spiral generates a geometric series with the common ratio of $\sqrt{2}\left(\sqrt{2}, \sqrt{4}=2, \sqrt{8}=2\sqrt{2}, \sqrt{16}=4, \sqrt{32}=4\sqrt{2},...\right)$.

Conclusion

In this chapter, we have shown that there are many places in which mathematical ideas are connected across domains. The concepts and skills discussed, together with confidence in thinking mathematically, will enable students to use these ideas in a range of vocations and in senior secondary mathematics. We have argued that knowing your students is important for developing a learning program that meets their needs, and using concrete materials, visual representations and estimation activities builds on students' understand-ings and addresses their misconceptions. We have also recommended using investigations that are interesting for students, and that engage them in reinventing algorithms, so that operating with numbers is meaningful for them.

REVIEW AND REFLECT : The learning cycle described by Frid (2000b) (see Figure 7.16) requires teachers to begin planning their teaching about a topic by finding out about their students' prior knowledge and using this information to plan a series of learning experiences that includes investigation, formalisation, practice and application.

Figure 7.16 **Learning cycle**

Prepare a learning cycle for proportion, exponential number, irrational number or integers.

Identify the formative assessment task.

Design an appropriate investigation that involves use of technology (calcula-tor, spreadsheet or CAS), a graphics organiser or concrete materials.

Explain how you would orchestrate discussion (Stein et al., 2008) to derive procedures or generalised laws with students following their investigation.

Find or design a game for students to practise operations with these numbers.

Include a real problem-solving or modelling task.

Recommended reading

Arnold, S. (2001). Thinking irrationally. *The Australian Mathematics Teacher*, 57(1), 38–41.

Beswick, K. (2011b). Positive experiences with negative numbers: Building on students' in and out of school experiences. *The Australian Mathematics Teacher*, 67(2), 31–40.

Kastberg, S., D'Ambrosio, B. & Lynch-Davis, K. (2012). Understanding proportional reasoning, *The Australian Mathematics Teacher*, 68(3), 32–40.

McIntosh, A. & Dole, S. (2004). *Mental computation: A strategies approach.* Hobart: Department of Education, Tasmania.

Stein, M., Engle, R., Smith, M. & Hughes, E. (2008). Orchestrating productive mathematical discussions: Five practices for helping teachers move beyond show and tell. *Thinking and Learning*, 10(4), 313–340.

Steinle, V., Stacey, K. & Chambers, D. (2002). *Teaching and learning about decimals* (Version 2.1). Melbourne: University of Melbourne. Sample materials available at extranet.edfac. unimelb.edu.au/DSME/decimals

Way, J. (n.d.). Fractions. In *Top Drawer Teachers: Resources for teachers of mathematics*. Adelaide: Australian Association of Mathematics Teachers. Retrieved from http://topdrawer.aamt. edu.au/Fractions

Young-Loveridge, J. (2005). Fostering multiplicative thinking using array-based materials. *The Australian Mathematics Teacher*, 61(3), 34–39.

CHAPTER 8
Teaching and learning algebra

Perhaps more than any other area of school mathematics, the study of algebra is bound to change dramatically with the infusion of currently available and emerging technology. What was once the inviolable domain of paper-and-pencil manipulative algebra is now within the easy reach of school-level computing technology. This technology demands new visions of school algebra that shift the emphasis away from symbolic manipulation toward conceptual understanding, symbol sense, and mathematical modelling. (Heid, 1995, p. 1)

Algebra—particularly when interpreted as symbolic manipulation—has an image problem in secondary schooling, with many adults seeing it as the beginning of their downward slide in school mathematics. Such mind-numbing activity is seen as having little relevance to everyday life, creating widespread disenchantment in mathematics classrooms in a context in which increasing numbers of students complete secondary schooling (Stacey & Chick, 2004, p. 2). To combat this, Stacey and Chick suggest that algebra needs to be reconceptualised as a topic of relevance to students in such a way that they are able to recognise this relevance and immediate purpose for themselves.

There is a plethora of current approaches to the teaching of algebra: the generalisation approach (Mason, 1996; Mason et al., 2005); the problem-solving approach involving word problems (Booker & Windsor, 2010); the functional approach (Ellis, 2011; Yerushalmy, 2000; Yerushalmy & Gilead, 1999); the language approach (Daroczy, Wolska, Meurers & Nuerk, 2015; Padula et al., 2001, 2002); the modelling of physical and mathematical phenomena approach (Cai et al., 2014; Galbraith, Stillman & Brown, 2010); and the historical approach (Puig, 2011; Radford & Grenier, 1996; V. Katz, 2006). However, it would be only fair to say that a balanced approach in the classroom would involve some of these approaches at different times since the various approaches highlight the fundamental concepts of algebra in different ways.

In this chapter, we provide an overview of how early algebraic ideas are fostered in primary school, illustrate approaches to facilitate students' transition from arithmetical to algebraic thinking, briefly highlight language difficulties associated with translation of word problems, and look at generational and transformational activities in algebra, and the use of CAS and algebra. This is followed by identification of the core ideas in school algebra with a major focus on functions.

Early algebraic ideas fostered in the primary years

Currently, in the majority of curriculum contexts, the introduction of algebra begins in late primary school or the beginning of secondary school. Warren (2003), in a study of Years 7 and 8 students in Queensland, reported that many students are leaving primary school with limited notions of mathematical structure and arithmetic operations as general processes—a dubious foundation for secondary school mathematics courses introducing algebra. Proponents of early algebra inclusion in the lower years of primary school (e.g., Carpenter & Levi, 2000; Lins & Kaput, 2004) do not see arithmetic and algebra as distinct, arguing that 'a deep understanding of arithmetic requires certain mathematical generalizations' (Schliemann et al., 2007, p. 8) and algebraic notation facilitates young children's expression of such mathematical generalisations—just as it does for adolescents and adults. In the particular approach adopted by Schliemann et al. (2007), for example, algebra is seen as a generalised arithmetic of numbers and quantities. Their approach encompasses 'a move from thinking about relations among particular numbers and measures toward thinking about relations among sets of numbers and measures, from computing numerical answers to describing relations among variables' (Schliemann et al., 2007, p. 10). Research findings from their studies in conjunction with other colleagues have shown that children as young as seven understand the equality principle of algebra, and children in Year 3 can develop consistent notations to deal with relationships involving known and unknown quantities.

The transition to algebraic thinking

Schliemann et al. (2007) see these difficulties for students beginning algebra as arising from teaching and learning experiences in arithmetic, not from students' cognitive development. For them, the sources are threefold:

1. the use of restricted arithmetical word-problem sets that focus on change problems (e.g., Joe has some marbles. He won three marbles. Now he has five marbles. How many marbles did Joe have at the start?); de-emphasising comparisons problems (e.g., There are eight riders but only three horses. How many riders won't get a horse?); and missing addend problems (e.g., Col has seven blue shirts and some brown shirts. She has 11 shirts altogether. How many brown shirts does Col have?)
2. the use of notation as a means of recording computation rather than as a description of what is known about the problem
3. focusing on computing particular values rather than on relations among sets.

As mentioned earlier, these researchers advocate the introduction of algebraic concepts and notation into the early primary years, and a different focus in arithmetic to address the difficulties highlighted above. However, several different approaches have been proposed by other researchers to facilitate students' transitions from arithmetical thinking to algebraic thinking.

A historical approach to transition

A historical analysis of Medieval Italian algebra by Radford (1995) has inspired a three-phase teaching sequence developed by Radford and Grenier (1996), designed to facilitate the difficult conceptual shift from solving concrete problems using words and numbers to the more abstract problem of using letters to designate unknown quantities. The fourteenth-century Italian mathematician Antonio de Mazzinghi explained the concept of unknown as a 'hidden' quantity, and it was thought that this notion would be a suitable means of helping students understand the role of letters as representing unknowns. In the first teaching phase, students were asked to solve word problems using manipulatives that embodied the notion of hidden quantity. A hidden number of lollies in a bag or a hidden number of hockey cards in an envelope were used to represent the unknowns in the problems. The teaching sequence was structured to allow students to master the two algebraic operations associated with the solutions of equations in a ninth-century Arab mathematician's text. *Al-jabr*, or *restoration*, was the operation of adding equal terms to both sides of an equation to remove negative quantities—or, less frequently, multiplying both sides of an equation by a particular number to remove fractions. *Al-muqabala*, or *reduction*, was the process of reducing positive quantities by subtracting equal quantities from both sides. In the second phase, manipulatives were replaced by drawings, while in the third phase, students used letters in place of the drawings of unknown quantities. Today, we know this method as the *balance method* of solving equations.

A generalisation and word–problem-solving approach to transition

Booker and Windsor (2010) used a problem-solving context involving word problems to stimulate the emergence and development of algebraic procedures with a Year 7 class (approximately 12 years old) working in small groups who presented their findings to the whole class for further discussion of the different approaches taken. They state that:

> Working on, representing and solving … problems in a variety of ways prepares students to think algebraically as they articulate and generalise their solutions. Initial verbal descriptions can give way to more mathematically based explanations, preparing for the more concise, symbolic arguments that will eventually develop into algebra as it is used in further mathematics. In particular, students can be helped to construct algebraic notation in a meaningful way through their representations using materials, diagrams, models, tables and graphs in their search for patterns and generalisations. An understanding of why and how the concepts of patterning and algebra have emerged in mathematics can then provide a richer background to algebraic thinking to teacher and students alike. (Booker & Windsor, 2010, p. 418)

Generalisation problem: dining chairs

To ensure there was enough room for all ten chairs of a dining table setting, a furniture maker needed to make a table that is 3 metres longer than its width. If the perimeter of the table is 14 metres, what is the width and length of the table? Can you draw a diagram of the table and its seating plan? (Booker & Windsor, 2010, p. 414)

When solving this problem, some children drew diagrams using 'w' to represent the width of the table (see Figure 8.1) and 'reasoned additively: w + w + w + w + 6 is 14, so w + w + w + w is 8 and w is 2' (Booker & Windsor, 2010, p. 415). Similarly, another group used 'w' to represent width, but reasoned multiplicatively 'I have four widths and six' (Booker & Windsor, 2010, p. 416).

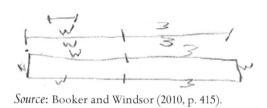

Source: Booker and Windsor (2010, p. 415).

Figure 8.1: **Spontaneous use of 'w' to represent the width of the table**

Classroom discussion and validation were also essential in explaining students' writing conventions and reasoning processes during this construction of meaning.

A quasi-variables approach to transition

A third contrasting approach is suggested by Fujii and Stephens (2001), who propose the use of quasi-variables as a bridge between arithmetical and algebraic thinking that students need to cross frequently during their formative algebra years. Quasi-variables appear 'in a number sentence or group of number sentences that indicate an underlying mathematical relationship which remains true whatever the numbers used are' (Fujii & Stephens, 2001, p. 259). The number sentence $57 - 47 + 47 = 57$ belongs to the class of algebraic equations of the type $a - b + b = a$, which is true for all values of a and b. Fujii and Stephens (2001) claim that working with quasi-variables assists students in identifying and discussing 'algebraic generalisations long before they learn formal algebraic notation' (p. 260). They focus on developing the concept of variable rather than the concept of an unknown. At the secondary level, there are many opportunities for using numerical expressions to signify variable quantitative relationships and to foster algebraic generalisation. We will look at an example from geometry involving shared diameters of semi-circles (see the box below).

Quasi-variable problem: shared diameters of semi-circles

Draw a semi-circle of diameter 60 units. Divide the diameter equally into three. Use these divisions to draw three touching semi-circles of diameter one-third of the diameter of the large semi-circle.

(a) Write a number sentence to compare the length of the arc of the large semi-circle to the total length of the arcs of the three smaller semi-circles.

(b) If there were four semi-circles dividing the diameter of the large semi-circle in the same manner, write a number sentence to show the total length of the arcs of the four smaller semi-circles.

If there were ten small semi-circles, write a number sentence to show the total length of the arcs of the ten smaller semi-circles.

Look at the four number sentences that you have written. In your own words, describe what you notice.

Now repeat for a large semi-circle with a diameter length of your own choosing. Compare your results with those of other students. In your own words, describe what you notice.

In part (a) of the above example, it is expected that students look across their series of number sentences to notice that, regardless of the number of identical smaller semi-circles used, the total sum of their arcs is equal to the length of the arc of the large semi-circle. These quasi-variable relationships allow students to understand the general relationship of the type $\frac{60}{n} \times 3.14 \div 2 \times n = 94.2$ where n is the number of identical small semi-circles. The immediate goal is not for students to write a formal expression, but to be able to articulate the relationship in their own words.

In part (b), by considering other cases for the length of the diameter of the large semi-circle, students are helped to see that the relationship holds for all semi-circles. There are many other topics (e.g., Pythagoras' Theorem with properties of tangents to circles) in which the use of uncalculated numerical expressions as quasi-variables in number sentences can be used in this way to discover underlying relationships.

Fujii and Stephens (2001) assert that the introduction of variables should not wait until students have been taught formal algebraic notation. Many situations in number and geometry are fertile ground for using quasi-variables to deepen students' understanding of algebraic thinking and to facilitate their transition from working with unknowns to variables. However, the emphasis needs to be on looking for the relationships rather than calculating.

REVIEW AND REFLECT : Design your own set of number tasks and one geometry task involving the use of quasi-variables to facilitate the understanding of variables by a lower secondary student.

Translation difficulties: reversal error

According to Drouhard and Teppo (2004), 'algebraic thought is made overt through the three components of natural language, symbolic writings, and compound representations' (p. 238). By 'compound representations', they mean illustrative elements of such classroom artefacts as textbooks that consist of symbolic writings such as numerals, drawings and natural language for labels and explanations. 'Acquiring a mastery of these components, however, is not straight forward' (p. 238), as the work of González-Calero et al. (2015) illustrates.

REVIEW AND REFLECT : Select a junior secondary mathematics textbook series and examine the approach taken to teaching algebra. In addition, examine the algebra component of your local mathematics curriculum for the junior secondary years.

Which approach to introducing algebra is taken in the series?

What assumptions about the transition to algebraic thinking are evident in the textbooks? In the curriculum document?

How do the textbooks deal with the concepts of equivalence and variable?

What methods of equation solving are included?

To what extent are the ideas and teaching sequences presented in the textbooks and curriculum document consistent with the research discussed above?

In what ways, and with what resources, would you need to modify or supplement the material in each textbook?

González-Calero et al. (2015) conducted a study of undergraduate students' ability to translate word-problem statements into equations. The results reported for 214 students show that this is not an easy task, echoing earlier research results from MacGregor (1990) and others. Some sentences that students had to translate were as follows:

B1 'There are five times less doctors than patients in a hospital.'

B2 'There are four times more boys than girls in a nursery.'

B3 'There are 13 waitresses more than cooks in a restaurant.'

B4 'There are six women less than men in a cinema.' (González-Calero et al., 2015, p. 140)

> **REVIEW AND REFLECT** : Examine these tasks in your groups. Give possible reasons B4 generated the most correct equations, while B2 generated the most reversal errors.

Generational activities: expressing generality

Generational activities involve forming expressions and equations—the objects of algebra. Algebraic notations can be used to express the generality we see in situations or to enable us, at times, to see generalisations we did not perceive before. Algebraic expressions (e.g., $5n + 3$) are the building blocks for the representations that we use for these expressions of generality. Students 'build on their understanding of the number system to describe relationships and formulate generalisations' (ACARA, 2015).

Generalisation as an approach to teaching algebra, particularly in the early phases in the lower secondary years, was given impetus in the Australasian region by the release of an Australian edition of the influential book *Routes to/Roots of Algebra* (Mason et al., 1987). Other resources have followed (e.g., 'Access to algebra', *Maths300*, n.d.), which have kept up the momentum of the approach. Mason et al.'s more recent book, *Developing Thinking in Algebra*, (2005) continues with this approach and is a ready source of activities.

One of the most important sources of generalisation is number patterns. The use of large numbers that are not easily computable (see the box below) is seen by Zazkis (2001) as a catalyst for young learners to become aware of generality, as it takes their focus away from completing operations.

> ## Generalising from number patterns
>
> Complete the next one in this sequence of number sentences:
>
> $(4 + 3) \times (4 - 3) = 4^2 - 9$
>
> $(5 + 3) \times (5 - 3) = 5^2 - 9$
>
> $(6 + 3) \times (6 - 3) = 6^2 - 9$
>
> $(7 + 3) \times \ldots\ldots\ldots\ldots$
>
> What would be the following number sentence?
>
> What would be the number sentence that starts $(59 \ldots?)$
>
> What would be the number sentence that starts $(1234567 \ldots?)$

Other sources of situations for expressing generality are diagrams and pictures. A typical introductory task would be to find a general rule for describing the number of tiles required to make cross patterns of varying size with tiles (see Figure 8.2). Students first express the rule in their own words (e.g., number of tiles is equal to four times arm length in tiles plus 1) and later in symbols (e.g., $n = 4l + 1$).

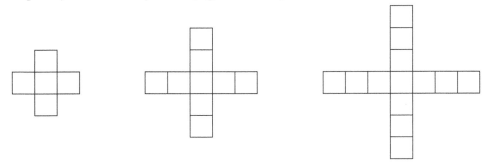

Figure 8.2 **Generalising from geometric patterns**

Transformational activities in algebra

Secondary school textbooks usually place their major emphasis in algebra chapters on explication of the process of using the rules for manipulating algebraic expressions (e.g., collecting like terms) and equations (e.g., backtracking, balancing and transposing), followed by practice to develop automaticity, rather than on conceptual notions underlying these rules or the structural features of the expressions and equations that are being manipulated (Kieran, 2004). Concrete materials can often be used to model these processes. For example, Algebra Experience Materials provide a geometric model for expansion and factorisation of related expressions (Lowe, 2013).

Rapid advances in technology have impinged on what students of today and in the future will need to know in order to live and work in a world equipped with technology. To be fully able to take their place as informed citizens of tomorrow 'students no longer need a high level of technical skill' with respect to algebraic methods, 'but the need for fundamental understanding is not diminished' (Ball & Stacey, 2001, p. 55). As an example, Pierce and Ball (2009) point out that the use of electronic technologies such as computers and handheld devices with spreadsheets and graphing software has already increased the efficacy of numerical and graphical techniques in secondary school classrooms (see Chapters 4 and 12 for examples), and also in workplaces. They list enablers of technology to be:

- motivating students to work on their mathematics
- helping students gain a deeper understanding of mathematics than is possible in a by-hand classroom
- making mathematics more enjoyable
- allowing students to engage in more real-world problems. (Pierce & Ball, 2009)

Using technologies with CAS, which allow symbolic manipulation, enables students to find exact solutions to algebraic equations, including those containing parameters (see Figure 8.3).

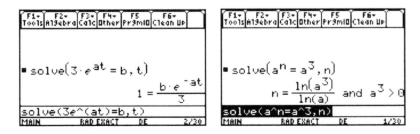

Figure 8.3 **Solving equations in a computer algebra environment**

Rather than such technologies signalling the demise of the teaching of algebra in secondary classrooms, Kenney (2014) points out that they necessitate the development of what is termed 'algebraic insight'. Algebraic knowledge is needed to decide which techniques are appropriate, to enter expressions in a form the system can handle, to monitor the solution process for possible errors and to interpret output in conventional format. The solving of the equation $a^n = a^3$ for n as shown in Figure 8.3 is correct, but requires a knowledge of logarithms to realise that the answer is $n = 3$ or what action you might take next with the calculator to arrive at such an answer.

Transformative activities in algebra are not being devalued by the use of CAS; however, the current emphases within the implemented curriculum certainly appear to need rethinking and refocusing to support our growing reliance on technological environments. As pointed out earlier, many transformational activities rely on, and promote, the development of notions of equivalent forms, and this development is as important today as it ever was.

CAS and algebra

CAS has untapped potential for changing the teaching and learning of algebra and the doing of mathematics, but what directions and forms this will take are still unanswered questions (Thomas et al., 2004). From an algebra perspective, the algebraic thinking that underpins the use of CAS in secondary classrooms is of paramount importance, whether CAS is a supplement to other technologies within the classroom or the first choice of technology. Pierce and Stacey's (2001) notion of algebraic insight being the subset of symbol sense that enables effective use of CAS in the solution of a formulated mathematical problem is useful in considering what the implications of the ready availability of CAS in secondary classrooms might be for change in emphasis in the implemented algebra curriculum. Algebraic insight has two components—algebraic expectation and the ability to link representations:

The term algebraic expectation is used here to name the thinking process which takes place when an experienced mathematician ponders the result they expect to obtain as the outcome of some algebraic process. Skill in algebraic expectation will allow a student to scan CAS output for likely errors, recognise equivalent expressions and make sense of long complicated results ... Algebraic expectation ... [involves] noticing conventions, structure and key features of an expression that determine features which may be expected in the solution. (Pierce & Stacey, 2001, p. 420)

Algebraic expectation involves: (1) recognition of conventions (e.g., the meaning of operators and letters for parameters and variable names and the order of operations) and basic properties (e.g., the non-commutativity of the division operation); (2) identification of structure (e.g., objects, strategic groups of components or simple factors); and (3) identification of key features (e.g., form, dominant term and being able to link form to type of solution). If a student is solving a quadratic equation of the form $ax^2 + bx + c = 0$, for example, an expected possible form of the answer should be $x = \dfrac{-b \pm \sqrt{b^2 - 4ac}}{2a}$. Thus, when an equation like the one shown in Figure 8.4 is solved, it should be expected that the non-standard form of the output (from a by-hand perspective) should be equivalent to $x = \dfrac{-5 - \sqrt{105}}{4}$ and $x = \dfrac{-5 + \sqrt{105}}{4}$.

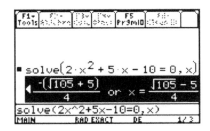

Figure 8.4 **Algebraic insight needed to interpret unfamiliar form of solution**

If algebraic expectation resulting from the symbolic representation is linked to the graphical and numerical representations, more algebraic insight is gained (Kenney, 2014). This ability to link representations in a CAS environment is two-fold. Firstly, it involves linking of symbolic and graphical representations (e.g., linking form to shape and linking key features to likely position and to intercepts and asymptotes). Secondly, there is the linking of symbolic and numerical representations (e.g., linking number patterns to form and linking key features to a suitable increment for the table or to critical intervals of the table).

The algebraic insight framework provides a structure for teachers to think about the algebraic thinking their students need to develop in using CAS. It highlights which areas of the algebra curriculum still need emphasis in teaching, and in formative and summative assessment.

Core algebraic ideas

Kieran (2004) presents a model for conceptualising what she sees as the principal activities of school algebra—namely, *generational, transformational* and *global/meta-level* activities. The generational activities involve forming expressions and equations—the objects of algebra. The underlying objects of equations and expressions are variables and unknowns. Transformational or rule-based activities include such processes as collecting like terms and solving equations. A great deal of transformational activity concerns equivalent forms and relies on well-developed notions of equivalence. Global/meta-level activities involve algebra as a tool in problem solving, modelling, noticing structure, studying change, generalising, analysing relationships, justifying and proving. According to Kieran (2004), these higher-level activities cannot be separated from the generational or transformational activities without the purpose of learning algebra being lost. Thus, algebraic thinking depends on the development of several core ideas, not the least of which are equivalence and variable (Knuth et al., 2005).

Equivalence

A deep understanding of the symmetric and transitive nature of equality is necessary in generating and interpreting symbolic representations such as equations. Godfrey and Thomas (2008) claim that 'understanding the mathematical equation object requires the formation and integration of individual properties from a number of areas' (p. 71), such as the notion of sameness or counting the total. 'If students have a view of the equals sign as signifying the result of a procedure ... and have not constructed the properties of an equivalence relation, they will not be able to interact fully with the mathematical equation object' (p. 89). They report that even university students still viewed the equals symbol as an operator rather than a relational symbol. Godfrey and Thomas (2008) assert that 'a teaching sequence influences the growth of the equation as a mathematical object for the learner ... [and] should make a deliberate effort to assist students to enrich their perspective on equation by paying explicit attention to the structure and properties of equations' (p. 90).

It is therefore imperative that students view the equals sign in algebra as a relational symbol indicating that two expressions are equivalent rather than as an operational symbol indicating that they should 'do the sum'. Typical responses to the question 'explain the meaning of the "=" sign', provided by a Victorian teacher's Year 8 class, illustrate these respective viewpoints: 'the = sign is used in equations to show that whatever is on the right of the "=" sign is equal to whatever is on the left of the "=" sign e.g., $49 = 7 \times 7$, $3 + 3 = 5 + 1$ etc.'

Much previous research has shown that many students hold an operational view of the equals sign (e.g., Falkner et al., 1999), and this extends well into secondary school (McNeil & Alibali, 2005). Knuth et al. (2005; Knuth, et al., 2006) investigated Years 6, 7 and 8 students' interpretations of the equals sign, their understanding of the preservation of the equivalence relation in the process of solving an equation, and the relation between students' equals sign understanding and their performance in solving algebraic equations. Three of the tasks used in these studies are shown in Figure 8.5.

Task 1: Interpreting the equals sign

The following question asks about this statement:

$$3 + 4 = 7$$

$$\uparrow$$

The arrow above points to a symbol. What is the name of the symbol?

What does the symbol mean?

Can the symbol mean anything else? If yes, please explain.

Task 2: Using the concept of mathematical equivalence

Is the number that goes in the $\square$ the same number in the following two equations? Explain your reasoning.

$$2 \times \square + 15 = 31 \qquad 2 \times \square + 15 - 9 = 31 - 9$$

Task 3: Equation solving

What value of m will make the following number sentence true?

$$4m + 10 = 70$$

Source: Knuth et al. (2005, p. 70; 2006, p. 301).

Figure 8.5 **Equivalence tasks**

Knuth et al. (2005) found that students' views of the equals sign increased in sophistication from operational to relational as they progressed through the middle years; however, the majority of students at each level held an operational view, with the percentage holding a relational view rising to only 46 per cent in Year 8. In Knuth et al. (2006), although there was a slight rise in the proportion of students exhibiting a relational view from Year 6 to Year 7, this fell away again in Year 8—this time to 31 per cent. These results are of some concern, as it was also reported by Knuth et al. (2005) that students who had a relational view of the equals sign outperformed their peers on the equation-solving task, which requires use of mathematical equivalence. Knuth et al. (2006) conclude that 'a relational view of the equal sign is necessary not only to meaningfully generate and interpret equations but also to meaningfully *operate* on equations' (p. 309). Thus, spending time in Years 7 and 8 on ensuring students develop understanding of the equals sign indicating equivalence rather than as an operational symbol may pay dividends in better algebra performance, particularly in transformational activities.

Variables

Variables provide the algebraic tool for expressing generalisations in mathematics. The notion of variable is fundamentally different from the concept of unknown. An unknown is a number that does not vary, whereas a variable denotes a quantity, the value of which can change. An often-quoted study in the UK by Küchemann (1978) highlights some of the many difficulties 13- to 15-year-old students have with the interpretation of literal symbols. Most students in Küchemann's study considered the symbols as objects. Few students considered them to be specific unknowns, and even fewer saw them as generalised numbers or variables. A worrying aspect of these students' misunderstanding of literal symbols was the potential for carryover of these misunderstandings into symbolising relationships in problems. When Knuth et al. (2005) investigated Years 6, 7 and 8 students' interpretation of a literal symbol and use of the concept of a variable, their results were not as pessimistic as those of Küchemann. The two tasks used in this part of their study are shown in Figure 8.6.

Task 4: Interpreting literal symbol

The following question is about this expression:

$$2n + 3$$

$$\uparrow$$

The arrow above points to a symbol. What does the symbol stand for?

Task 5: Using the variable concept

Can you tell which is larger, $3n$ or $n + 6$? Please explain your answer.

Source: Knuth et al. (2005, p. 70).

Figure 8.6 **Tasks investigating the notion of variable**

The most common meaning that students at all year levels provided for n in Task 4 was that it was a variable, with the percentages of correct responses ranging from just below 50 per cent in Year 6 to just over 75 per cent in Year 8.

In Task 5, an acceptable justified response would be: 'No, because n is not a definite number. If n was 1, $3n$ would be 3 and $n + 6$ would be 7 so $3n < n + 6$. Conversely, if n was 10, $3n$ would be 30 and $n + 6$ would be 16 so $3n > n + 6$. You cannot tell which is larger unless you know the value of n.' Only 11 per cent of Year 6, 51 per cent of Year 7 and 60 per cent of Year 8 students provided a justification based on a variable interpretation of the literal symbol. Students who provided a variable interpretation were more likely than their peers to correctly respond 'can't tell', and to provide correct justifications for Task 5 with the proportion who responded correctly increasing with the year levels. Different problem contexts did appear to activate different aspects of students' knowledge, as 20 per cent of the students who provided a correct justification on Task 5 did not provide a variable interpretation for Task 4. Overall, it would appear that students' improved understanding of variable (as compared with Küchemann's results) was associated with better performance on the problem task. However, students' knowledge of the concept of variable may take some time to stabilise, as the Year 6 students' increased performance on Task 4 was not matched on Task 5.

> **REVIEW AND REFLECT :** Use the five tasks from these last two sections to investigate the understanding of equivalence and variables of a lower secondary student. Compare your findings with those obtained by other pre-service teachers in your class.

Rate

'Rate is an important mathematical concept that is often poorly understood by many people. It is a complicated concept comprising many interwoven ideas' (Herbert, 2013, p. 173) (see Figure 8.7). Lamon (1999) refers to a diversity of ideas about rate, stating 'it is difficult to arrive at a single definition of rate on which everyone will agree. If we look at the ways rates are used in everyday contexts and the ways in which the word rate is used mathematically and colloquially, we see a complex web of meanings' (p. 204).

Rate is a ratio comparing two different numeric, measurable quantities; for example, density is a rate that compares a measure of mass with a measure of volume. It expresses the change in the dependent variable resulting from a unit change in the independent variable, and involves the ideas of change in a quantity, coordination of two quantities, and the simultaneous co-variation of the quantities. Carlson et al. (2002) refer to co-variational reasoning as 'the cognitive activities involved in coordinating two varying

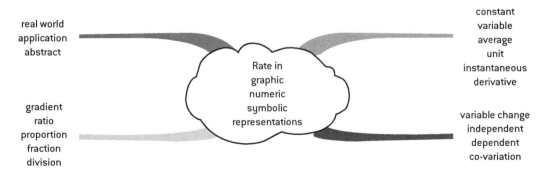

Figure 8.7 **Multiple views of the mathematical concept of rate**

quantities while attending to the way in which they change in relation to each other' (p. 354).

Common examples of rates are price per gram, speed, cost per kilometre travelled in a taxi, acceleration, volume per second in filling a tank, density, and variable rate distribution of fertiliser and seed. Some of these rates, such as speed and volume per second, display a temporal aspect, whereas other rates, such as price per gram and cost per kilometre, do not. Some rates are said to be constant (they do not change) and others are said to be variable (they do change). For example, speed expresses a constant relationship between the distance and time, while this relationship varies when speed is changing (i.e., acceleration). 'Speed has typically been used as a foundational example for rate, on the premise that speed is familiar to all students' (Herbert & Pierce, 2011, p. 456). However, students may not link an understanding of speed with the concept of rate. Herbert and Pierce (2011) report that an understanding of speed 'was not necessarily helpful to these participants in understanding rate in a context not involving speed ... [and] symbolic representations held almost no rate-related meaning' (p. 476).

Some strategies related to proportion and constant rate problems are not helpful in solving problems where the rate is variable. De Bock et al. (2005) warn that the application of proportional reasoning to non-proportional problems is prevalent and difficult to change even when strong contradictory evidence is provided. An over-reliance on proportional reasoning may inhibit the development of an understanding of variable rate. Herbert and Pierce (2012) determined four educationally critical aspects of the concept of rate and advised:

- Emphasising the relationship between the changes in two quantities when students first meet the notion of rate may facilitate rate being seen as a relationship and not a single value resulting from some calculation.
- Expressing the relationship, in descriptive words, between the variables in a variety of familiar rate contexts, including both constant and variable rate contexts.
- Attending to quantification of rate during the teaching of linear functions, linking rate to gradient, and also consideration of average rate in functions where the rate varies.

- Studying rate in a variety of contexts, including contexts involving distance and time; other temporal rates; and rates that do not involve time as one of the variables would emphasise rate, independent of context. (p. 99)

REVIEW AND REFLECT : Consider these four educationally critical aspects of the concept of rate and formulate a teaching sequence for rate that fosters a co-variational view of rate for a lower secondary student. Compare your sequence with those obtained by other pre-service teachers in your class.

Gradient is a manifestation of rate in the graphic representations of functions. Sometimes gradient is referred to as slope. A broader understanding of rate may be restricted by the view of slope as a measure of steepness rather than seeing slope as a rate that measures the relationship between two variables. Walter and Gerson (2007) describe the emergence of connections in 25 practising teachers' thinking between the notions of 'additive structure, recursive linear equations, proportional relationships in discrete measurements, graphing, rise-over-run, data tables, and an embodied sense of slope as steepness of a mountainside' (p. 227). They claim that an understanding of slope based on the calculation of the ratio 'rise over run' limits the development of connections between slope and rate. Van Dooren, De Bock, Janssens and Verschaffel (2005) report that even after lessons emphasising the limitations of linear functions in describing non-linear situations, students still experienced difficulties in deciding when it was appropriate to use proportional reasoning.

The ascendency of the function concept

Kieran and Yerushalmy (2004) point out that 'construction of the function concept … is now widely considered to be part of the knowledge of algebra' (p. 115). Many researchers espouse the importance of the function concept in secondary school mathematics (Coe, 2007; Ubuz, 2007). Yerushalmy and Schwartz (1993) see function as 'the fundamental object of algebra', imploring that 'it ought to be present in a variety of representations in algebra teaching and learning from the outset' (p. 41) in the instructional sequence of algebra.

The co-variational approach to functions places emphasis on the relationship between the variables and the way they change in relation to each other. The correspondence approach emphasises building a rule of correspondence between the values of the variable, matching the value of one variable with the corresponding value of the other variable. The initial steps to a co-variational view of function may be taken when rate and proportion are first presented to students to avoid the ambiguous and overly narrow treatment of rate in many mathematics textbooks (Herbert, 2010). Difficulties with the concept of rate mentioned in the previous section suggest that when teaching about linear functions, a teacher needs to make explicit connections between gradient and rate and a discussion of rate is also important when studying functions in which rate varies. Perhaps the average rate is a bridge between constant rate and the concept of instantaneous rate required in differentiation.

Doorman et al. (2012) pointed out that the concept of function is a central but difficult topic in secondary school mathematics curricula, which encompasses a transition from an operational to a structural view. *They claimed that* 'the learning arrangement with a computer tool helped students to overcome the difficulty of integrating operational and structural aspects of the concept function [and] ... the use of the tool supported explorative activities for coordinating and investigating the dynamics of co-variation'. (p. 1262)

CAS facilitates access to the numeric, graphic and symbolic representations of functions (Kaput et al., 2008; Zbiek & Heid, 2008) and hence may foster a deeper understanding of mathematical concepts through numerous experiences of exploring a concept such as function in a variety of representations (Zbiek & Heid, 2008). CAS affords automatic connections between representations, assisting students to notice common features across representations more easily:

> The facility of CAS to perform procedures accurately and quickly, challenges a mathematics curricula with an emphasis on procedures ... The sharing of the representational burden and automation of routine symbolic manipulations frees cognitive load for deeper conceptual understanding of mathematical ideas unhampered by tedious, error-prone calculations and procedures. (Herbert, 2013, p. 177)

REVIEW AND REFLECT : Here are a number of solutions to the equation $x^2 - 2x = -1$ presented by Year 10 learners to their class.

Explain clearly which of these solutions is correct/incorrect and why.

Explain how you would communicate the strengths, limitations or errors in each of these solutions to the students.

What questions could you ask Clara to assist her to understand and be able to formulate a more general response?

Jim: $x = 1$ because $x^2 - 2x = -1$, then $x^2 = 2x - 1$ and $x = \sqrt{(2x-1)}$

x can't be 0 because we get $0 = \sqrt{-1}$

x can't be negative because we get the square root of a negative

$x = 1$ works because we get $1 = 1$ and no other number bigger than 1 works.

Jen: $x = 1$ because if $x^2 - 2x = -1$, then $x(x - 2) = -1$ and so $x = -1$ or $x - 2$ $= -1$, which leaves us with $x = 1$ (because $x = -1$ does not hold true)

Mark: $x = 1$ because if $x^2 - 2x = -1$, then $x^2 - 2x + 1 = 0$ and this factorises to get $(x - 1)(x - 1) = 0$; so $x = 1$.

Val: $x = 1$. I drew graphs $y = -1$ and $y = x^2 - 2x$. They intersect in only one place, at $x = 1$.

Carla: $x = 1$. I substituted a range of values for x in the equation and 1 is the only one that works.

Source: Adapted from Adler and Davis (2006, p. 273).

Zbiek and Heid (2008) called for teachers to consider different ways of learning about algebra, with tasks that encourage students to look for patterns and allow different problem-solving strategies using technology to connect representations and encourage deeper thinking and appreciate the power of symbols. The language of technology quite naturally depends on the concepts of variable and function. However, the concepts of variable and function in a technological world are much richer than those found in current school textbooks or in the minds of today's students. The search for variable values that satisfy equations need no longer be the unquestioned and primary goals of beginning algebra. In a technological world, both functions and variables take on new meanings, as they are no longer seen as mere abstract notions in the classroom—especially in the context of exploring real-world phenomena. 'Variables represent quantities that change, and algebra is the study of relationships among these changing quantities. What was the search for fixed values that fit statically defined relationships is now the dynamic exploration of mathematical relationships' (Heid, 1995, p. 1).

REVIEW AND REFLECT : With other students, draw a concept map of your understanding of function as developed through secondary school. Make sure you label object and procedural links (e.g., a link from 'graph of function' to 'gradient' could be labelled as 'graph of function' HAS A 'gradient' [an object link] or IS USED TO FIND a 'gradient at a point' [a procedural link]). Compare your completed maps to those of two secondary students and an expert in the paper by Williams (1998).

The availability of these CAS calculators challenges much of what was seen as essential mathematical knowledge (Tall, 1996), but may facilitate new ways of studying mathematics. For example, the graphing facility of a calculator may be used to examine common features of families of graphs by enabling students to experience many correct graphs quickly—many more correct examples than they could prepare accurately by hand. Other concepts could also be explored in this way 'making numeric, symbolic or graphic patterns explicit for students to consider, thus exposing them to variation in critical aspects of the function concept' (Herbert, 2010, pp. 2–33).

A CAS calculator may facilitate students' problem solving about some authentic data collection in which symbolic representation is more complicated than sanitised functions usually appearing in textbooks, affording a focus on the underlying mathematics without the burden of accurate, tedious, by-hand calculations or manipulations (Tall, 1994). For example, calculator-generated regression equations approximating real-world data afford a focus on interpretation of the results and problem solving (Nolan & Herbert, 2015). Nolan and Herbert introduced linear functions using statistical data from the Australian Bureau of Statistics relevant to middle secondary students, such as reaction time versus sleep time, where the relationship is roughly linear. 'This alternative approach was relevant to the students; they were familiar with the data; and there was a recognition that

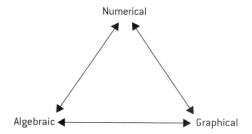

Figure 8.8 **Function representations and the links to be developed between these**

least squares lines are genuinely used for problem solving in a real-life context' (Nolan & Herbert, 2015, p. 418).

Functions are 'multifaceted' (Lloyd & Wilson, 1998, p. 250) and cannot be fully understood within a single representation environment. Being able to make links between representations (see Figure 8.8) is crucial to the underlying concepts of functions (Kaput et al., 2008).

Function-graphing technologies provide students with the opportunities to make these links and to develop rich conceptual schemas but students do not necessarily make the links merely by using technology. When taking a multiple-representation approach to teaching function, it is often pointed out that students should do tasks or teachers provide teaching examples that necessitate linking across the various representations of functions. However, even with the best-designed tasks, there is always a place for teacher monitoring and possible intervention. Arcarvi (2003) reports how students do not always notice what an expert would expect in a graphing software environment, such as what the multiplier does or the common y-intercept in a series of graphs of the form $y = ax + 1$ (see Figure 8.9). Instead, they can notice irrelevancies (such as the graphs starting at the bottom of the screen) that are 'automatically dismissed or unnoticed by the expert's vision' (Arcarvi, 2003, p. 232).

REVIEW AND REFLECT : Design three tasks that would require students to link across representations. Set at least one of these tasks in a real-world context. The tasks should be targeted for Year 11 where the students have covered the appropriate topics on functions. See *Graphic algebra: Explorations with a function grapher* (Asp et al., 1995), *Navigating through algebra in grades 9–12* (Burke et al., 2001), or *Developing thinking in algebra* (Mason et al., 2005) for ideas.

> For each task, indicate which representations (e.g., algebraic [A] and graphical [G]) and links (e.g., from algebraic to graphical [A → G]) you would expect solvers of the task to be able to use.

Families of functions

Functions-based approaches to algebra in secondary schools often focus on polynomial and rational functions, but emphasis is placed on the explicit study of just a few of the families within these, such as linear, quadratic, cubic and, to a lesser extent, quartic functions.

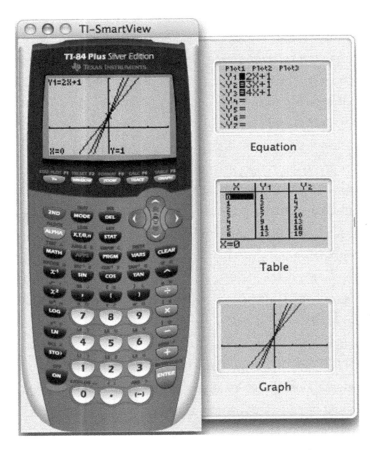

Figure 8.9 **The family of graphs $y = ax + 1$ in the multiple representation environment (View^3) of the TI-SmartView**

In addition, exponential and trigonometric functions are often studied. From time to time, other functions are suggested (e.g., the Lambert W function; see Stewart, 2006), but these are yet to gain a toehold in Australian curricula. Unfortunately, this study of functions is fragmented over the years of secondary schooling, and often students develop deep understanding of particular aspects of one family of functions but fail to transfer this knowledge to other families.

The easy creation of graphs in a technological environment allows a large number to be observed and provides easy access to myriad function types. In addition, observing multiple views of a single function can—though does not necessarily—add to the development of a broad 'concept image' (Vinner & Dreyfus, 1989, p. 356) of the prototypical graphical representation of a particular function type (Brown & Stillman, 2006). For example, the graphical representation of a cubic function has three possible 'shapes' (see Figure 8.10), based on the number of stationary points.

However, when using graphing technology, only a portion of the graph can be seen— hence, a cubic function can also appear linear (with positive, negative or zero gradient) if the viewing window of the technological tool is focused 'closely' on a part of the graph. Examining 'locally straight' sections (Tall, 1994) of a non-linear function thus provides a link between constant rate and variable rate, which is an important consideration for students preparing to move on to calculus. As Hershkowitz and Kieran (2001) point out in their report of a case study of Year 10 students using multiple representation and regression tools on graphing calculators, the crucial pedagogical questions for teachers are 'how much should the tool be used' and 'in what way should the tool be used'. The properties of the software and the tasks for which they are used can lead to the formation of different knowledge.

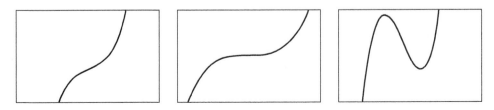

Figure 8.10 **Possible shapes for graphs of cubic functions**

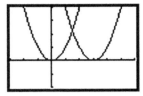

Figure 8.11 **Algebraic transformation with resulting numerical and graphical representations using a function grapher**

Manipulating functions as entities

Handheld function graphers allow the function to be entered and modified via the algebraic (e.g., $y = f(x) = x^2$) and numeric (e.g., a list of ordered pairs) representations, and the resultant graphical representation can be viewed but not directly manipulated (see Figure 8.11).

However, on computer programs such as GeoGebra (Hohenwarter & Preiner, 2007), TI–Nspire CAS (Texas-Instruments, 2016), and new generation calculators, graphs can also be manipulated enactively by translating, stretching and reflecting (see Figure 8.12). The graph is treated as a single object to be transformed.

> **REVIEW AND REFLECT:** Examine the chapters on functions in a senior secondary mathematics textbook series. To what extent are the ideas and tasks presented in these chapters consistent with the research findings discussed above?

Conclusion

As Kendal and Stacey (2004) point out, algebra is a very large content area—far too large to fit into the school curriculum or a chapter in a text such as this, for that matter. Choices and focuses have had to be made selectively. However, we must not lose sight of the fact that algebra is 'a rich field with many possibilities for applications and for addressing meta-mathematical goals, such as learning about problem solving, or axiomatics, or mathematical structure, or the benefits of an organised approach. Again this means that choices can and must be made' (Kendal & Stacey, 2004, p. 345).

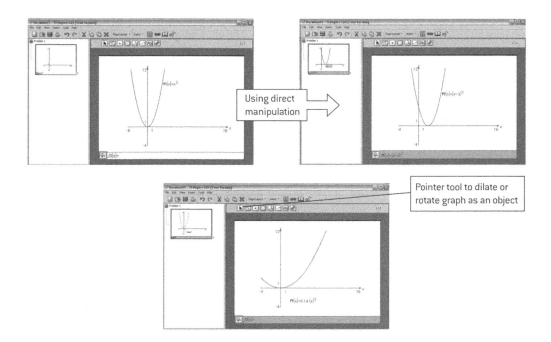

Figure 8.12 **Translation and dilation of function using a function as object manipulator**

Recommended reading

Asp, G., Dowsey, J., Stacey, K. & Tynan, D. (1995). *Graphic algebra: Explorations with a function grapher.* Melbourne: Curriculum Corporation.

Brendefur, J., Hughes, G. & Ely, R. (2015). A glimpse into secondary students' understanding of functions. *International Journal for Mathematics Teaching and Learning* (January), 1–22.

Ellis, A. B. (2011). Algebra in the middle school: Developing functional relationships through quantitative reasoning. In J. Cai & E. Knuth (Eds), *Early algebraization* (pp. 215–238). Berlin: Springer.

Heid, M. K., Thomas, M. O. & Zbiek, R. M. (2013). How might computer algebra systems change the role of algebra in the school curriculum?. In M. A. (Ken) Clements, C. Keitel, K. S. Leung, A. Bishop & J. Kilpatrick (Eds), *Third international handbook of mathematics education* (pp. 597–641). New York: Springer.

Herbert, S. & Pierce, R. (2011). What is rate? Does context or representation matter? *Mathematics Education Research Journal*, 23(4), 455–477.

Leung, F. K., Park, K., Holton, D. & Clarke, D. (Eds). (2014). *Algebra teaching around the world*. New York: Springer.

Mason, J., Graham, A. & Johnston-Wilder, S. (2005). *Developing thinking in algebra.* London: Sage.

Stacey, K., Chick, H. & Kendal, M. (Eds). (2004). *The teaching and learning of algebra: The 12th ICMI study.* Norwell, MA: Kluwer.

CHAPTER 9
Teaching and learning geometry and spatial concepts

Geometry is regarded as one of the original and essential aspects of mathematics. Kline (1979) has argued that knowledge and understanding of geometry—especially shape and location—is more fundamental than number. Objects in our world cannot exist without shape, but can be described without number. Without spatial sense, we would become lost all the time. The place of geometry in the school curriculum is more profound. As Johnston-Wilder and Mason (2005) argue, the development of geometric reasoning is important for all learners:

> The significant contribution to all learners' development available through geometrical thinking is to develop the power to imagine, to discern elements that are not shown, to 'see' a dynamic, as something is permitted to change, and to recognise that there are facts which must be true, relationships which may sometimes hold and relationships which can never hold. These facts and relationships are encountered and justified in the Spartan world of geometrical diagrams, but apply to the material world. Architects, engineers, scientists and artists must have taken them into account in their professional activities. (p. 211)

Geometry is included in the current *Australian Curriculum: Mathematics* as one of the content domains along with measurement. It includes the concepts and skills of location, such as maps, as well as two- and three-dimensional shapes and geometric reasoning. Measurement concepts traditionally associated with geometry, such as Pythagoras' Theorem and trigo-nometry, are discussed in Chapter 10. In recent years, there has been a shift of emphasis in geometry curricula around the world from Euclidean geometry to transformation geom-etry (Johnston-Wilder & Mason, 2005), though the amount of classroom time devoted to geometry varies within Australia and elsewhere. There are many connections between geometry and other mathematics concepts, so there are opportunities for teachers and learners to integrate thinking and learning through the study of geometry. For example, proportion is central to the concept of similarity, and understanding the properties of geometric shapes underpins learning about area and volume. Geometry also provides an important context in which to develop visual thinking and reasoning used in all fields of mathematics.

A constructivist approach in which students solve problems, explore geometric figures and reflect upon changes made to figures enables them to experience the surprising rela-tionships in geometry and discover these relationships for themselves (Johnston-Wilder & Mason, 2005). Geometry provides a language to describe and interpret reality and a struc-ture to organise it, and teachers should use real contexts to motivate students and estab-lish a link between school learning and everyday learning. Students should be encouraged to use a range of geometric tools for drawing, making, constructing, problem solving and communicating. These should include pen-and-paper drawing tools, virtual tools such as dynamic geometry software and programming environments (e.g., MicroWorlds), and the vocational tools of carpenters, dress designers, cartographers, draftspersons and surveyors.

In this chapter, we will discuss some of the difficulties encountered by students in this area of mathematics, as well as teaching and learning geometric and spatial concepts, pro-cedures and reasoning.

Geometry and spatial concepts in the primary years

For children, spatial sense and visual skills are needed for mathematics learning. Touching and visualising objects and materials often arranged in basic geometric structures are powerful means of establishing the abstract concepts of number in the primary years of schooling. For example, children see 'three' in the triangular arrangement of object and the rectangular arrangement of materials is used to model numbers to ten, multiplica-tion and fractions. Visual-spatial skills are needed for measurement, including estimation.

Geometry learning in the primary years includes the development of language for shape and location that is essential for modelling, visualising and communicating in all areas of mathematics—for example, in, between, under, beside, near, left and right. Geometric learning also involves visualising, drawing, making, communicating, problem solving and reasoning about two- and three-dimensional shapes.

Van Hiele's (1986) levels of geometric thinking have informed the development of the geometry curriculum, especially as it relates to Euclidean geometry, throughout the primary and secondary years of schooling:

- *Level 1: recognition*—children are able to recognise and name basic shapes.
- *Level 2: analysis*—children are able to describe attributes or properties of the basic shapes and sort, classify and make them.
- *Level 3: ordering*—students begin to establish relationships between the properties of shapes. They are able to identify families of shapes, and make conjectures and simple deductions.
- *Level 4: deduction*—the idea of a minimum number of properties for definitions is grasped. Students recognise relationships between properties and make logical arguments about properties.
- *Level 5: rigour*—students form chains of reasoning and justify their thinking.

The *Australian Curriculum* expects that students entering secondary school are thinking geometrically at level 3 of Van Hiele's framework. However, many of these students will not have established the analysis level of thinking (level 2). Some will not recognise the congruency of shapes that have been rotated. For example, students may call a square a 'diamond' if it is oriented so that the vertices point to the top and bottom of a page rather than one side being horizontal with the base of the page (Pugalee et al., 2002).

Geometry and spatial concepts in the secondary years

In secondary school, students continue to develop their spatial sense, geometric concepts, visual thinking and reasoning. They investigate properties and relationships of two-dimensional and three-dimensional shapes including closed curves, polygons and polyhedra. They also interpret and draw plans of three-dimensional objects using multiple points of view. *Invariance*—that is, a relationship or property that stays the same and does not change when some change is permitted—is a key aspect of geometric thinking (Johnston-Wilder & Mason, 2005). Students explore and apply the concepts of similarity by scaling, stretching and shrinking shapes. This understanding is then drawn upon and applied when students learn trigonometry (see Chapter 10). When students investigate *transformations,* they deepen their understanding of *congruence* and *symmetry* and the *properties* of shapes. They solve problems by constructing, referring to known properties and using chains of logical argument, that is, reasoning. In the senior secondary years, depending on the subject, students use vectors and matrices to describe and investigate transformations (see the classroom scenario in Chapter 2).

Students' spatial sense and location skills are further developed through the interpretation of maps, the use of scales, giving and following directions using compass points, and in some senior secondary subjects, drawing and analysing network diagrams to determine critical paths and shortest routes. Depending on the senior secondary or vocational mathematics subject, they may use Cartesian, polar, spherical and navigational systems to investigate and solve problems about location and spatial relationships. In vocational post-compulsory mathematics subjects, these concepts are explored through practical applications and problems.

In international studies, over the last decade, Australian students have consistently recorded poor performance in geometry and spatial reasoning (Thomson, Hillman et al., 2013). It is the weakest mathematics content area for Australian students. Time devoted to teaching and learning of geometry varies from state to state, but the teaching of spatial reasoning and properties of shapes is typically under-emphasised. Herskowitz (1998) was concerned that some teachers may be assuming that people are born with visual thinking skills; however, other teachers may believe the popular theory that some students are visual learners, even when confronted with the evidence of poor performance on spatial and geometric tasks. Students in the junior secondary years will often identify shapes and solids according to what they look like rather than checking properties (Ubuz & Üstün, 2004). The use of technology—in particular, dynamic geometry software—has drawn attention to visual spatial skills, including the need to make the distinction between a drawing and a figure, as some students are inclined to draw and check shapes with these tools by eye rather than construct and check using properties (Hoyles & Jones, 1998; Mackrell & Johnston-Wilder, 2005). Further, students' experiences have often been limited to regular polygons, so they do not recognise irregular or concave polygons or irregular or oblique prisms, pyramids and cylinders. For example, students do not recognise that this shape ⬡ is a hexagon. For many students, defining shapes does not mean that they understand the properties or can classify shapes and solids by property (e.g., 90° rotational symmetry). Visualisation of transformations, especially rotational symmetry, is difficult for some students (Wesslen & Fernanez, 2005).

Geometric reasoning

Reasoning is one of four proficiencies in the *Australian Curriculum: Mathematics* that is defined as the 'capacity for logical thought and actions, such as analysing, proving, evaluating, explaining, inferring, justifying and generalising' (ACARA, 2012b). Reasoning actions are used when learning and communicating mathematics and enable students to make sense of mathematics (Kilpatrick et al. 2001). Reasoning is a proficiency for all content domains in the *Australian Curriculum*. In addition, geometric reasoning is assigned a separate content description for all secondary years consistent with levels 3–5 in Van Hiele's (1986) framework.

Duval (1998) claims that geometric thinking involves three kinds of cognitive processes: visualisation, construction and reasoning. Visualising, the capacity to see, notice and compare and contrast are important components of learning, and for forming conjectures, generalising and justifying generalisations Lo (2012). Johnston-Wilder and Mason (2005) and Brodie (2010) also give importance to language and points of view, that is, the importance of talking and communicating to establish meaning and to develop reasoning, in particular when forming conjectures and expressing generalities. Construction is also used when searching for common properties or relations to form conjectures, to justify and prove, as well as when applying geometry to real-world problems (Battista, 2007; Lannin, Ellis & Elliott, 2011). Geometric modelling involves creating spatial representations to model mathematical situations, draws on problem-solving proficiencies and requires visualisation, construction and geometric reasoning (Pugalee et al., 2002).

Visualising, comparing and contrasting

Visualisation is a two-way process between a person's mind and an external medium (Borba & Villarreal, 2005). Conversely, it involves the ability to notice, interpret and understand figural information, such as seeing and recognising a shape or attribute of a figure or object (e.g., parallel lines), and interpreting a map or graph of a function. It also involves the creation of images from abstract ideas, such as imagining an ellipse or imagining how objects appear from different perspectives, how objects are positioned in relation to each other, how two-dimensional representations are related to three-dimensional objects, and predicting appearances of objects after transformations (Pugalee et al., 2002). These images can be mental or created with the aid of paper and pencil, or technology.

Noticing, that is, attending to the elements of an object that matter, whether it be elements of a diagram or geometric figure, or even a symbolic expression, is needed if one is to learn Lo (2012). Students need to see or visualise 'what is the same' and 'what is different'. When students are looking for similarities (invariance) or differences (variance) they are actually comparing and contrasting. For example, students need to be able to see similarities and differences between different polygons, that is, 'families' of two dimensional shapes, to identify and generalise the common properties of one 'family' of polygons, such as the set of quadrilaterals. Only then, can they attend to the differences or variation within the family of shapes, for example, the set of quadrilaterals.

In everyday life, we visualise three-dimensional objects and spatial relationships—for example, we interpret diagrams for assembling furniture and machines, we interpret plans and elevations for constructing buildings, we interpret patterns for constructing garments and packages, we use visual projection to create some styles of visual art, and we interpret maps to travel to particular locations. These are excellent contexts for student activities.

Difficulties in visual interpretation begin for students when they do not discern the same details of figures and objects as the teacher or other students (Battista, 2007). Johnston-Wilder and Mason (2005) recommend three teaching strategies to enhance students' visualisation and geometric reasoning: 'say what you see', 'same and different' and 'how many different'. For the first strategy, when considering a figure, the teacher invites each student to 'say what you see' (see Figure 9.1). As each student says what he or she sees, the student or the teacher points to this detail. By sharing these multiple views of the figure, attention will be drawn to details that may otherwise be overlooked and discussion can then proceed to focus on the detail that is most important for the problem being considered.

There are many contexts in which 'what is the same and what is different' is useful for developing visualisation and geometric reasoning—for example, establishing relationships between shapes, such as the families of quadrilaterals, and establishing the concepts of similarity and congruence through transformation. Figure 9.2 includes two-dimensional representations of a three-dimensional shape. You might ask 'what is the same and what is different' about these representations. In Figure 9.2, a cube is represented using different drawing techniques such as isometric and perspective drawing. Some of these depictions preserve parallel lines and some preserve equal lengths. Some are more familiar, yet we need to be able to recognise each as a representation of a cube. 'What else belongs' and 'what doesn't belong' (Small, 2011) are two other problem types for developing students' skills of comparing and contrasting. The images included in Figure 9.2 could be used for either of these problem types.

'How many different figures' tasks are also useful for establishing attributes of geometric shapes and concepts of congruence. Students can be challenged to find different shapes within shapes or to make shapes with given shapes. A range of materials is useful for these

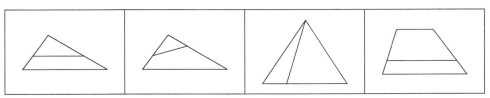

Source: Johnston-Wilder and Mason (2005, p. 36).

Figure 9.1 **'Say what you see'**

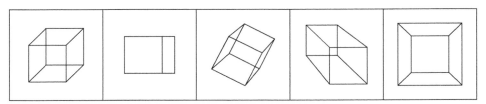

Source: Adapted from Johnston-Wilder and Mason (2005, p. 59).

Figure 9.2 **Many cubes**

types of activities—for example, commercially produced geometric materials such as Multi-link blocks, Polyhedra, commercial games such as *Tangrams, Soma Cubes, Kaleidoscope,* as well as geo-boards and paper for paper folding (Burnett, Irons & Turton, 2005). Objects in everyday life are also very useful for these types of tasks.

REVIEW AND REFLECT : The design on the walls of the buildings in Federation Square is made using right-angled triangles with sides in the ratio $1:2:\sqrt{5}$

Figure 9.3 **Federation Square, Melbourne**

- How many different shapes are made using two congruent right-angled triangles on the Federation Square buildings? Describe the properties of the shapes formed.

  Cut out any two congruent right-angled triangles and place them together along equal edges.

- Are different shapes possible other than those shown on the buildings in Federation Square?

- What if the triangles are scalene, isosceles, equilateral or obtuse-angled?

- Discuss with colleagues the possible learning outcomes for students.

Actually handling objects and materials is very helpful for developing geometric ideas and visualisation (Battista, 2007). In these tasks, students are challenged to determine what is the same and different about the possibilities that they consider. Teachers should encourage students to articulate these differences and justify their findings.

Imagining, that is, creating mental images and predicting change, is an important component of visualisation, and necessary for the development of reasoning (Borba & Villarreal, 2005; Johnston-Wilder & Mason, 2005; Takahashi, 2012). Imagining a construction, for

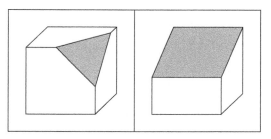

Source: Johnston-Wilder and Mason (2005, p. 63).

Figure 9.4 **Slicing a cube**

example, requires you to use relationships and properties, and hence to reason using this knowledge. Imagining cross-sections of three-dimensional shapes and transformations of three-dimensional shapes is especially challenging, and students need to test their conjectures using actual materials. In the problem depicted in Figure 9.4, plasticine or cheese would be useful.

Imagine slicing through a cube with a single plane. Two of the many possibilities are shown in Figure 9.4, giving cut faces of a triangle and a rectangle. What would the missing portions of the cubes look like? How many different polygons can be formed as the cut face of a cube?

REVIEW AND REFLECT : The *Maths300* website (membership required for access) includes a number of problems that support the development of visualisation and geometric reasoning in three dimensions—for example, 'Four-cube houses' (rotation of solids and isometric drawing), 'Cube nets' (nets of solids and congruence of nets) and 'Building views' (lateral, front and plan views of solids).

- Complete and analyse each of these (or similar) tasks: How do these tasks enhance students' visualisation skills? What geometric concepts are explored in these tasks? How do these tasks enhance the geometric reasoning of students?
- Prepare an assessment rubric for these tasks (see Chapter 6).
- Plan a follow-up task that would enable students to establish further the relevance of geometric concepts in a real-world context.

Language and communicating reasoning

Learning and using mathematical language, terms and representations is not only important for mathematical learning but is required when forming conjectures, generalising, justifying and communicating reasoning (Johnston-Wilder & Mason, 2005; Lannin et al., 2011). There are many new terms for students to learn and use in geometry—not only the names of shapes, but also terms given to properties in geometry (see Figure 9.5). One amusing example of confusion for some students is the idea of 'left-angled' triangles that some teachers have observed among secondary students. However, students can be confused about the meaning of mathematical language—especially when the same term is used to mean something different in another context. For example, what does 'corresponding' mean in a definition of 'congruent' shapes?

Failure to pay attention to language in geometry can obstruct students' learning, as the following example shows. Students in a Year 9 class were given a worksheet that required them to use dynamic geometry to investigate exterior angles of polygons. A series of instructions on a worksheet (Rasmussen et al. 1995) required students to begin this task by constructing a pentagon with exterior angles (see Figure 9.6), then asked them to measure

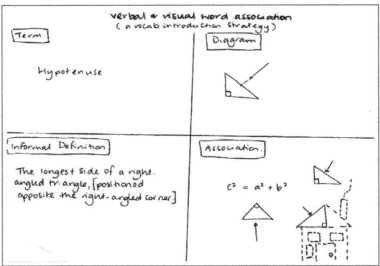

Source: Quinlan (2004, p. 23).

Figure 9.5 **Verbal and visual word association diagram showing the concept of hypotenuse**

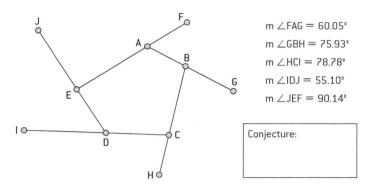

Figure 9.6 **Sum of exterior angles in a polygon**

the angles before dragging the figure and forming a conjecture about the sum of exterior angles.

Ellen, one of the high-achieving students in the class, expressed her frustration:

> Are you enjoying this maths thing? [Reads from the sheet] 'Move parts of the pentagon to see if the sum changes. Make sure the pentagon remains convex.' How are we meant to know what to do when we don't even know what the words mean? Convex? Conjucture [conjecture]?

It is important to ensure that students know the meaning of terms in an investigation. In this case, the teacher needed to use a task that would enable students to develop an understanding of convex and concave, perhaps, a task like 'what doesn't belong' before they began the investigation. The teacher also needed to explain the meaning and purpose of forming conjectures. 'Conjecturing involves reasoning about mathematical relationships to develop statements that are tentatively thought to be true but not known to be true' (Lannin et al., 2011, p. 12).

Johnston-Wilder and Mason (2005) argue that a transmission model of teaching leaves some students trying to guess what is in the teacher's mind. A more successful approach involves collaboration among students about problems and discussion with students of the mathematical ideas and the terms used to describe, explain or make conjectures about these ideas, where students are challenged to consider alternative points of view in order to clarify the meaning of their ideas and to reach a common and shared meaning of these ideas.

The following episode, taken from the same lesson about exterior angles (see Figure 9.6), illustrates collaboration among students and the importance of language. It also illustrates the strategy 'do–talk–record' for investigations. In this episode, three boys discussed their results. Che had completed the task for homework and Lawrie asked him what he had written for the conjecture:

Lawrie: What do you do here, what did you write?

Che: Um, I wrote, um, I found out that all the angles equal up to 360 degrees.

Lawrie: Not matter what shape as long as its perimeter … [interrupted].

Che: I found it for all pol, polygons or something like that equals up to … [interrupted].

Darren: The hexagon equals up to … [interrupted].

Che: It's not a hexagon. Do control later on. No, no you don't. You go to calculator. Where's your calculator?

Lawrie: I already calculated it. [Points to the result on the screen.]

Che: Yeah, well there you go. You done it all. Now you just write there [points to the screen] that all the angles equal up to 360 degrees. That's your conjuncture. [Waves his hands as if to say 'or whatever it is'.]

In their discussion, the boys argued about the name of the shape and the constraints to be included in their conjecture. They also use 'equal up' instead of 'equals' or 'adds up' (see Chapter 8 for more about this inaccurate use of language). This example further illustrates a learning objective for geometric reasoning, that is, students are able to use and clarify the meaning of terms, symbols and representations, when conjecturing and generalising, in addition to identifying the boundaries or constraints of their generalisation (Lannin et al., 2011).

Constructing

Constructing in geometry is used to experience and explore geometric concepts, to form conjectures or verify properties and relations as well as to model situations and solve problems. It involves using tools, both real and virtual (or digital), to build figures or models so that actions on these figures or models result in expected observations or enable generalising about properties or relations (Battista, 2007; Duval, 1998). It is important to distinguish between sketching or drawing and constructing. In the real world, it is not sufficient for builders to construct by eye or use insufficient criteria. For example, to ensure that their building is rectangular, builders do not rely on measuring the length of the sides only; they also measure the length of diagonals. They do this because they need to check that the walls will be constructed at right angles. Instead of measuring the angles, they measure the diagonals, because having diagonals of equal length is the necessary criterion to ensure that the parallelogram is a rectangle. The same requirements apply in geometry. A drawing or sketch might be used to communicate an idea, but a figure is need to explore, form and test conjectures about properties or when modelling to solve a problem. A figure is constructed using known geometric properties (Hoyles & Jones, 1998).

Compass and straightedge are the traditional tools for constructing in geometry. Paper folding is also useful (Burnett et al., 2005; Coad, 2006; Dickerson & Doerr, 2014). Students enjoy paper folding, and there are many resources for constructing three-dimensional shapes. Experienced secondary teachers acknowledge paper folding as an efficient strategy for many geometric proofs (Dickerson & Doerr, 2014).

Dynamic geometry software (DGS) (e.g., Autograph, Geometer's Sketchpad, GeoGebra, Cabri Geometre II, Cabri 3D and Cinderella) and programming languages such as Logo and MicroWorlds are powerful tools for construction and geometric reasoning. Use of technology, especially dynamic geometry systems, enhances students' visualisation and mental models for geometric reasoning and improves students' use of the property-based concepts when analysing geometric shapes (Ubuz et al., 2009). Many other applets and interactive tools are available that are useful for demonstrating concepts and properties, or for structured exploration by students of properties using these ready-made interactive construction devices. When using software tools, teachers need to be aware that students sometimes use the software tool to draw shapes by eye or freehand rather than using the Geometer's tools embedded in the menus of the soft-ware to construct shapes (Hoyles & Jones, 1998; Mackrell & Johnston-Wilder, 2005). The shapes that they create by eye do not stay intact as squares or rectangles—for example, when sides or vertices are dragged. Their learning may therefore be limited because they can't then use the dynamic aspects of the software for investigating these shapes. See, for example, the isosceles triangles constructed using Geometer's Sketchpad shown in Figure 9.7.

Ian was satisfied when his drawing looked like an isosceles triangle. The teacher had to remind him to check by measuring the angles or sides. Ben used the grid tool, and so used line symmetry in the construction of his isosceles triangle. In the third example, the 'perpendicular bisector' tool was used. To construct the isosceles triangle, a perpendicular line is constructed (CE) after drawing a line for one side of the triangle (AB). Then two equal sides can be constructed from the ends of side AB to the perpendicular line (AD and BD) to form the third vertex.

These examples show that teachers need to assist students to distinguish between drawing by eye and constructing shapes using the software tools. Familiarity with the software will improve students' knowledge and skills with using the software, but these

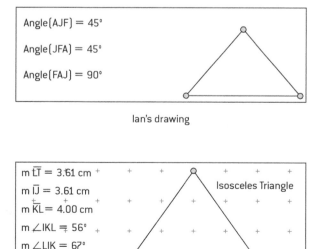

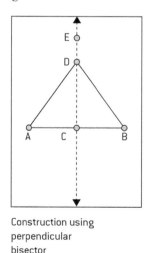

Ian's drawing

Ben's drawing

Construction using perpendicular bisector

Figure 9.7 **Isosceles triangles constructed using *Geometer's Sketchpad***

may need to be built up over a period. Initially, teachers could provide pre-constructed shapes for students to use for investigations (Mackrell & Johnston-Wilder, 2005; Ubuz et al., 2009). Mackrell and Johnston-Wilder (2005) recommend using pre-constructed shapes or templates:

- in the early stages of learning to use the software
- when students are at the stage of recognising the shapes and properties but do not understand the relationship between them (van Hiele (1986), level 2—see earlier in chapter)
- when it is not necessary for students to be able to construct the shape for the learning objectives to be achieved
- when the construction is too complex for students to achieve in the context of the lesson.

Serow and Inglis (2010) provide an example of using templates to engage students through construction in forming conjectures regarding the properties of circles. Students should be encouraged to construct independently:

- when they have skills and confidence with the software
- when they have some idea of the relationship between properties and understand that these properties need to be embedded in the shape
- when the process of constructing the shape leads to learning outcomes planned for the lesson
- when the construction provides a worthwhile challenge for the student
- for open-ended tasks.

Generalising, justifying and proving

The preceding sections have described processes that all contribute to learning geometry and to developing geometric reasoning. Forming conjectures, generalising, justifying (or refuting) and proving are actions associated with reasoning (Brodie, 2010; Kaput, 1999; Kilpatrick et al., 2001; Lannin et al., 2011). Explaining what a conjecture means or how it works might be considered informal reasoning, whereas justification is used to verify that the proposition holds for all cases and is convincing (Pedemonte, 2007; Vale et al., 2016). Lannin et al. (2011) describe reasoning as an 'evolving process' as students move backwards and forwards between conjecturing, generalising and investigating why to evaluate and justify their findings or generalisations arising from investigations. Using truth propositions, for example, 'do quadrilaterals tessellate' (Mousley, n.d.), encourages students to evaluate conjectures and develop convincing arguments. Developing convincing arguments in geometry involves constructing figures and using visual, symbolic and/or worded arguments communicated on paper, digitally or on video.

REVIEW AND REFLECT : Find and analyse the following examples of proof from the Web:

- Do quadrilaterals tessellate? www.topdrawer.aamt.edu.au/Reasoning/Big-ideas/ Mathematical-truth/Truth-of-propositions/Do-quadrilaterals-tessellate
- Pythagorean theorem water demo: www.youtube.com/watch?v=CAkMUdeBO6o
- Pythagoras in 60 seconds: www.youtube.com/watch?v=pVo6szYE13Y

What methods do these people use to communicate their reasoning and proof?
Are these proofs convincing arguments? Why or why not?
What other arguments might you use for these proofs?

Using dynamic geometry software deepens students' understanding of geometry because the learning environment involves 'students as conceptualising participants, rather than spectators in the geometry classroom' (Ubuz et al., 2009, p. 158). If geometric reasoning is to flourish in classrooms, teachers need to create a safe and supportive environment for talking, listening, questioning, challenging and convincing (Johnston-Wilder & Mason, 2005). Students need to be invited to make conjectures—that is, to make an assertion that something is reasonable or that a property holds. They also need to be praised for making modifications to their conjectures following experimentation or challenging questions from peers.

Problems and investigations using paper folding, compass and rulers or dynamic geometry provide opportunities for students to visualise, compare and contrast, conjecture, generalise, justify, refute and prove. Group work and whole-class discussions of their investigations provide opportunities for students to communicate their reasoning and to convince. Further examples of investigations that aim to develop geometric reasoning and specific geometric knowledge and understanding are discussed in the following sections.

Topics in secondary geometry curricula

In the following sections, we discuss geometry content included in the *Australian Curriculum* from Years 7 to 10 that build on earlier understandings developed in the primary years. Connections are made between location and algebra using the Cartesian plane in the junior secondary years, otherwise location is revisited in some senior secondary curricula. Likewise, vectors is a topic normally associated with senior secondary mathematics.

Lines, planes and angles

In Chapter 10, we discuss the difficulties that some junior secondary students have with identifying and measuring angles. With respect to geometry, the concepts of co-interior, alternate and corresponding angles are often introduced to secondary students in textbooks using a series of definitions and exercises. A more engaging way to introduce students to these concepts is through investigation. Paper folding, pen-and-paper constructions and dynamic geometry are suitable tools for these constructions and suitable investigations can be found in various resources (e.g., Dickerson & Doerr, 2014; Rasmussen et al., 1995).

> **REVIEW AND REFLECT :** Develop or find an activity for investigating angle properties involving parallel lines using a dynamic geometry tool. Consider whether this task should involve construction by the students or use of a template and what you would do if the technology fails on the day.

The difficulties that some students have with angles became clear to James, a preservice teacher, in a geometry lesson when students investigated the geometry of the Leaning Tower of Pisa. The students had been learning about the different types of angles.

He decided that an investigation would show how these angles and relationships could be used for a real-world problem. For this task, the students worked in small groups and had to predict how long it would take the Leaning Tower of Pisa to fall over after measuring the current lean of the tower from a photograph. James had gathered information from the internet to develop the materials for the task. He prepared a sequence of questions and an assessment rubric to guide the students in their investigation. He demonstrated and explained an experiment for them to find the angle at which a cylinder would topple over (see Figure 9.8), though he did not include a diagram of this experiment on the worksheet.

Afterwards, James and his university supervisor discussed the lesson, and the students' problem-solving strategies and understanding of angle. They noted that some students had difficulty identifying the angle to measure on the photograph of the Leaning Tower of Pisa. The students did not know whether to measure the deviation from the vertical or the horizontal. It was also not clear that they understood why the angle that they were measuring in the experiment related to the angle at which the tower leaned. Duncan (2011) describes a similar investigation about measuring the slope of ramps to ensure construction conforms to building and safety regulations, which provides strategies for developing understanding of the relations between angles involved in the problem.

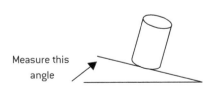

Place a cylinder on an inclined plane with a rough surface and measure the angle at which it topples over.

Measure this angle

Source: <www.kidsnewsroom.org/images/100299/pisa.jpg>.

Figure 9.8 **Leaning Tower of Pisa and experiment**

REVIEW AND REFLECT : What would you prepare for students if you were going to use the Leaning Tower of Pisa investigation in your classroom? Why? (See Duncan, 2011.)

Design an assessment rubric for an authentic geometric problem that you would use in the classroom. Explain and justify your rubric according to the curriculum and models for assessing performance tasks in mathematics. (Refer to Chapter 6.)

Visualising angles between planes in three-dimensional objects is especially difficult for students. Students need experiences of physical models of these objects and virtual models are also useful (e.g., 3D-XplorMath, Autograph, Cabri 3D and various dynamic digital resources).

Properties and relationships of two- and three-dimensional figures

The features of the boundaries, surfaces and interiors of two- and three-dimensional figures (closed curves, polygons and polyhedra) constitute a body of mathematical knowledge. Through investigation of these features and classification of shapes, students develop an understanding of uniqueness—that is, the information that is necessary and adequate for defining a shape and a class of shapes. It is in this part of the curriculum that students learn about Pythagoras' Theorem, the Golden Ratio and other 'classics' of mathematics. Students will apply their knowledge of shapes in many other areas of mathematics and in practical situations. In this section, we have chosen to focus on quadrilaterals to illustrate three different approaches that can be used for investigating properties of figures.

The first approach is *analytic* and uses dynamic geometry software. Students are provided with a pre-constructed file of a shape and asked to analyse the features by 'dragging' vertices or sides (Rasmussen et al., 1995). In the example shown in Figure 9.9, students use a pre-constructed file of a quadrilateral (ABDC) with an interior quadrilateral constructed from midpoints. They drag the vertices or sides (in Figure 9.9, vertex A) to investigate the interior quadrilateral. They use the measure tools to form and test a conjecture about the interior quadrilateral.

Mackrell and Johnston-Wilder (2005) stress the importance of the questions that teachers use to scaffold students' learning for such investigations. Well-designed activities will

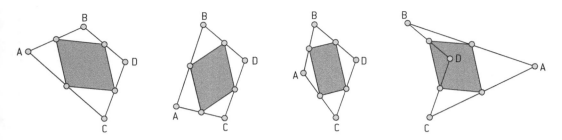

Figure 9.9 **Midpoint quadrilaterals formed by dragging vertex A of a quadrilateral**

draw students' attention to the aspects of the figure that are invariant (do not change) when it is 'dragged'. The question 'what is happening to the figure' may produce descriptive accounts of a figure, but teachers need to use a more concise question: 'what stays the same and what changes?' This question leads students to look carefully to find the elements of the figure that do not change—that is, the essential properties of the shape. It also invites students to explain and perhaps to justify. 'If I change this, what else changes and what stays the same' questions are useful for whole-class discussions of a projected figure.

Using dynamic geometry software or an expressive tool (such as turtle geometry with Logo or MicroWorlds) to construct shapes is the second approach for investigating the properties of shapes. D. Leigh-Lancaster (2004) provides a dynamic geometry example. For this investigation, students start with two line segments of different lengths that intersect and construct a quadrilateral using these line segments as the diagonals (see Figure 9.10).

Students are then invited to move one of these line segments and observe 'what stays the same and what changes'. A more structured approach to this task would be to pose a series of 'what if …' questions—for example, 'what if the intersecting lines bisect each other' or 'what if the intersecting lines are at right angles?' Teachers are not limited to computer-based activities for these investigations; for example, A. Leigh-Lancaster (2004) suggests paper-folding activities. For one of these activities, students predict how to cut a piece of paper that has been folded twice at right angles in order to produce a rhombus, square, rectangle and octagon (see Figure 9.11).

A third approach is using rich tasks or real problems. Rich tasks represent the ways in which the knowledge and skills are used in the real world, address a range of outcomes in the one task, are open ended, and encourage students to disclose their own understanding of what they have learned (Clarke, D.M. n.d.). 'The shape we're in' (Department of

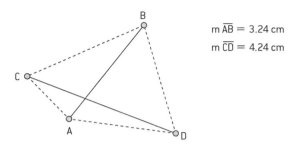

$$m\ \overline{AB} = 3.24\ \text{cm}$$
$$m\ \overline{CD} = 4.24\ \text{cm}$$

Figure 9.10 **Constructing quadrilaterals from diagonals**

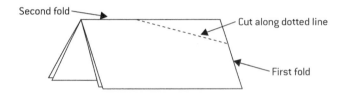

Figure 9.11 **Cutting folded paper to construct quadrilaterals**

Education, Queensland, 2004) is one example. For this task, students investigate the mathematical concepts of one container, one domestic object, one mechanical device and an object from nature. They also have to investigate the consequences—mathematically and practically—of changing the object in some way. Vincent (2005) includes structured investigations of the geometric features of buildings and objects in central Melbourne. Various places that provide a context for investigations are regularly described in *The Australian Mathematics Teacher*—for example, the parabola formed by the cable on the Golden Gate Bridge (Brinkworth & Scott, 2002). Pierce et al. (2005) describe how students can use digital photography in combination with dynamic geometry software to investigate geometric features of nature, architectural features and mechanical objects.

Each of these approaches use an inquiry or problem-solving approach to teaching. Such an approach is commonly used in Japan for individual and sequences of lessons (Shimizu, 2010; A. Takahashi, 2006; T. Takahashi, 2012), where the first few lessons in a sequence of lessons involves students working on carefully selected problems or investigations. The problems for these lessons are ordered and build to a key mathematical learning goal for the sequence. For each of the problem-solving lessons, students' conjectures and reasoning are displayed and discussed in a whole-class discussion and the teacher elicits the 'big ideas' from students during discussion. Later lessons in the sequence involve exercises to develop fluency with procedures and applying concepts and procedures, for example, to real-world problems. Serow and Inglis (2010) describe a similar approach to sequencing lessons using dynamic geometry software templates. The steps in their approach also involve students developing ownership of the ideas generated and identifying and pursuing further investigations.

REVIEW AND REFLECT : Work in a team to prepare a sequence of lessons for a topic concerning the characteristics and properties of a specified polygon, closed curve, polyhedron or conic section.

Document learning objectives (or learning intentions) for the sequence and each lesson.

Include investigations and problems that enable students to generate ownership for the concepts explored.

Include a 'truth statement' (Mousley, n.d.) as one of the tasks for a lesson to encourage students to justify using a convincing argument.

Use technology, compass and ruler, or concrete materials for investigations and proofs.

Explain how you would formalise geometric properties with students that arise from their investigation and conjectures.

Isometric transformations

Transformations that keep shapes the same—that is, congruent—are called 'isometries'. (Isometric means same measure.) They include reflecting (flipping), translating (sliding) and rotating (turning). According to Wesslen and Fernanez (2005), there are two key ideas about isometric transformations. The first is that only one transformation is needed to map one shape onto a congruent figure. The second is that two transformations are the same as making one other transformation. (A reflection followed by a translation is sometimes called a glide.) Some students have difficulties or misconceptions regarding transformations (Wesslen & Fernanez, 2005). They don't realise that translating a shape with reference to one point (e.g., on a point on one edge) is the same as translating with

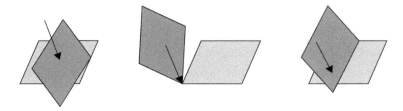

Figure 9.12 **Rotation about the centre, the edge and a point not on the centre**

reference to another point (on another edge). In addition, students are not confident rotating a shape where the centre of the rotation is not on the edge of the shape or in the centre of the shape—that is, they think that for all rotations, the figure must stay on the same spot (see Figure 9.12).

Paper folding and digital technology resources are useful media for developing students' understanding of transformations and addressing these misconceptions. Students can use technology to move (transform) one shape on to another (Johnston-Wilder & Mason, 2005; Vincent, 2000; Wesslen & Fernanez, 2005). Some computer-based games and interactive learning materials are available and drawing tools in word processing software or pre-constructed figures using dynamic geometry software may be used. Students will need instruction on how to use the transformation menu in these tools. You may notice differences in learning preferences for media or resources among your students. Since different students will find different media easier to use, it is important to incorporate a variety of media and tools to support their learning.

REVIEW AND REFLECT : An A4 sheet of paper has been folded in half lengthways and in half again. A hole-punch tool has been used to make two holes in the folded piece of paper, as shown below.

Second fold ⟶   ⟵ First fold

- Draw a rectangle to represent the unfolded piece of A4 paper and mark where you predict the holes will be when the paper is unfolded.
- Work with a partner and fold the paper in different ways; use a hole punch and predict the placement of the holes.
- Design some other paper-folding tasks about reflection (see Johnston-Wilder & Mason, 2005).
- Investigate technology-based activities for isometric transformations.

Students learn about the properties of shapes through investigations of transformations. They also develop further understanding of congruence through explorations of tiling patterns and tessellation. They can be challenged to find which two-dimensional and three-dimensional shapes tessellate or find different tiling patterns using one or two shapes (see Figure 9.13).

For these explorations, students should be encouraged to describe the transformation used to create the tessellation and form conjectures about the properties of shapes, including irregular shapes that tessellate. Simply working with physical objects may not achieve the desired learning; students need to imagine in order to understand (Johnston-Wilder & Mason, 2005). Again, drawing attention to particular features and relationships is important for developing students' thinking.

Investigations involving analysis, construction or rich tasks are suitable activities for teachers to use. Famous or well-known buildings may be the source of rich tasks, especially for unusual tiling patterns (see Eppstein, 2007 for examples). The particular

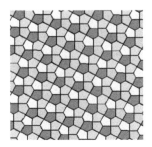

Figure 9.13 **Cairo tiling (pentagons)**

tiling pattern of an irregular pentagon, shown in Figure 9.13, is called the Cairo tessellation because it appears in a famous mosque in Cairo. The buildings in Federation Square in Melbourne (see Figure 9.3 above) show an example of Pinwheel Aperiodic Tiling (Bourke, 2002; Vincent, 2005). This tiling pattern, made with right-angled triangles in the ratio $1:2:\sqrt{5}$, can be constructed through a series of iterations. Five triangles can be transformed to make a similar right-angled triangle, and so on. It is aperiodic tiling because the tiling pattern is not repeated within a region bounded by a parallelogram.

Non-isometric transformations

The tracking of transformations through a third dimension illustrates a different relationship between two-dimensional and three-dimensional shapes from that developed through a study of boundaries and nets. Drawing tools in word processing software can be used to show translations of two-dimensional shapes in the plane (x,y) through a third dimension (z) to create prisms and cylinders (see Figure 9.2). Three-dimensional dynamic geometry (Autograph, Cabri 3D) permits rotations, reflections and translations to create three-dimensional figures. For example, a right cylinder can be constructed by rotating a rectangle around a line in the plane of the rectangle (see Figure 9.14).

Transformations that change distances between points in a shape produce different figures. Such transformations include dilation (or scaling), squeezing and stretching, and shearing. Dilating (or scaling) a figure uniformly in all directions produces similar figures. For dilations, students need to use proportional thinking and explore the ratios of the distances between points on the projected figures to answer the questions and determine scale factors. Contexts for studying dilation or scaling and similar figures include scale drawings and perspective drawing. These understandings are used for understanding Pythagoras' Theorem and trigonometry.

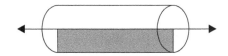

Figure 9.14 **Cylinder created by rotation of rectangle**

Location and spatial reasoning

Rich tasks are suitable contexts for location and spatial reasoning. 'To find their way, people have to take on board the links between the orientation of their body, that of the plan and that of real space, and sometimes the orientation of persons they are asking their way from' (Berthelot & Salin, 1998, p. 75). To extend students' spatial reasoning, using either coordinate systems or directions and distances, they need to have experiences of unfamiliar situations—that is, places and buildings of which they have no mental images. Excursions and school camps are therefore ideal contexts for developing spatial reasoning. With the required levels of legal supervision, students should be provided opportunities to plan and/or follow routes on various types of maps: road maps using Cartesian systems, public transport maps using network diagrams, and topographical maps showing geographical features. See, for example, an investigation of the London Underground (Brinkworth & Scott, 2001). These tasks also involve the development of location and spatial language and communication skills.

In the junior secondary years, Cartesian coordinate systems are the focus of students' learning and included in the algebra curriculum for Years 8–10 as students explore functions and relations. In the senior secondary years and vocational programs, polar, spherical and navigational systems of spatial reasoning may be developed. Spherical geometry begins with a study of coordinates using latitude and longitude. Using a cross-sectional view of the globe, students consider problems of calculating distances on the surface of the globe between places with the same longitude and then places with the same latitude. Hence, students encounter the meaning of a nautical mile and apply the spherical cosine rule to solve these problems (see Figure 9.15).

Interesting investigations of invariance are encountered when considering the mapping of the globe (map projection)—that is, transforming a spherical grid onto a rectangular grid for two-dimensional maps. Wilkins and Hicks (2001) discuss three commonly used map projections: Mercator, Robinson and Mollweide. Each of these map projections has preserved or distorted different properties of the spherical geometry: the distance between points, area, direction or shape or a combination of these. The problem that they pose for students is to calculate the area of the oceans on the Earth for each of these maps. Johnston-Wilder and Mason (2005) propose a different approach to the

Carnarvon (24°53′S, 113°40′E), in Western Australia and Bundaberg (24°52′S, 152°21′E) in Queensland are on opposite sides of Australia at approximately the same latitude (25°S). How far west is Carnarvon from Bundaberg?

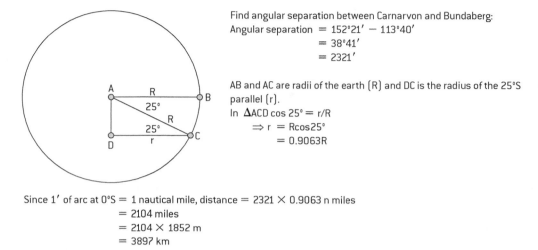

Find angular separation between Carnarvon and Bundaberg:

$$\begin{aligned} \text{Angular separation} &= 152°21' - 113°40' \\ &= 38°41' \\ &= 2321' \end{aligned}$$

AB and AC are radii of the earth (R) and DC is the radius of the 25°S parallel (r).

In $\triangle$ACD $\cos 25° = r/R$

$$\Rightarrow r = R\cos25°$$
$$= 0.9063R$$

Since 1′ of arc at 0°S = 1 nautical mile, distance = 2321 × 0.9063 n miles

$$\begin{aligned} &= 2104 \text{ miles} \\ &= 2104 \times 1852 \text{ m} \\ &= 3897 \text{ km} \end{aligned}$$

Source: Hodgson and Leigh-Lancaster (1990, p. 471).

Figure 9.15 **Distance at the same latitude**

problem of map projection with the following problem: Carnarvon (24°53′S, 113°40′E) in Western Australia and Bundaberg (24°52′S, 152°21′E) in Queensland are on opposite sides of Australia at approximately the same latitude (25°S). How far west is Carnarvon from Bundaberg? They use a longitudinal cross-section of a globe to illustrate three possible projections of the globe: gnomic, stereographic and orthographic. The implications of map projections for navigation provide a rich source for problem-solving tasks. For example, Hodgson and Leigh-Lancaster (1990) pose a series of navigation problems to illustrate the difficulties of charting a course using Mercator maps.

Network analysis is a relatively new area of school mathematics arising from the field of operations research and graph theory. A network diagram or graph is shown in Figure 9.16. Network diagrams are constructed of points (vertices or nodes) and links (lines or edges) used to show connections between places on a map or nodes on a network (such as the internet). Network analysis is included in some senior and vocational mathematics subjects.

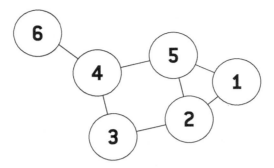

Figure 9.16 Network diagram or graph

REVIEW AND REFLECT : Use a map on the internet or *Google Earth* to find your home.

Investigate different coordinate systems, map projections or network diagrams, and teaching materials for these location contexts and topics for learning (see Farmer, 2005; Faulkner, 2004; Hekimoglu, 2005 for senior secondary examples).

Find or develop a rich task using topographical maps, network diagrams or Cartesian, polar or spherical coordinate systems.

Identify the learning outcomes for mathematics and for other disciplines and generic skills in the case of junior secondary mathematics.

Vectors

Vectors and matrices can be used to describe and investigate transformations and to solve a range of geometric problems involving magnitude and direction, with particular application to physics. These topics are included in some senior secondary curricula. It is often assumed that students taking advanced senior mathematics subjects understand mathematics, but this is not necessarily the case, and the study of vectors illustrates some of the difficulties students encounter when their mathematical thinking is based on operational or procedural knowledge (Forster, 2000a, 2000b). Researchers and experienced mathematics teachers agree that vectors are best introduced through problems based in a real-world context (Forster, 2000a, 2000b; McMullin, 1999). Certainly, there are some interesting contexts that could replace the more predictable physics and navigation examples, including amusement park rides (McMullin, 1999) and cave exploration (Vacher & Mylroie, 2001). However, there is no consistent view on whether students are likely to have more success in solving problems using geometric, trigonometric or Cartesian component methods. There is some evidence that the use of digital technology can support the development of students' conceptual understanding of vectors (Forster, 2000b; Goos, Galbraith, Renshaw & Geiger et al., 2000). In both these studies, students also worked with concrete aids and/or pen-and-paper diagrams. In the problem illustrated in Figure 4.11 in Chapter 4, the teacher used transparent grid paper, cut-out polygons and an overhead projector to demonstrate the matrix transformation problem. The students used the same materials to explore the problem and a graphics calculator for the matrix calculations.

Conclusion

Too often, the geometry curriculum in secondary schools is implemented as a series of disconnected topics or 'fun' activities without careful consideration of the need to develop geometric reasoning, connection of geometric ideas, and application to real problems for students. Geometry lessons should connect with prior knowledge and engage students in creative thinking and problem solving. Visualising and predicting are important to this process. Students should be encouraged to imagine or visualise a figure before sketching it, to sketch it before constructing it, and to make a conjecture about what will happen to it when investigating or problem solving.

Recommended reading

Coad, L. (**2006**). Paper folding in the middle school classroom and beyond. *The Australian Mathematics Teacher*, 62(1), 6–13.

Johnston-Wilder, S. & Mason, J. (**2005**). *Developing thinking in geometry.* London: Open University.

Mackrell, K. & Johnston-Wilder, P. (**2005**). Thinking geometrically: Dynamic geometry. In S. Johnston-Wilder & D. Pimm (Eds), *Teaching secondary mathematics with ICT* (pp. 81–100). Maidenhead Berkshire, UK: Open University Press.

Serow, P. & Inglis, M. (2010). Templates in action. *The Australian Mathematics Teacher,* 66(4), 10–16.

Takahashi, T. (2012). Let's investigate solid figures (IMPLUS Lesson Plan). Tokyo: Gakugei University. Retrieved from www.impuls-tgu.org/en/library/geometry/page-87.html

Software and interactive manipulatives

3D-XplorMath: http://3d-xplormath.org vmm.math.uci.edu/3D-XplorMath/index.html

Autograph: www.autograph-maths.com

Cabri 3D: www.cabri.com

Cabri Geometre II: www.cabri.com

Cinderella: www.cinderella.de/tiki-index.php

GeoGebra: www.geogebra.org/cms/en/

Geometer's Sketchpad, Key Curriculum Press: www.dynamicgeometry.com

MicroWorlds: el.media.mit.edu/Logo-foundation/products/software.html mia openworldlearning.org

National Library of Virtual Manipulatives: www.matti.usu.edu/nlvm

CHAPTER 10
Teaching and learning measurement

Teaching and learning measurement concepts and skills provide opportunities for learners to connect mathematics concepts and skills across all fields of mathematics. Further, investigating, solving and modelling measurement problems derived from real-world situations that connect with students' interests enables mathematics to 'come to life' for students. It is essential that students go outside the classroom and into the community to develop the 'hands on' and practical skills of estimating and measuring, as well as the social skills for working cooperatively on problems and projects with real applications.

As is the case for number, students entering secondary school will have developed many of the underpinning concepts and skills in measurement; however, these understandings may not be secure for many students. In this chapter, we will focus on the measurement concepts and skills in the *Australian Curriculum: Mathematics* (ACARA, 2015) that are important for mathematical literacy and success in senior secondary mathematics, and that are routinely included in vocational mathematics subjects in senior secondary mathematics. We begin the chapter by providing some background on children's measurement learning in primary school and outlining the challenges for students in secondary school. The particular topics and skills discussed include estimating and measuring, relating measurement units, perimeter and circumference, area and volume, Pythagoras' Theorem and trigonometry. Finally, we discuss teaching approaches for integrating and making connections between measurement and other fields of mathematics, and provide examples of problems to engage students in problem solving and working mathematically on measurement.

Measurement in the primary years

Young children begin to understand length, mass and capacity through play and measure using direct comparison. They place objects beside each other to see which is longer or taller, they heft objects to see which is heavier and pour substances from one container to another to find out which holds the most. When objects cannot be compared directly, they use informal units to compare the length, mass and capacity of objects. They cover surfaces with informal units or squares to develop a sense of area and construct objects using interlocking blocks or play dough to develop a sense of volume. Situations are created to show that informal units cannot be used for accurate or reliable comparisons, and so the idea of standard units is introduced. The experience of estimating and measuring enables primary students to develop a sense of the size of metric units of measurement and the relationship between them. Primary students begin to appreciate the importance of accuracy when measuring, though reading measures on scaled instruments remains a challenge for some students entering secondary school.

Understanding the area of rectangles is enhanced when children use concrete materials arranged in arrays to discover the rule for finding area and when they make layers of arrays to discover the formula for volume of rectangular prisms (Mitchelmore, 1995). However, some junior secondary students may have been taught these formulae too soon and rely too much on remembering them. These students need further experiences of measuring area and volume using arrays of materials to make sense of these formulae and to form and test conjectures about calculating the area and volume of other shapes.

Understanding angle is challenging for primary children, since many students do not identify the physical experience of turning—the definition used in mathematics—as an angle (Mitchelmore & White, 1998, 2000). Further, they do not recognise the similarity of the everyday representations of angles they observe: turn, meeting, slope, corner, bend, direction and an opening. Understanding of angle is not established for many students entering secondary school.

Measurement in the secondary years

Measurement enables us to describe and compare attributes of objects or events in space and time, and to use measurement to solve real problems—including the design and construction of objects and events. In the secondary years, students continue to develop measurement sense for the attributes of length, angle, mass, capacity, temperature, time, area and volume and explore relations between these attributes to make sense of surface area, speed and density in real contexts. They develop more sophisticated strategies for estimating measurements, and appreciate that, in real contexts, units of measurement are selected for the purpose of the measurement and not just according to the relative size of the unit. They explore and use relationships between attributes and units of measurement to calculate measurements of regular and irregular shapes and objects, and composite attributes such as speed and density. The shapes and objects include polygons, regular curved regions, irregular closed shapes and regions, regular and irregular polyhedra, regular solids with curved faces (spheres and cones) and irregular solids. In Years 9 and 10, measurement connects concepts in number and geometry in the study of triangles through Pythagoras' Theorem and trigonometry. These understandings are developed further in algebra in senior secondary years. Measurement therefore provides the opportunity not only to problem solve but to reason mathematically by comparing and contrasting, generalising and proving.

Developing measurement sense also means that students understand the structure of the system of measurement units and can move flexibly between these units when solving problems in context; a strong foundation in the decimal place value system and multiplication and division by multiples of 10 is therefore needed. You can expect that, given the number of students who continue to make errors when comparing and calculating with decimals, finding equivalent units of measurement—especially for area and volume—will need attention in the secondary classroom. You will find that the real context of measurement will assist students to make sense of decimals, but you need to focus on the structure of number and units of measurement and on developing students' multiplicative thinking rather than teaching tricks for converting units, such as moving decimal points (see Chapter 7).

Performance by 15-year-old Australian students on problems concerning space and shape (measurement and geometry) for the 2012 PISA was weaker than performance in other content areas (Thomson, Hillman et al., 2013). Other studies report secondary students at various year levels find it difficult to: identify an angle when one or both of the lines had to be imagined (White & Mitchelmore, 2000); measure the weight of objects using analogue kitchen scales (Drake, 2013); interpret scales on maps (Livy & Vale, 2011) and area and perimeter of polygons (Thomson, Hillman et al., 2013); and scale up area and volume (Van Dooren et al., 2005). Student misconceptions and effective teaching approaches are discussed in the sections that follow. Emphasis is placed on the connections between measurement and number and between geometry and algebra to be made, consolidated and applied to enhance students' mathematical learning during the secondary school years. Making sense of measurement through problem-solving approaches is emphasised. Online interactive tools as well as digital tools and software can be useful learning environments and aids for students.

Estimating and measuring

Measuring involves counting units that uniquely apply to particular attributes (e.g., degrees to measure temperature), using tools of measurement (e.g., scales, stopwatch, clinometers and thermometers) and using relationships to calculate measurements of other attributes such as speed, volume, circumference and density. Estimating is more than guessing. It supports students to make sense of units of measurement, but is ultimately a skill that people use to test the accuracy of measurement (Muir, 2005). Estimation plays a key role in the working lives of people in a range of occupations. We estimate measurements as part of everyday activities, such as how long it will take to reach a destination and how much butter to use in a recipe.

Students need many different experiences of estimating and measuring in order to internalise benchmarks for measuring length, mass, angle, time, area, volume and temperature, to improve the accuracy of their estimations, as well as to develop strategies for making estimates (see Lovitt & Clarke, 1988). They should gather and analyse data about their estimations. Teachers can help scaffold this skill by asking students to explain and justify the strategies and benchmarks that they use when estimating. Particular strategies for estimating can be explored through problem-solving tasks, especially for irregular shapes and solids.

Students need to explore measurement in context, to estimate and to measure. The estimation and measurement problems need to be meaningful for students. For example, students could investigate the accuracy of digital pedometers and design their own (Peters et al., 2012). Meaningful contexts enable students to consider whether an under-estimate or over-estimate is appropriate and the degree of accuracy needed (Muir, 2005). They should conduct investigations to find out how various tradespeople and professionals make estimations of things that are central to their work. These may include 'rules of thumb' that involve more than one step to arrive at an estimate.

Students should use analogue and digital measurement tools of the real world, not just those in the mathematics and science classroom (e.g., tape measures, pressure gauges and theodolites); however, attention needs to be given to reading measures from scales on analogue instruments. Secondary students' understanding of the conventions of linear and circular scales, that is, the role of the marks and spaces on the scaled instrument, cannot be assumed (Drake, 2013). Especially difficult are scales that show both metric and imperial measures and scales with iterations other than multiples of 10 or 100 (e.g., Figure 10.1). Drake (2013) found that less than 30 per cent of a small sample of Year 9 students could correctly read the gross weight of an object on a 1 kilogram kitchen scale with 25 gram iterations on the circular scale. Fewer students accurately reported the net weight, as the scale had not been accurately zeroed before measurement. Hands-on experience with measuring tools where measurements are compared through whole-class discussion is needed. Online interactive tools may also be used to support accurate reading of various scales and scaled instruments.

Particular attention needs to be given to developing estimating and measuring angles. Mitchelmore and White (2000) recommend that students develop a language to describe various real situations of angles, have experiences where they compare and discover the similarity of these situations, and use informal units to estimate and measure angles. As well as providing experiences in the real world with tools, there are various internet sites and digital interactive learning objects that enable students to practise estimating angles (see 'Estimating angles', Maths300, n.d.). Duncan (2011) describes an interesting investigation concerning measurement of slope to ensure that construction conforms to safety regulations.

Relating metric units of measurement

During the primary school years, students become familiar with the most common units of measurement and recall the relationship between them as isolated facts without necessarily appreciating the base 10 structure of the prefix in the metric system (see Table 10.1). Comparing measurements using an analogue and digital measurement tool will assist students to apply the decimal place value structure to conversion of units. Investigating measurement and solving problems in familiar contexts, such as costs of digital products, water and solar power, will assist students to make sense of the prefixes for very large units and very small units, and relations between all units. For example, how much water is needed to be stored in the dams to supply Melbourne (or another city or town in Australia)? Understanding of the structure of the metric system will enable students to transfer this knowledge to the metric measurement units for attributes studied in other

Table 10.1 Common prefixes in the metric

Prefix	Symbol	Meaning	Power of 10
giga	G	1 000 000 000	10^9
mega	M	1 000 000	10^6
kilo	K	1 000	10^3
hecta	H	100	10^2
deka	D	10	10^1
		1	10^0
deci	D	1/10	10^{-1}
centi	C	1/100	10^{-2}
milli	M	1/1000	10^{-3}
micro	μ	1/1 000 000	10^{-6}
nano	N	1/1 000 000 000	10^{-9}

subjects, such as energy (joules), electricity (watts), computer memory (bytes) and digital picture elements (pixels).

The difficulty that students have in making sense of measurement units is illustrated by errors made by beginning undergraduate students when responding to a test item concerning interpreting scales and measurements on maps (e.g., Figure 10.1). As noted in Chapter 7, secondary students find problems involving proportion difficult. In the case below, the student had more difficulty with making sense of the solution that they calculated. This student did not correctly interpret the scale for the map, did not think about the measurements shown on the map and did not check that the answer made sense by thinking about a distance of 7,500,000 g/centimetre^3.

Digital tools such as *Google Earth* and *Google Maps* can be used to make sense of large units of measurement for length and area along with tasks concerning spatial relations and location (Fitzgerald, 2012). Students who have consolidated these skills of converting units of measurement, and who demonstrate an understanding of attributes measured using composite units—such as speed and density—will be able to demonstrate flexible thinking when solving problems and choosing and calculating equivalent composite units (e.g., g/centimetre^3/hour and g/centimetre^3). Understanding the relationship between units of measurement in the metric system (i.e., 1 litre of water at sea level on the equator has a mass of 1 g/centimetre^3 and takes up 1000 cubic metres of space) is also necessary for solving a range of real problems. For example, how much water does it take to fill an Olympic swimming pool?

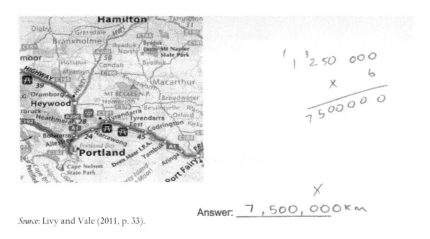

Source: Livy and Vale (2011, p. 33).

Figure 10.1 **Incorrect response to scale item**

Online tools for converting units between systems of measurement can be investigated and used by students, particularly vocational students who may regularly need to work with imperial units.

Perimeter and circumference

Using open-ended tasks and real problems, students can develop strategies for finding the perimeter of closed regions, and thus avoid confusing formulae for perimeter and area. Consider, for example, the following problems:

1. Draw, or make on a geo-board, a rectangle with a perimeter of 18 units. Draw and make another rectangle with a perimeter of 18 units. Find a third rectangle.
   - How did you find these rectangles?
   - Formalise the process you followed by writing a rule to find the dimensions of a rectangle with a given perimeter.
2. You need to build a new pen for a pet lamb. You have 12 metres of fencing wire and you can use the side of a building as one side of the pen. What are the dimensions of the pen that will give us the largest area for the pet lamb?

When these problems include calculating perimeter and area, students can also form and test conjectures about the relation between perimeter and area. For example, 'if you increase the perimeter of a shape you will increase the area of this shape'. As this problem infers, learning about perimeter should not be restricted to rectangles, but rather include problems with other polygons and irregular shapes.

Investigating the circumference of a circle provides a mathematical and real context for learning about proportion, and is often students' first formal encounter with an irrational number, that is, pi (π). A teacher with whom one of the authors of this book recently worked decided to conduct a whole-class investigation into circumference using technology when she discovered that students in her Year 9 class still had many misconceptions about pi (one being that pis were round and that was the association between pi and circles). For this investigation, the students measured many different circular objects and entered the measurements of the diameter and circumference into a spreadsheet. The teacher had prepared the spreadsheet with the whole class using her laptop computer and a data projector, and included a column for the ratio of the circumference and diameter. As each student entered the measurements of their circular object, the ratio was displayed for all to see. The average ratio was also displayed. The teacher explained that 'the big drama was when you hit enter and it calculated pi [the ratio]. There was lots of competition. Could they get to 3.14? How close?' She noticed that the students appreciated the need for accurate measurements and went back and measured their object again. They 'would come back and say, "my diameter was out by about 2 millimetre, will that change my answer?"' The use of technology in this lesson enabled these students to learn from instant feedback (see Chapter 4). They appreciated the need for accuracy when measuring, as they observed the dynamic calculations of the ratio of circumference and diameter and convergence of the average ratio to a constant. By making this investigation a whole-class activity rather than an individual one, the teacher ensured that the inaccuracies in measurement that often occur in an investigation like this did not hinder students' 'discovery' of this constant. Pre-service teachers who have taught this lesson reported that measuring the circumference of circles using string and identifying the diameter are non-trivial tasks for students. These pre-service teachers used an interactive whiteboard.

Area and volume

Students can develop a sense of ownership or agency if they reinvent formulae for finding the area of different polygons. Paper folding, drawings or construction of triangles circumscribed in rectangles on paper or on geo-boards, or online interactive learning objects and drawings using dynamic geometry software, are alternate environments for investigating the area of triangles (Burns & Brade, 2003; Frid, 2000a). Similarly, students can derive the formulae for finding the area of quadrilaterals, other polygons and circles (Stacey & Vincent, 2009a).

Understanding of volume should be developed using materials to derive the formulae of prisms and cylinders as area of the base times the height (see Figure 10.2). Further, students can discover the relationship between the volume of a prism and a pyramid with the same base (and similarly a cylinder and cone with the same base) to derive the formulae for the volume of pyramids and cones. Following this investigative approach, along with solving problems about irregular shapes and solids, develops students' skills in reasoning and their capacity for flexible thinking and problem solving.

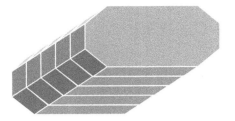

Figure 10.2 **Representing volume (volume = area × height)**

REVIEW AND REFLECT : Derive the formula for the area of a triangle. Show that the formula works for a right-angled triangle, an acute-angled triangle and an obtuse-angled triangle (see Frid, 2001).

Research and discuss different methods for deriving the area of a circle, quadrilaterals, other polygons and the volume of a pyramid and a cone. Include review of online learning objects in your research.

When solving area and volume problems, students often make errors when the dimensions of shapes and solids are given in different units. They need to pay attention to the units given in each problem, and to think about which unit to use when calculating. Students who do not have a strong conceptual understanding of area and volume and the units of measurement, or who have poor multiplicative skills, will have difficulty converting square units (such as 40 000 centimetre² = 4 metre²) and cubic units of measurement. They need visual models of units for area and volume for successful conversion. Use problems that enable students to make sense of hectares. For example:

- How long does it take to run around a hectare?
- A bushfire burnt 40,000 hectares of farm and bush land. Make sense of the extent of the damage by mapping this area onto a map of a region with which you are familiar.

Problems based on a real context may also reveal poor conceptual understanding or reliance on applying a given formula without due consideration of the context. Consider the following problem in Figure 10.3.

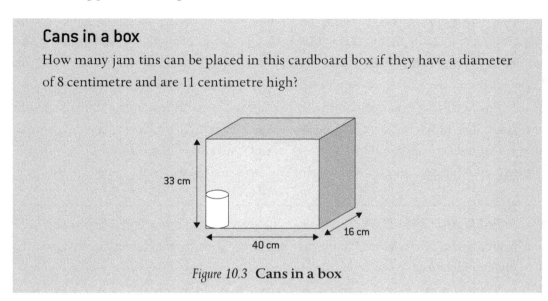

Cans in a box

How many jam tins can be placed in this cardboard box if they have a diameter of 8 centimetre and are 11 centimetre high?

33 cm

16 cm

40 cm

Figure 10.3 **Cans in a box**

Procedural thinkers—that is, students who rely on applying procedures rather than using conceptual understanding—are likely to solve this problem by calculating the volume of a can using the formula for volume of a cylinder and calculating the volume of the box, then dividing the volume of the box by the volume of the can to find the answer. There are many opportunities for students to make errors following this method, and their final solution may not make sense. Redefining the dimensions of the box according to the diameter and height of the can is a more efficient method of solving this problem. Using concrete materials may assist some students to solve these types of authentic problems for area and volume, and strengthen their understanding.

Problems of scale in area and volume are a persistent difficulty for students throughout secondary schooling. Consider the following multiple-choice question:

Which of these pipes will fill a pool the fastest?
- one pipe with a diameter of 60 centimetre
- two pipes each of diameter 30 centimetre

- three pipes each of diameter 20 centimetre
- all the same.

Students assume a linear relationship in scaling up problems—that is, if you double the dimensions of a rectangle or rectangular prism, or radius of a circle or sphere, then you will double the area and double the volume (Van Dooren et al., 2005). To overcome this misconception, students need experiences with concrete materials, along with diagrams, to be convinced that area and volume increase exponentially when the dimensions are increased. Real problems (e.g., *Baby and the heatwave*, Chapter 15) provide a personal connection for students to appreciate the concept of scale and enlargement and the effect on the dimensions, area, surface area and volume of shapes and solids.

REVIEW AND REFLECT : Try the two problems *Cans in a box* and *Filling a pool* with a group of students. Ask them to explain and justify their solution. Analyse their responses.

What strategy did they use to solve each of these problems?

What conceptual understanding of area, volume and scaling was evident or not evident?

How did they justify their solution?

How could you improve their understanding, problem solving or reasoning proficiencies?

Pythagoras' Theorem

The study of triangles led to many significant theorems in mathematics. These theorems and their applications constitute an important part of secondary school measurement, geometry and algebra learning. As for other content in measurement, students should rediscover this theorem and generate informal proofs through guided investigations (see Chapter 9 and Lovitt & Clarke, 1988). Concrete materials and various technologies, including Java applets and templates available on the internet and dynamic geometry software, are useful (see Chapters 4 and 9).

REVIEW AND REFLECT : A Victorian secondary teacher asked his students to use different technologies to solve the following problem involving the application of Pythagoras' Theorem (Noura, 2005). They used a spreadsheet (Excel), a graphics calculator, dynamic geometry software (Geometer's Sketchpad) and classical geometry.

We need to establish a single exit (E) from a freeway to serve two towns, Melford (M) and Extown (C), which are located 1 kilometre and 2 kilometre from the freeway respectively (see Figure 10.4). The road MEC should be of minimum length in order to minimise the cost of the roads. Where should the exit (E) be placed?

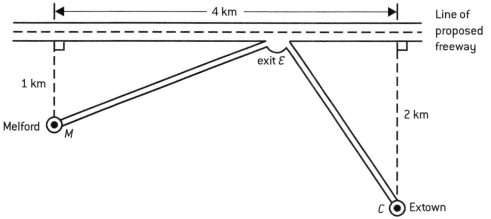

Source: Swan, K. et al. (1997, p. 86).

Figure 10.4 **The freeway problem**

Work in a small group to solve this problem using one of the technologies listed above. Each group should choose a different technology.

Present and discuss the solution processes used in each group. What understandings of the problem and mathematics are demonstrated? In what way may the skills and understandings vary according to the technology used, if at all?

Prepare an assessment rubric to use with students for this task. (Refer to Chapter 6 for examples.)

Trigonometry

Quinlan (2004) recommends that students should work from the concrete to the abstract and from the particular to the general when being introduced to new topics in mathematics. He illustrated this approach to learning about trigonometry by setting Year 10 students the task of finding the height of the wall in a classroom using a 45–45–90 set square, straws and rulers, and tape measures initially, and then challenging them to find it using the 30–60–90 set square. The teacher guided the students' investigation by drawing their attention to similar triangles and ratios of the sides. The teacher generated interest in trigonometry by involving students physically in this measurement problem.

Moving from the particular to the general can be achieved through further investigation of similar triangles. Lo and Marton (2012) argue that students learn by paying attention to what remains unchanged when other elements of the object vary. Noticing the unchanged elements enables students to generalise properties or relations. Dynamic geometry software provides instant and accurate feedback for students as they conduct an investigation of similar triangles or the unit circle to identify what stays the same. Such software makes it possible to keep one element constant (the angles) and to vary the length of one side, so that students notice that the ratio of side length that is the focus of attention remains unchanged. Figure 10.5 shows the results of an investigation of similar triangles using Geometer's Sketchpad. Students need to be able to construct shapes that retain their geometric properties when sides and vertices are dragged (drag-resistant constructions), and to use the calculation and tabulation features of the software. Cavanagh (2008) agrees that it is important to engage students in a particular example through problem solving and modelling before investigating similar triangles to discern the general properties.

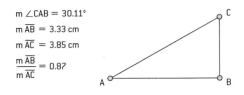

m ∠CAB	m $\overline{AB}$	m $\overline{AC}$	$\dfrac{m\,\overline{AB}}{m\,\overline{AC}}$
30.11°	2.75 cm	3.18 cm	0.87
30.11°	2.96 cm	3.43 cm	0.87
30.11°	3.10 cm	3.58 cm	0.87
30.11°	3.33 cm	3.85 cm	0.87

m ∠CAB = 30.11°

m $\overline{AB}$ = 3.33 cm

m $\overline{AC}$ = 3.85 cm

$\dfrac{m\,\overline{AB}}{m\,\overline{AC}}$ = 0.87

Figure 10.5 **Noticing constant ratio in similar triangles for cosine 30.11°**

However, some teachers prefer the unit circle in order to introduce trigonometric functions. 'Trigonometry walk' (Lovitt & Clarke, 1988) is a physical outdoor investigation of the unit circle, and the dynamic software products include a pre-made sketch for a unit circle investigation. This investigation holds the hypotenuse of a right-angled triangle constant while varying the angle from $0°$ to $360°$. Students will notice the range in the value of ratios, and a pattern in the variation of ratios. Kendal and Stacey (1999) conducted a teaching experiment to compare the unit circle method with the similar triangles method to establish the trigonometric ratios—that is, the sine, cosine and tangent of an angle in a right-angled triangle—with Year 10 students. They found that the students in the class who learned trigonometry using similar triangles and the SOHCAHTOA mnemonic for the ratios to calculate sine, cosine and tangent achieved better results in the topic test than did students in the unit circle class.

Making connections, problem solving and modelling

Solving measurement problems provides a context for many number concepts and skills, including multiplicative, proportional and exponential thinking. It consolidates geometric understanding of shapes and objects, and provides an opportunity to use algebraic thinking. An approach to teaching that makes these connections explicit and integrates learning in these fields will enable students to make connections with and between mathematics and real-world problems (see Chapter 3). Teachers need to solve the problem themselves and prepare enabling and challenging prompts that support students to collaborate and persist (Clarke, Roche, Cheeseman & Sullivan, 2014). These problems should be drawn from contexts that are familiar to students. These contexts will vary according to opportunities for learning in their community setting. Rather than working through exercises of routine applied measurement problems, teachers should design lessons in which students work collaboratively in groups on investigations, problem solving and modelling tasks such as:

- How far away does an on-coming car have to be for it to be safe to cross the road? (Lovitt & Clarke, 1988).
- How much rainwater could you collect from the roof of your home if you have 4 millimetres of rainfall? How much could you collect in a year? What size water tank should you purchase for your home?

- Make a cylinder by rolling an A4 piece of paper lengthways. Make another by rolling an A4 piece of paper widthways. Which cylinder has the largest volume? Which cylinder has the largest surface area? Why? (See 'Measuring cylinders', *Maths300*, n.d.)
- Which one of these sportspeople needs the fastest reaction time: cricketer (batting), baseball player (batting) or tennis player (receiving serves)? (See Barrow, 2013.) What is your reaction time?

Problems that involve exploring relationships between distance and time, area and perimeter, and surface area and volume lay the foundations for applications of rates of change and differential and integral calculus in later secondary years.

> **REVIEW AND REFLECT :** In a group, choose an object, setting, context or theme of interest to you. Brainstorm issues and problems related to this theme. Identify the issues or problems that could be investigated or modelled using mathematics. Conduct a search for information to use for an investigation and create a video (or still pictures) to use when setting and launching the problem. Consider the modelling processes discussed in Chapter 3 and design a lesson, series of lessons or a project on learning and applying measurement concepts and skills using this video.

Conclusion

In this chapter, we have discussed some of the issues and approaches for teaching measurement in secondary schools. We have seen there are many places in which mathematical ideas are connected across fields and are applied in a range of vocations and real settings. Estimating and measuring in real settings, and investigating and problem solving in groups, will enhance engagement and learning in this field of mathematics.

Recommended reading

Barrow, J. (2013). *Mathletics.* London: Random House.

Clements, D. H. (Ed.) (2003). *Learning and teaching measurement: 2003 Yearbook.* Reston, VA: NCTM.

Noura, K. (2005). How to teach mathematics using technology. In J. Mousley, L. Bragg & C. Campbell (Eds), *Mathematics: Celebrating achievement* (pp. 256–64). Brunswick: MAV.

Quinlan, C. (2004). Sparking interest in trigonometry. *The Australian Mathematics Teacher,* 60(3), 17–20.

Van Dooren, W., De Bock, D., Janssens, D. & Verschaffel, L. (2005). Students' over reliance on linearity: An effect of school-like word problems? In H. Chick & J. Vincent (Eds), *Proceedings of the 29th conference of the International Group for the Psychology of Mathematics Education* (Vol. 4, pp. 265–272). Retrieved from http://onlinedb.terc.edu

CHAPTER 11
Teaching and learning chance and data

Citizens today need to negotiate immense amounts of information and uncertainty in a complex society. Over the past decade, the school curriculum has responded dramatically to educate the public to be more critical consumers of information. Other factors have also played a part. Economic demands require workers to manage data-intensive problems in probabilistic processes. New technologies increase accessibility to large data sets and complex situations. Technology enables visual approaches to analysis that were once too technical for non-statisticians. Finally, an explosion of cognitive and classroom research in probability and statistics is changing the school curriculum (Batanero, Burrill & Reading, 2011; Biehler, Ben-Zvi, Bakker & Makar, 2013; Chernoff & Sriraman, 2014; Garfield & Ben-Zvi, 2007, 2008; J. Watson, 2006). Overwhelmingly, these influences argue for new learning environments emphasising a more inquiry-based approach to teaching chance and data.

Both chance and data use uncertainty and randomness. Chance topics—or probability—measure uncertainty, while data-based or statistical processes examine patterns of variability in aggregating uncertain outcomes. Chance and data topics are usually taught in mathematics, but there are important differences that teachers need to know between mathematics and statistics. In mathematics, a major goal is to identify invariant properties and processes that generalise across different contexts, whereas statistics is consumed with understanding variation within a context. Statistical conclusions can be contentious and interpretive, requiring students to develop skills of building convincing arguments supported with data-based evidence.

Chance and data provide natural opportunities to integrate other topics in mathematics. For example, data can be used to teach functions through modelling and curve fitting, or areas of unusual shapes can be estimated using randomisation (see Figure 11.1). Using contexts from science, social studies or health and physical education are ideal ways to integrate chance and data with other subject areas (Day, 2013).

In primary school, learners examine randomness using coins and dice, organise data with graphs, and engage in statistical investigations by surveying classmates. During the secondary school years, students extend these experiences with more complex situations, and link probabilistic and data-based phenomena via simulations.

At the heart of statistics is the ability to investigate: to formulate and test hunches, justify conjectures with evidence, and make inferences with a convincing argument. In this chapter, we discuss how teachers can develop students' probabilistic and statistical reasoning by engaging them in investigations using chance and data. We introduce a framework—statistical literacy, statistical reasoning and statistical thinking—for building different types of statistical skills (Garfield & Ben-Zvi, 2008). Briefly, statistical literacy is aimed at creating 'statistically literate consumers of claims made in wider social contexts' (J. Watson, 2006, p. 23), developing students' ability to critique statistical information encountered in the media and daily life. Statistical reasoning is the ways in which people use statistical tools in the context of understanding a particular phenomenon, and statistical thinking includes the dispositions and skills needed to undertake statistical investigations in order to create, question and evaluate the processes and evidence used to make statistical claims. These levels are hierarchical in their increasing cognitive

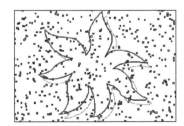

Figure 11.1 **Estimating unusual areas—the area of the figure can be estimated by calculating the proportion of random points in the rectangle that fall inside the figure**

demands; however, rather than teaching them in this order, we recommend embedding teaching statistical literacy and reasoning in the process of developing students' statistical thinking.

Chance and data in the primary years

Until recently, students' main exposure to chance and data in the primary years was through collecting data on coin tosses or rolls of the dice to introduce tallying and elementary probability; data handling was limited to constructing bar charts or reading information in a graph or table. Over the past decade, however, shifts towards teaching higher-order thinking have inspired the introduction of concepts like informal inference and statistical investigations early in the primary years, embedding skills like graph interpretation and invention within mathematical investigations, and postponing teaching formal calculations of averages (mean, median and mode) to the beginning of the secondary years (J. Watson, 2014).

Chance in the primary years

Research documents that young children often hold deterministic beliefs about randomness. For example, children's ideas of destiny, personal preference, 'lucky' outcomes or unrealistic causes can influence their ability to conceptualise chance events. Teaching basic combinatorics (counting)—generating all possible outcomes of throwing two dice, or listing ways that three shirts and four pants can be combined into an outfit—can build ideas about sample space and promote systematic thinking vital for theoretical probability. The language of chance developed in primary school also assists students to begin to connect measures of likelihood with events—for example, expressing a certain outcome as having a probability of 1.

Data in the primary years

An important shift in the primary years is moving students from a focus on individual data points ('Kym watches 16 hours of TV per week') towards holistic descriptions of data to characterise a group ('most of my classmates watch between 10 and 15 hours of TV per week'). While most children have been exposed to averages by the end of primary school, few consider applying an average as representative of a group in problem situations. Konold and his colleagues (Konold & Pollatsek, 2002; Konold & Harradine, 2014) argue that, by focusing early on the concept of data as a combination of signal (central tendency) and noise (variation), students build more robust understandings of average. In contrast, if students learn algorithmic approaches early ('add 'em up and divide'), they see average only procedurally, and this can hinder later development (J. Watson, 2014).

Connecting chance and data in the primary years

One of the main connections made between chance and data concepts in the primary grades is through linking experimental and theoretical probabilities. Although this is possible through games of chance, students also need experience with outcomes that are not equally likely to counter an intuition that if an event has two outcomes then the probability of each must be one-half. Concepts can be developed informally by performing empirical probability experiments, such as observing that in repeating independent trials, the relative frequency of an event approaches its theoretical probability. Connecting relative frequency proportions with likelihood in real-world events is another approach that provides opportunities to build informal concepts of sampling—for example, how would one estimate the proportion of the population with Type O blood, and how might hospitals use this information to plan surgery?

Challenges in understanding chance and data

Research has identified a range of challenges that students experience in understanding and using chance and data concepts as they move into the secondary years. These are related to ideas about randomness, the relative emphasis on conceptual versus procedural knowledge, the nature of statistical investigations, the need to view data from a global perspective rather than as a set of points, difficulties with graph interpretation, and the effect of real-life contexts on students' understanding of variability. Each of these challenges is discussed below.

Intuition about randomness

Students' thinking about randomness is initially deterministic in nature; intuitions about randomness develop through experience and instruction. Misconceptions decrease with age in problems that are clearly probabilistic in nature (e.g., outcome of rolling a die), but not necessarily in problems that are set in a real-life context. Even with statistical training, students have difficulty letting go of strong personal beliefs; they frequently respond to teaching that opposes their intuition by holding dual beliefs (Jones, Langrall & Mooney, 2007; Greer, 2014). A significant challenge in teaching probability and statistics is to pursue a gradual process of building students' intuition before jumping to what may appear to be more efficient formulae and procedures (J. Watson, 2005). Even in secondary school, students should be encouraged to work with manipulatives or concrete materials to assist them in building intuitions about probability and data. Collecting their own data and working with simulations may also help with personal beliefs about randomness that run counter to statistical principles—for example, the belief in a 'hot hand' in basketball (Pfannkuch & Ziedins, 2014).

Procedural knowledge

Research has clearly shown that computational and rote learning can shut down learners' meaning making in mathematics (e.g., Shaughnessy, 2007). For example, teaching students how to calculate a probability or a mean before they build a strong and flexible conceptual understanding of these ideas can undermine their ability to build these notions later on. Plenty of research has pointed out that, while students are able to calculate averages, few students choose to make use of an average in applied problems—for example, when comparing two data sets (Stack & Watson, 2013; J. Watson, 2014). For this reason, teachers are encouraged to hold off teaching students to calculate the mean until after primary school, after they have had multiple informal experiences developing and utilising concepts of central tendency. Similarly, if students are taught specific graphing skills before they develop a conceptual understanding of the purpose and utility of graphs to organise and display information, this conceptual understanding can be impaired. Instead, teachers are encouraged to provide students with repeated experiences handling and organising raw data to invent ways to represent data by making meaning through sorting, stacking and ordering the data to communicate patterns. (A wonderful and classic publication that showcases humorous examples of misuses of averages and graphs is Darrell Huff's 1954 book *How to lie with statistics.*)

Statistical investigations

Most students manage concepts of statistical literacy fairly well, but struggle with the higher-order thinking skills required for statistical thinking. For example, several researchers have noted the difficulty students have in creating questions that can be addressed using statistics (e.g., Arnold, 2013; Makar & Fielding-Wells, 2011). Often, questions are too narrow—requiring simple yes/no answers—or too broad—not measurable or requiring data that are impractical to collect. Frequently, once they do collect and analyse their data, students find it challenging to connect their conclusions back to the question being investigated (Hancock et al., 1992). Most of the time in school is spent in the analysis phase, which is the least challenging part of the statistical investigation cycle (described in more detail later in the chapter). If students' experiences with statistics are imbalanced towards too much time on the calculations and procedures in the data analysis stage, this can reinforce a simplistic view of statistics rather than one in which responses to questions must be tempered by uncertainty. Developing students' statistical thinking involves a concerted effort by teachers to counter this simplistic perspective of statistics by immersing students in all aspects of the statistical investigation cycle.

Aggregate perspective of data

A major goal in elementary data analysis is to encourage students to move from a focus on seeing data as a set of individual points towards a more global, aggregate view of data in which students see a distribution as an entire entity. For example, in Figure 11.2, the quality of 36 brands of peanut butter was rated. By comparing the ratings of 'natural' peanut butter (no additional additives or sugar) with 'regular' peanut butter, students who see data as individual points may conclude that regular brands of peanut butter are higher in quality because the highest rating went to a regular brand. However, a student with a global perspective of data would likely conclude that natural brands are higher in quality because the bulk of the natural brand data is higher than the bulk of the data in the regular brands. Additionally, students may discuss whether natural or regular brands are more consistent in quality.

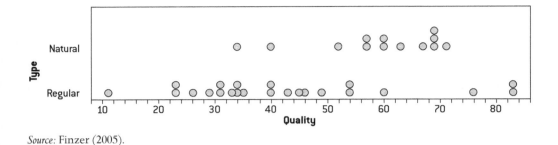

Source: Finzer (2005).

Figure 11.2 **Comparing the quality of brands of peanut butter**

Interpreting graphs

In interpreting graphs, one of the major difficulties students have is that, rather than see-ing a graph as a purposeful tool, they focus too much on learning the features of graphs. In research with secondary students in New Zealand, Pfannkuch et al. (2004) found that students frequently struggled to use appropriate features of graphs as evidence. Although students did learn the properties well, they lacked understanding of the *utility* of the graphs in applications. For example, when comparing the weather in two cities, many of the students in their study used the range of the data or compared upper with lower quar-tiles when this information was not evidence for which city had the hotter weather. The problem was that when students were learning about box plots, instruction focused on learning *features* of a box plot (e.g., locating the median, quartiles, five-number summary or interquartile range) rather than using the box plots as evidence for a particular *reason.* As a result, students didn't see the purpose in using the box plot and focused on what they presumed their teacher wanted—identifying properties.

Similar problems are often evident when students use software to generate different types of graph without making considered decisions about which type of graph is most appropriate for representing the data.

Integrating contextual knowledge

The context within which chance and data topics are taught can affect students' under-standing. There is a misconception that learning about randomness within probabilistic contexts (e.g., dice, coins) will transfer to understanding of randomness in real-life con-texts. Research has shown that tolerating randomness in deeply contextual problems is much more challenging. (See the following Review and Reflect task.)

REVIEW AND REFLECT : Discuss the following two problems with a partner. Compare your solutions with others in the class.

Dice problem

A six-sided die is thrown seven times, resulting in the following outcome: 3, 3, 3, 4, 4, 5, 5 (order is not important). Do you think there is evidence to suspect that the die is unfair? Why or why not?

Media problem

Every year in New Zealand, approximately seven children are born with a limb missing. Last year, the children born with this abnormality were located in New Zealand as shown on the map. Note that the population in each region is approximately equal. A group of families in the central regions has filed a legal case claiming the incidence in their region is unusually high. Do the data support their claim? Why or why not?

Several studies have shown that students interpret the variability in two problems very differently, even though the two situations are structurally identical (Pfannkuch & Brown, 1996). In the first example above, most students interpret the lack of 1s, 2s and 6s as expected in such a small number of rolls of the die. However, the interpretation of randomness in the second problem was quite different. Note that, in the New Zealand context, the presence of abnormalities in six regions with similar populations is equivalent to six possible outcomes of the die, if each region is thought of as representing a particular outcome on the roll of a die. In this context, however, students often interpreted the unevenness of the random outcomes to be influenced by contextual information, such as a potential chemical plant in one of the central regions of the country. One research project reported that two-thirds of the participants in the study (secondary pre-service mathematics and science teachers) attributed the imbalance of outcomes in the dice problem to the randomness expected from such a small sample whereas only one-sixth of them recognised this in the New Zealand problem (Makar, 2004).

Chance and data in the secondary curriculum

Although chance and data are generally taught as distinct topics in primary school, by the secondary years, they begin to merge with only some computational aspects of probability taught independently of statistics. For this reason, the emphasis in this chapter is on statistical reasoning processes, all of which involve basic concepts in probability and probabilistic thinking.

Probability

Only a few topics in the chance strand are taught separately from data at the secondary level. All of these have to do with deductive probability concepts—that is, those in which probabilities are calculated based on other known values. In particular, topics such as calculations of compound events (addition, multiplication and conditional rules), marginal probabilities in two-way tables, independent and mutually exclusive events, expected values, counting principles, and sampling with and without replacement are taught at the secondary level. Students find probability a challenging topic, particularly if the emphasis is on isolating the calculations from a meaningful purpose. Therefore, when teaching probability, it is important for students to try to visualise the sample space. Graphic organisers such as a Venn diagram, tree diagram or two-way table often help clarify subtleties in problems.

REVIEW AND REFLECT : How could you use a graphic organiser to help students see the difference between the probabilities of the following two events (adapted from J. Watson, 2006):

- A randomly chosen man is a 'lefty' (left-handed).
- A randomly chosen 'lefty' is male.

With a partner, come up with three more examples of events that can be represented with graphic organisers. What observations do you have about the kinds of problems that lend themselves to using graphic organisers?

***

Investigate the meaning of the term 'false positive' in medical screening tests in relation to probability. (E.g., even if a person has a positive result in an HIV screening test, it's unlikely that they have the virus.)

***

Punnett squares (see Figure 11.3) are used in genetics to determine the likelihood of genetic traits being passed from one generation to the next.

Do some internet research to find out how Punnett squares work.

Discuss with a partner the appropriateness of this application for junior and senior secondary school students.

Create a set of questions using Punnett squares to help students understand the list of probability topics in the paragraph above. (You may want to include grandparents in your example for looking at compound events.)

Male parent

		T	t
	T	*TT*	*Tt*
Female parent	t	*tT*	*tt*

Source: Australian Bureau of Statistics (2006).

Figure 11.3 **A Punnett square**

This Punnett square shows the offspring of two pea plants, both heterozygous ([defn]), with one dominant gene for tall plants.

It is always important to let students know how the mathematics they are learning is used in people's work. Make a list of professions that rely heavily on probability (an actuary is one example) and find out how they use probability in their work. Give an example and/or explain how these professions use probability in ways that secondary students will understand.

Although probability is an important concept in the chance and data strand, at the secondary level most chance concepts are taught in conjunction with statistics. This is largely because, while probability can help us to work out *expected* values and *theoretical* probabilities, most applications of probability involve empirical probabilities with data. One reason for this is that, in using probability, we may not pick up the subtleties in the context that could influence whether variables are independent. We need statistics to help us decide whether our theoretical probabilities appear to be justified in an applied setting in which variation adds a new level of uncertainty to problems. In new areas, in which probabilities are often changing or unknown (such as in weather forecasting), data are used to generate probability estimates from empirical results. The real power of statistics lies in making predictions through statistical inference, a concept heavily influenced by probability theory.

Statistical literacy: consuming statistical information

> Statistical literacy, although based on concepts currently in the school curriculum, goes beyond them to be embodied in a complex construct that weaves together literacy skills, critical thinking, contextual understanding, and motivation to be involved in decision-making. (J. Watson, 2006, p. 3)

Statistical literacy is a powerful tool that citizens need to interpret statistical information and critique statistical claims found in everyday contexts. At a content level, statistical literacy may only require that students be able to read and interpret graphs and tables, and understand fundamental ideas of average, variation, sampling and chance. However, statistical literacy extends beyond content to include the ability to debate and judge data-based information and claims (Ridgway, Nicholson & McCusker, 2011). J. Watson (2006) proposes six levels of appreciation of aspects of statistical literacy, from idiosyncratic to critical mathematical. These levels incorporate increasingly complex understandings of the context of a problem, sampling issues, data representations, the meanings of average, variation, chance and inference, and mathematical content.

Students have developed most of their basic statistical literacy skills during primary school, and this understanding is refined and deepened as they apply these skills to increasingly complex real-life problems. With the advent of powerful technologies accessible to school-age students for visualising and calculating large data sets (such as Excel, *Fathom, Tinkerplots*), less time need be spent on specifics of graph construction and hence the focus can be shifted to the more challenging and engaging process of utilising graphs, tables and summary values. The difficulties students have with constructing and interpreting graphs or calculating appropriate summary statistics and probabilities are likely due to teaching that focuses more on attributes and calculations than on their utility and purpose as tools for communication and analysis (Shaughnessy, 2007). We therefore move on to the teaching of statistical reasoning.

REVIEW AND REFLECT : Numeracy in the News (www.mercurynie.com.au/mathguys/mercury.htm) is a website for teachers in Australia, with newspaper articles and discussion questions that integrate literacy and numeracy with everyday events reported in the media. Most of the articles involve statistical literacy. Explore the site and some of the articles and activities posted. Work with a partner to find a current article that could be used to discuss statistical ideas with students and write a set of discussion questions for the article like those posted on the website.

Statistical reasoning: utilising statistical information

Statistical reasoning is 'the way people reason with statistical ideas and make sense of statistical information' (Ben-Zvi & Garfield, 2004, p. 7). The classroom environment needed to develop statistical reasoning is more student-centred to enable students to explain their reasoning (Garfield & Ben-Zvi, 2009). The focus is on how people *utilise* data, graphs and statistical information in the context of understanding a particular phenomenon, and how they integrate and explain probabilistic and statistical ideas within applied situations. This involves a higher level of cognitive demand and mathematical content than statistical literacy. This section focuses on Exploratory Data Analysis (EDA), developed by Tukey (1977) to support 'data sleuthing' in descriptive statistics. Within this tradition, we will look at univariate (one variable) statistics as well as bivariate (two variables) statistics. Finally, we will look at the reasoning students use to integrate probability and statistics.

EDA is 'about looking at data to see what it seems to say' (Tukey, 1977, p. v). It focuses on visual impressions of data as partial descriptions and supports attempts to 'look beneath them for new insights. Its concern is with appearance, not with confirmation' (ibid.).

Using an EDA approach to learning statistics encourages students to be 'statistical sleuths'. Although the processes involved in EDA support the development of statistical thinking, it is primarily a tool of descriptive statistics (describing the data in front of you), not statistical inference (making generalisations from a sample of data to a larger population or process). Here, we will look at univariate statistics (one variable) and the more powerful tools of co-variation (relationships between two or more variables).

Univariate descriptive statistics

Univariate statistics is the description of the distribution of a single variable. The important aspects of working with univariate statistics are to focus on qualities of the distribution as an entity and the potential of descriptive statistics to describe and predict phenomena. In the CensusAtSchool archive (Australian Bureau of Statistics, 2006), school students are surveyed and data are collected about multiple variables—for example, the number of hours of sleep. Figure 11.4 shows a box plot with a sample of 50 students from across Australia in Years 7–10 showing the number of hours of sleep they reported getting on a typical school night. In this sample, we can say that about half of the students slept between 8.5 and 10 hours at night and only a few of these students got less than 6.5 hours of sleep.

The limitation of working with a single variable is that it is very easy to lose sight of the *purpose* of describing the particular variable. In this example, it may be of use to have a feel for the number of hours of sleep students report, but if too much time is spent decoding graphs, students lose sight of the purpose of their interpretation and which statistics might be useful to report as evidence in a particular situation. It would be more useful to look

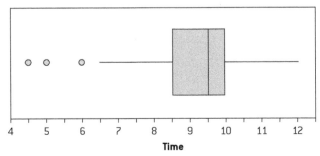

Source: Australian Bureau of Statistics (2006).

Figure 11.4 **Hours of sleep of Years 7–10 students**

at relationships in variables by investigating their *co-variation*. For example, did the amount of sleep differ between students in Year 7 compared with students in Year 10? Or is there a relationship between the amount of sleep students have and how much exercise they do?

Co-variation

Co-variation is the comparison of relationships between variables. There are three distinct kinds of co-variation that secondary students study depending on the types of variables being compared. When two numerical variables are compared (such as height versus arm-span), students work with scatterplots to observe trends in the data. When two categorical variables are compared (such as whether accident victims wore seatbelts compared with the category of their injury), then association of the variables is investigated using two-way tables and comparison of marginal probabilities. Finally, when one variable is numerical and the other is categorical, students use stacked graphs such as box plots or dot plots to compare the numerical data for each group. An example of each of these is given below.

The first type of co-variation compares numerical data. For example, when archaeologists find human bones at a dig site, they use data that compares the length of the bones to heights of individuals, thus allowing them to predict the likely height of the person. The *Comparing numerical data* task below shows how students can participate in such an investigation by collecting this data themselves.

Another type of co-variation examines associations between two categorical variables. Students will likely have some experience with representing categorical variables in two-way tables in primary school (see Table 11.1), but may not have used them as a tool to investigate association. Since secondary students are more comfortable with proportional reasoning, they can begin to appreciate the power of proportions in working with unequal-sized groups.

Comparing numerical data

Secondary school students were told that an archaeologist had discovered a human femur measuring 41 centimetres. They collected data on the lengths of their own femurs (measuring from the middle of the kneecap to the hipbone) and their heights, and produced the graph shown in Figure 11.5. Identify different techniques for predicting the height of the individual whose femur was found by the archaeologist. Discuss the appropriateness of using regression versus non-regression techniques with junior and senior secondary students.

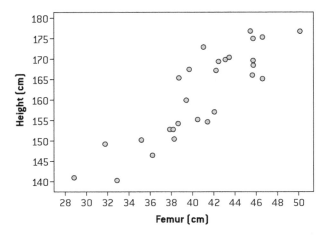

Figure 11.5 **Association between lengths of femur bones and heights in humans**

Categorical variables

A random sample of 300 residents from the UK is split according to whether or not they smoke (see Table 11.1).

Table 11.1 Association between where UK citizens live and whether they smoke

		Smoke?		
		Yes	No	Overall
Live in London?	Yes	5	25	30
	No	64	206	270
	Overall	69	231	300

Do the data suggest that those living in London are more or less likely to smoke than those who live elsewhere in the UK? How is this problem similar to and different from establishing the concept of independence in probability? Can we use the formula for independence to help us decide whether these variables appear to be related?

Comparing groups has been found to be one of the most effective approaches for sup-porting students' understanding of statistical concepts. As some learners find it difficult to move away from focusing on individual points in a distribution, group comparisons can be a way to support aggregate thinking.

Comparing groups

With a partner, discuss the usefulness of the two graphs in Figure 11.6 for describ-ing spring temperatures in Perth.

Discuss your interpretations of the graph on the right, noting any compari-sons you make between the months. In which graph were your interpretations richer? Why?

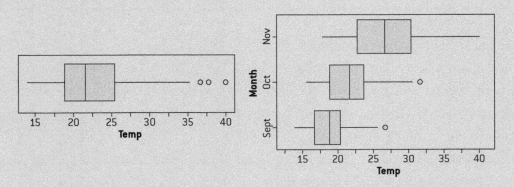

Figure 11.6 **Spring temperatures for Perth (left), disaggregated by month (right)**

The graph of the spring temperature for Perth (see Figure 11.6, left) may be useful to describe the overall spring temperature, but it may also be difficult for students to know which aspects of the distribution to focus on. Conversely, by comparing the tem-peratures by month (see Figure 11.6, right), there is a more natural tendency to focus on interesting aspects of the distribution—for example, comparing how the median temperature changes in each month, giving a reason to look at the *centre*. There is a distinctive difference in variability as well, noting that the temperatures in September appear to be more consistent than in November, giving a reason to focus on aspects of *variability*. Finally, in all three distributions, the data appear to be skewed to the right somewhat, meaning that higher temperatures are less likely within the monthly range than lower temperatures.

Duncan and Fitzallen (2013) discuss the advantages of using multiple representations of data sets when comparing groups. They describe an experiment carried out by the class to collect two sets of data for comparison and then provide a learning sequence that enables students to work from ordered sets of data to stem-and-leaf plots to box plots in order to compare the two sets of data.

REVIEW AND REFLECT : Use data from the CensusAtSchool archive website to devise a set of tasks that illustrates each of the types of comparison outlined above (comparing numerical data, co-variation between two categorical variables and comparing groups).

Combining probability and statistics

Simulations, whether manual or technological, are wonderful teaching tools that can be used to build reasoning about the interaction between probability and statistics. For example, theoretical probabilities can be estimated by conducting empirical experiments, as illustrated by the example below.

Conducting empirical experiments

A family has four children. What is the probability that all four are girls? How common is it to have an even split between boys and girls?

Work in groups of three to tackle this task by generating a list of 100 families with four children. One person generates random four-digit numbers using a graphics calculator and the second person by using the list provided below, where an even number represents a female and an odd number represents a male. The third person conducts a manual simulation using coins, with heads representing females and tails representing males. Each person is to predict, then calculate, the empirical and theoretical probabilities of a family having two girls and two boys.

8964 8980 8093 5700 9531	6583 2680 2073 9480 2305	7918 8964 9808 9357 0845
2857 3513 8492 7826 7205	8671 4737 1522 0245 8697	2817 5221 9933 3689 2857
1407 3219 7389 9449 3743	3869 9967 6561 8568 8246	3115 4476 3475 9677 5654
7708 3740 0808 1393 7186	9698 9039 3129 4005 5176	9320 5222 7934 3220 5584
2277 3114 6911 3179 5559	0602 2061 2463 9173 2185	0966 6861 6773 9625 2628
2498 0654 4380 6536 0104	3471 7474 0857 0411 8525	7688 3727 7891 4668 8287
1263 4538 0665 6872 5648	5747 3306 8592 4513 5373	5235 6948 6462 6011 0108
4117 1872 7920 5900 3837	9097 0148 6428 0367 6885	7459 2427 9355 2148 3253
7589 2333 7300 6848 6784	4262 6646 5005 1333 6718	4138 4289 8752 9920 5183
2626 1016 1979 6818 2007	1471 3374 4121 0389 2012	7549 6245 0253 6509 9555

- How close were your empirical and theoretical results?
- How can you explain the difference in outcomes?
- In your group, compare the three approaches used to generate the list of families and discuss the advantages and disadvantages of each for classroom implementation.
- Extend this investigation to generate and answer your own questions (e.g., 'What is the likelihood of getting consecutive siblings with the same gender').
- How could you adapt this problem for different age levels?

Statistical thinking: creating statistical information

Statistical thinking involves an understanding of why and how statistical investigations are conducted and the 'big ideas' that underlie statistical investigations. (Ben-Zvi & Garfield, 2004, p. 7)

The 'big ideas' of statistics involve understanding the nature and omnipresence of variation, how samples can be used to make inferences to populations, the utility of models to simulate random processes, the centrality of the context of a problem in drawing conclusions, and engagement with the process and limitations of a statistical investigation from problem conception to critiquing of outcomes. Statistical thinking entails a higher level of cognitive demand than either statistical reasoning or statistical literacy; it also requires greater support from teachers to develop a supportive classroom environment to allow students to explore statistical ideas and make connections (Garfield & Ben-Zvi, 2009). It includes both the knowledge and dispositions required to understand the underlying concepts of statistical investigations (Hannigan et al., 2013). A model of statistical thinking developed by researchers in NZ looks at both the investigative processes that engender statistical thinking and the kinds of dispositions critical to statistical thinking (Shaughnessy & Pfannkuch, 2002; Wild & Pfannkuch, 1999). Three critical aspects of statistical thinking will be discussed in this section: *dispositions* critical to statistical thinking; the *statistical investigation cycle;* and *informal statistical inference.* In the process of developing and supporting students' statistical thinking, there is an added benefit of simultaneously deepening their statistical reasoning and literacy skills.

Dispositions for statistical thinking

Developing students' statistical thinking is an ongoing process requiring a classroom culture that encourages risk, collaboration, reflection and open debate (Garfield & Ben-Zvi, 2009; Hannigan et al., 2013). Wild and Pfannkuch (1999) list the following dispositions as critical to statistical thinking:

- *scepticism*—the ability to worry about potential pitfalls in reasoning or lack of justification for assumptions and claims
- *imagination*—the realisation that statistics requires creativity often surprises people; however, understanding core leverages and dynamics of a problem, seeing a problem from multiple viewpoints, and generating possible explanations are deeply imaginative processes
- *curiosity and awareness*—triggering and reacting to internal questions asking why
- *openness to ideas that challenge preconceptions*—registering and considering new ideas, particularly when information conflicts with assumptions
- *a propensity to seek deeper meaning*—looking beyond initial and superficial impressions, being prepared to dig deeper

- *being logical*—being able to construct a valid, coherent and reasoned argument
- *engagement*—intense interest heightens sensitivity and observation skills, and background knowledge of the context is one key to eliciting engagement
- *perseverance*—trying new approaches, sticking with a problem when obstacles are encountered, and managing ambiguity all require perseverance. (p. 233)

In addition, Wild and Pfannkuch (1999) identify several types of thinking unique to statistics:

- *Recognition of the need for data*—this is the understanding that personal experience and anecdotal evidence are inadequate for making decisions, leading to a desire to collect data.
- *Transnumeration*—this is the process of finding ways of collecting data (through measurement or categories) that capture meaningful aspects of the phenomenon under investigation (in mathematics, this is similar to the process of mathematising a problem). Transnumeration also includes looking at several different representations or ways of organising and summarising data that give insight. This aspect of statistical thinking is where the creative insight lends particular power to statistics as a tool to understand the world.
- *Consideration of variation*—statistics empowers one to be able to manage uncertainty, or variation, in data. If there were no variation, there would be no reason to do statistics. The ability to 'see variation' is fundamental to understanding statistics.
- *Reasoning with statistical models (aggregate-based reasoning)*—this process develops more formally when students look at experimental design, but it can be included early when thinking about distributions as representing outputs of a process, as in considering why heights tend to follow a bell-shaped distribution. Models are used to think about statistical processes.
- *Integrating the statistical and contextual*—statistics is meant to be strongly grounded in the context of the problem under investigation. Therefore, students need to be able to integrate statistical and contextual information fluidly. The context allows the data analysis to constantly check with meaning making to generate conjectures, explain causes and seek further meaning in the data.

Statistical investigation cycle

Statistical investigation is part of an information gathering and learning process which is undertaken to seek meaning from and to learn more about observed phenomena as well as to inform decisions and actions. The ultimate goal of statistical investigation is to learn more about a real world situation and to expand the body of contextual knowledge. (Australian Bureau of Statistics, 2006)

The statistical investigation cycle (see Figure 11.7) is a process for using data to learn about phenomena that models the practice used by statisticians in solving problems (McQuade, 2013; Wild & Pfannkuch, 1999). Statistical investigations require a different kind of thinking from that used in mathematical problem solving, and are 'mastered only over an extended period and depend on thoughtful instructional support and repeated opportunities for practice and use' (Lehrer & Schauble, 2000, p. 114). In

Source: www.censusatschool.org.nz.

Figure 11.7 **The statistical investigation cycle**

addition, the first experience with statistical investigations can be frustrating for students and teachers alike because of the difficulty dealing with the multiple uncertainties encountered (Makar & Fielding-Wells, 2011; Makar, 2013). Since statistical inquiry often generates more questions than it answers as the inquirer digs deeper into the data, the process is considered cyclic, sometimes stopping in the middle of a cycle to generate a new one if new questions are deemed to provide better insight or answer questions more efficiently as the inquirer gains deeper understanding of the problem at hand. This process can leave the learner a little overwhelmed at times, particularly in early experiences.

There are five steps involved in a statistical investigation: *problem, plan, data, analysis* and *conclusion.*

- *Problem.* What is the question that is being investigated? Students must have an understanding of the context being investigated in order to come up with questions that make sense. Although it may seem trivial, generating measurable questions and conjectures has been shown to be extremely challenging for students. Unfortunately, most learning experiences in statistics skip this critical step. Until students conduct their own investigations and gain multiple experiences working with data, they tend to either ask questions that are too simple (requiring a 'yes' or 'no' answer), questions for which data are not available or practical, or questions that are not measurable and generally cannot be answered using data. Those inexperienced with data often believe that data analysis provides definitive answers to complex questions. Therefore, one of the goals of the statistical investigation cycle is to provide students with sufficient experience with data to break the black-and-white or overly relativistic mentality that statistics can 'prove' anything.

- *Plan.* Once the question under investigation is generated, students plan their investigation. This includes deciding what data will be useful in answering the question, finding out more about the situation to aid in understanding the data, and logistics of how the data will be collected and recorded. It is advisable for students to collect pilot data to check whether their data collection plan is viable. In upper secondary school, this may also include an appropriate experimental design needed to answer the question. Again, this step is often overlooked in school but includes multiple opportunities to discuss and gain insight into statistics.

- *Data.* Once a plan is in place, students collect and record their data. Often, during this process, there are ambiguities in data categories that require further discussion. For example, in an investigation of whether older classmates had healthier lunches at school than did younger ones, the student investigators realised there was a dispute as to whether items such as cheese and muesli bars should be classified as 'healthy'. The data stage also includes clarifying and 'cleaning' data to prepare it for analysis, such as deciding what to do about missing or omitted values as well as scrutinising outliers in the data.

- *Analysis.* The stage most commonly practised in school is the analysis phase of an investigation. This involves organising and summarising the data, and digging into the data to find meaning, investigating hunches that develop, following leads, and seeking to explain patterns observed. A goal during this stage is to answer the question being posed but often, new questions arise during this process and the inquirer goes back to collect more data or refine their original question based on a better understanding of the data. Often, students realise that the data they have collected doesn't actually answer their original question. There is also difficulty—particularly if the investigation cycle is taught as a rote process—in seeing the analysis stage as an opportunity to generate evidence for the question under investigation (Hancock et al., 1992; Pfannkuch & Ben-Zvi, 2011).

- *Conclusion.* The conclusion stage is where final decisions are made about the interpretation of results and analysis, including inferences made to larger processes. Although this will often include communicating findings to address (completely or partially) the question initially posed, it is meant to include other elements, such as additional information that was learned during the investigation cycle about the phenomenon under study, limitations to the conclusions drawn, and ideas for potential further investigations based on what was learned.

REVIEW AND REFLECT : Design a statistical investigation that engages students in all five steps as outlined in this section; discuss potential areas of difficulty students may have and how these difficulties can be addressed. Use the *Australian Curriculum: Mathematics* to identify strands or topics to be incorporated in your investigation (see McQuade, 2013). Here are some ideas:

- Find out the school's 'top ten' most popular songs.
- Investigate the sleep habits of students in your school—compare weekdays and weekends.
- How many hours in a week do students study (or watch TV, use a computer, spend outdoors)? *Note:* look at the data distributions, not just averages!
- Check online for ideas from your local curriculum authority or mathematics teachers' organisations.

Informal statistical inference

Statistical inference is the process of making a generalisation, with a level of uncertainty, to a population or mechanism based on sample data. Although formal hypothesis testing is not always part of the secondary mathematics curriculum, basic concepts of statistical inference should be taught during the school years. Research has shown that formal statistical inference (such as tests of significance and confidence intervals for population parameters) is very difficult for students to learn when they first encounter the topic in university (Castro Sotos, Vanhoof, Van den Noortgate & Onghena, 2007; Harradine et al., 2011). Developing a strong conceptual basis behind inference in school through less formal means can ease this difficulty. Simulations are often used to help build students' experiences in visualising inferential relationships such as variability in sampling (see www.rossmanchance.com/applets/index.html for examples). Building on work from primary school, junior secondary mathematics can use data with greater sophistication to make subtler interpretations about the population from which data were collected. A benefit of focusing on inference is that it helps to refocus learning on the *purpose* behind the statistics—understanding the context. It makes little sense to draw inferences without bringing in the context. After all, statistics is meant to provide a set of tools to gain insight into a context using data.

Statistical inference differs from *descriptive statistics,* where the given data are described and interpretations are made only about the given data. The following example (see the box below) demonstrates the distinction. Remember that, ideally, informal statistical inferences should be carried out within a statistical investigation cycle.

Statistical inference

Data were collected on the ages of 100 couples when they got married. A graph is given in Figure 11.8 with the means of each group marked.

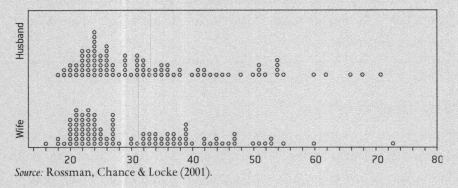

Source: Rossman, Chance & Locke (2001).

Figure 11.8 **A comparison of the ages of 100 couples getting married**

One could use descriptive statistics to describe this sample by making statements such as 'the mean age of the husbands was 33 years and the mean age of the wives was 31'. Or, 'on average, the men in the sample were two years older than their wives', or '4 per cent of the women in this sample getting married were in their teens'. However, any generalisation made to all married couples (assuming this is a representative sample)—such as 'from this sample, one can hypothesise that husbands tend to be older than their wives'—would be an inferential statement, the validity of which would likely need further analysis (formally or informally).

This is an important distinction, as the power of statistics lies in the ability to draw inferences about phenomena based on appropriately drawn samples. Descriptive statistics, while it does allow us to gain insight into patterns found in data, is limited to describing the data at hand. Of course, descriptive statistics forms the backbone of inferential statistics. For further discussion of the role of informal inferential statistics in the curriculum, see Makar (2013). Chick (2013) and Stack and Watson (2013) provide examples of statistical investigations and discuss key concepts regarding inferential statistics.

REVIEW AND REFLECT : A poll of 50 randomly chosen voters is taken to estimate how a local community will vote on a referendum to ban pets from the local park. Results of the poll are given in Figure 11.9.

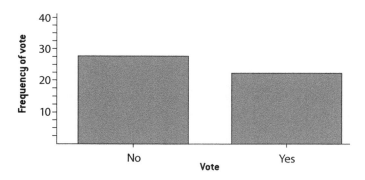

Figure 11.9 **Random sample of 50 voters—22 in favour, 28 against**

- Do you think the referendum will pass based on this poll? How sure are you?
- Use your response and the poll results to write one descriptive and one inferential statement about the data.
- Make up your own example to help students understand the difference between descriptive and inferential statistics.
- Compare your responses with a partner.

In the later secondary years, sampling distributions are introduced to give an idea of how multiple samples can be used to provide more precise estimates of population parameters (Harradine et al., 2011). The tools for inference draw on and extend earlier concepts related to probability, centre, variability and distribution. Sampling distributions present a number of conceptual challenges for students. In descriptive statistics, the data points generally represent a single measurement. However, sampling distributions are made up of a collection of data in which each point represents an entire sample.

> **REVIEW AND REFLECT** : Analyse the curriculum of the senior secondary mathematics subjects in the *Australian Curriculum* (ACARA, 2014) and compare these with those (current or previous versions) in your state or territory. Identify and document the statistical literacy and the statistical inference understanding and reasoning that are included in each of these documents.

Conclusion

Teaching chance and data provides many opportunities to implement the mathematical pedagogies and practices outlined in the chapters in Part II of this book. Problems in probability and statistics naturally incorporate context and hence provide teachers with multiple opportunities to connect students' learning in school with problems encountered in real life. By simultaneously developing students' statistical literacy, reasoning and thinking practices, teachers can prepare students for consuming, utilising and creating statistical information throughout their lives.

Recommended reading

The Australian Mathematics Teacher, **69**(4)—Special issue on statistics, 2013.

Mathematics Teaching in the Middle School, **21**(6)—Focus issue on probability and statistics, 2016.

Batanero, C., Burrill, G. & Reading, C. (2011). *Teaching statistics in school mathematics— Challenges for teaching and teacher education.* Dordrecht, the Netherlands: Springer.

Garfield, J. & Ben-Zvi, D. (2008). *Developing students' statistical reasoning: Connecting research and teaching practice.* New York: Springer.

Makar, K. (2013). Predict! Teaching statistics using informal statistical inference. *The Australian Mathematics Teacher,* 69(4), 34–40.

Shaughnessy, M. & Pfannkuch, M. (2002). How faithful is Old Faithful? *The Mathematics Teacher,* 95(4), 252–259.

Strayer, J. & Matuszewski, A. (2016). Statistical literacy: Simulations with dolphins. *The Mathematics Teacher,* 109(8), 606-611.

Watson, J. (2005). Is statistical literacy relevant for middle school students? *Vinculum,* 42(1), 3–10.

Watson, J. (2006). *Statistical literacy at school: Growth and goals.* Mahwah, NJ: Lawrence Erlbaum.

Watson, J., Fitzallen, N. & Carter, P. (2013). *Top drawer: Statistics.* Melbourne: Education Services Australia. Retrieved from http://topdrawer.aamt.edu.au/Statistics

CHAPTER 12
Teaching and learning calculus

Calculus plays an important role in secondary and tertiary education. Future teachers, engineers, doctors, economists, scientists, and, of course, mathematicians undertake the effort of learning and understanding calculus concepts and techniques ... Calculus is used as ... a fundamental preparation to take on applied problems in partner disciplines, preparing students to bring an understanding of rates, concavity, functional relationships, among other topics to bring to bear on multi-disciplinary problems. (Rasmussen, Marrongelle & Borba, 2014, p. 507)

Derivatives and differentiation, and integrals and integration are central concepts and procedures, with very important underpinning notions that begin their formation in other areas of mathematics studied earlier. Derivatives and differentiation link to the notion of gradient of a function, which links to rate of change and so ratios of differences. Integrals and integration link to the notion of area under a curve, which links to summation. According to Tall (1996), there is a:

Spectrum of possible approaches to the calculus, from real-world calculus in which intuitions can be built enactively using visuo-spatial representations, through the numeric, symbolic and graphic representations in elementary calculus and on to the formal definition–theorem–proof–illustration approach of analysis which is as much concerned with existence of solutions as their actual construction. (p. 294)

'Procept' is the term used by Gray and Tall (1994) to name the combination of process (e.g., differentiation) and the concept (e.g., derivative) produced by the process that may both be evoked by the same symbol $\left(e.g. \frac{dy}{dx}\right)$. In introductory calculus, three procepts are important: *function*, incorporating the notion of *change*; *derivative*, incorporating the notion of *rate of change*; and *integral*, which incorporates the notion of *cumulative growth*. This area of study, then, becomes a study of the *doing* and *undoing* of the processes involved with these procepts.

Tall (1996) suggests there is 'a fundamental fault-line in "calculus" courses which attempt to build on formal definitions and theorems from the beginning' (p. 293). Researchers such as Vinner (1989) claim that students tend to use algebraic representations and methods when solving calculus questions, avoiding the visual methods

The ever-present temptation in teaching and learning mathematics is to succumb to the pressure to develop fluency and automaticity in techniques over the building of relational understanding of the underlying concepts and procedures and the ability to apply these concepts in a wide variety of task contexts (both real world and purely mathematical). Nowhere is it more important to resist this temptation than in introductory calculus.

In this chapter, we will examine the background students require for successful learning of calculus, key steps in the introduction of the calculus, and obstacles to learning that have been identified by researchers. Key notions in the teaching of differentiation and integration will then be addressed. In keeping with the tremendous changes that have been enabled in teaching by the advent of electronic technologies in classrooms of high mathematical capability, there is a focus on technology in addressing difficulties and introducing key ideas.

Background required for successful learning of the calculus

Representational diversity, fluency and versatility

Modern technology supports three (external) representations of function: the numerical representation (through tables of values), the graphical representation, and the algebraic representation (especially with the capability of CAS). Image digitisers such as GridPic (Visser, 2004) and Plot Digitizer (Huwaldt, 2015) allow photographs to be imported into the program then overlaid with a Cartesian coordinate grid where students can click on points on the image to collect a set of points (listed in a table of values) that form the basis for generating a graphical model of things of interest in the image such as the Gothic arch in Figure 12.1. By selecting a quartic polynomial from a number of options, students can use various strategies to refine the parameters of an algebraic model for the graph that best fits the points. The ability to establish meaningful links between representational forms and the concept being represented, which Moore et al. (2013) call representational fluency, involves connecting 'data, real-world situations, theories and mathematical models that allow students to demonstrate understanding in different representations' (p. 144). Also useful in providing these links is video analysis and modelling software such as Tracker (Brown, 2015), an open-source program enabling tracking of the path of a moving object in a video with a data analysis tool affording powerful automatic and manual curve fitting. The Digital Library Browser available at the site provides easy access to online collections of videos and Tracker resources along with video tutorials in the use of Tracker.

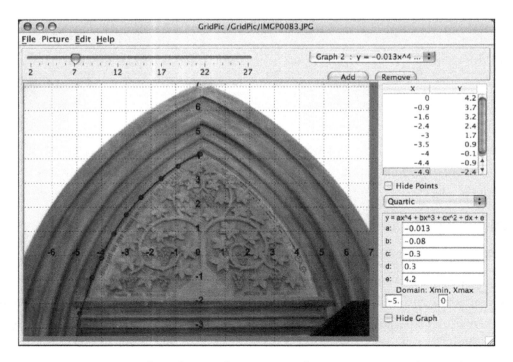

Figure 12.1 **Use of GridPic to fit a quartic function to a Gothic arch**

Multiple representations may enable students to view and explore mathematical concepts in a variety of ways. A function may be expressed in words, represented in a table of values, depicted as a graph or in a symbolic representation:

> Different representations clarify different aspects of the concept by providing complementary roles where understandings in one representation are reinforced in another, for example, seeing the pattern in the numbers in a table may give greater meaning to the slope of a line in the corresponding graph. Interpretation of meaning is constrained by comparisons of representations and connections between representations facilitate the construction of deeper understanding (Herbert, 2013, p. 175).

Different features of functions are more apparent in graphic representations than they are in symbolic representations. The symbolic representations of two different quadratic functions may appear very similar, for example, $f(x) = x^2 - 2x + 1$ and $g(x) = -x^2 + 2x + 1$, but their graphic representations (see Figure 12.2) look very different.

In particular, the graphic representation better illustrates the zeroes and turning points of these functions, while the numeric representation (see Figure 12.3) may emphasise the relationship between the variables. Multiple representations may provide complementary roles assisting students to notice different aspects of concepts and so construct deeper understanding of them.

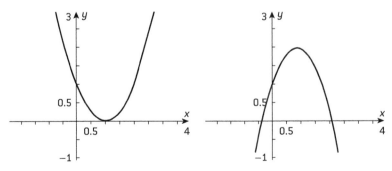

Source: Herbert (2013).

Figure 12.2 **Graphic representations of $f(x) = x^2 - 2x + 1$ (left) and $g(x) = -x^2 + 2x + 1$ (right)**

x	y		X	y
−2	9		−2	−7
−1	4		−1	−2
0	1		0	1
1	0		1	2
2	1		2	1
3	4		3	−2

Source: Herbert (2013).

Figure 12.3 **Numeric representations of f(x) = x² − 2x + 1 (left) and g(x) = −x² + 2x + 1 (right)**

However, others such as Kendal and Stacey (2003) question the wisdom of such an approach, in the limited time available in an upper secondary curriculum. Drawing on the results from a teaching experiment involving the introduction of differential calculus over 22 lessons of 45 minutes each to two Year 11 classes, they found that even though both classes had access to CAS calculators, 'only the most capable students developed an appreciation of the concept of derivative across all three representations and about half demonstrated mastery in at least two representations' (Kendal & Stacey, 2003, p. 38).

A major claim made by advocates of the use of technology for building students' understanding of mathematics (Kaput et al. 2008) is the ease of students' access to multiple representations of functions, perhaps assisting students to develop a deeper understanding of mathematical concepts, such as function and derivative by providing numerous experiences in which a particular concept can be explored in a variety of representations (Zbiek & Heid, 2008). In their book chapter, Kaput et al. (2008) show how the development of technology-supported external manifestations of knowledge and transformational skill, thus changing representational infrastructure, have the potential to make this intellectual power available to all, not just the mathematically privileged elite. They assert that employing multiple representations extends students' processing power and ability to solve a wider range of problems that were previously only accessible to the intellectual elite.

Figure 12.4 **Students linking numerical and diagrammatic representations in a real-world task**

Stewart and Thomas (2006) use the more expansive term 'representational versatility' to encompass 'the need for both conceptual and procedural interactions with any given representation and the power of visualisation in the use of representations' (p. 488), as well as representational fluency. Thus, one of the first prerequisites for the beginning calculus student is a well-rounded multiple representation understanding of function (see Chapter 8)—which will, of course, be developed further as elementary calculus ideas are developed. These understandings can, and should, be developed through tasks at the lower secondary level encompassing the use of multiple representations (see Figure 12.4).

The study of change

From an early age, children recognise examples of change in their environment and describe change in qualitative terms such as a bucket becoming heavier as it fills with water or a pile of potato chips becoming smaller as more are eaten. They notice that the bucket becomes heavier more quickly as the tap is turned on further, but that the rate at which vegetables disappear from a dinner plate does not change if you continue to eat just as slowly as you can. They become aware that some changes are increasing, others decreasing and yet others change at the same rate. By measuring and comparing quantities, children learn to quantify change and about the predictability of some change situations and the randomness of others.

Several important ideas are embedded in change situations, and these need to be addressed explicitly in the lower secondary years to lay a foundation for calculus:

- Change in one quantity may or may not be related to change in another quantity.
- The rate at which the change is occurring may be constant or may vary.
- How quickly or slowly the variation in the change occurs is important.

Change can be represented in many different ways (e.g., verbally, in a table, diagramatically, graphically). A graph shows the relationship of one quantity to another with the shape giving insight into the nature of the change. Although there have been many opportunities in the pre-calculus years to focus on interpreting and drawing graphs, as preparation for calculus, it is important that students focus on global properties of graphs, such as whether a graph is increasing or decreasing, whether this is happening steadily, or whether it is increasing/decreasing quickly or slowly. Sketching graphs of relationships (especially motion) described in words, interpreting graphs in this global sense, and matching graphs to situations described in words (see the box below for examples)— *not* just drawing graphs from algebraic formulae—provide a firm foundation for rate of change. Excellent examples of such tasks are provided in Barnes (1991) and *The language of functions and graphs* from the Shell Centre for Mathematical Education (1985).

Examples of describing relationships verbally and graphically
From graph to story

Australian Open Tennis Championship

Centre Court in Rod Laver Arena is used for important matches that draw large crowds during the Open. Draw a graph showing how the number of people in the stands varies during the evening program. Write a brief story explaining the changes shown in the graph.

From story to graphs

Petrol theft

Non-payments by customers at self-service petrol bowsers were high at the beginning of the year when petrol prices were high because of high consumption overseas in a cold northern winter. These thefts dropped slowly during the next six months, then a fire at the refinery resulted in a 20-cent increase in petrol prices overnight and the thefts suddenly doubled. They stayed high for the next three months as petrol prices remained high while the refinery was repaired and after that gradually decreased as petrol prices dropped again. Draw a graph to illustrate the number of these petrol thefts over the period. If you drew a graph of the petrol prices over the same period, how would it differ?

Matching graphs to story

Aeroplane landing

An aeroplane coming in to land at an airport has been put into a holding pattern while it waits to land. It is circling at a constant height and a fixed speed. At a particular moment, the pilot is informed it is safe to land and to taxi some distance to the allocated airport terminal gate to disembark passengers. Which graph in Figure 12.5 models most realistically the relationship between the speed of the aeroplane and the distance it travels to the disembarkment bridge at the terminal gate? Explain your choice. If none of the models is realistic, draw your own version and explain it fully.

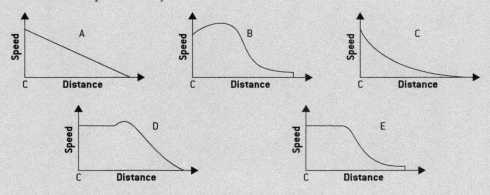

Figure 12.5 **Aeroplane speed models**

Slope of secants

Technology can help students interiorise these actions to a single process as the points get 'closer and closer' together—the second step on the graphical path to derivative. For example, it can be used to draw the tangent at a point, say (−3.5, 0.9) (see Figure 12.6), by selecting the straight line option and refining choices of values for parameters in $y = ax + b$. The picture can then be turned off to remove extraneous detail and the slopes of secants to a series of points approaching (−3.5, 0.9) can be computed and used to draw secants through these points and (−3.5, 0.9) as shown in Figure 12.7.

Rate of change

One of the key properties of any function is how it changes (or even whether it changes). This is referred to as the rate of change of the function. It provides an important underpinning for the concept of derivative (Herbert, 2010). As we have seen in Chapter 8, middle secondary students have difficulties with the concept of rate, so it is not surprising that students of calculus have difficulties with differentiation that have persisted over many years (Hashemi, Abu, Kashefi & Rahimi, 2014; Orton, 1983).

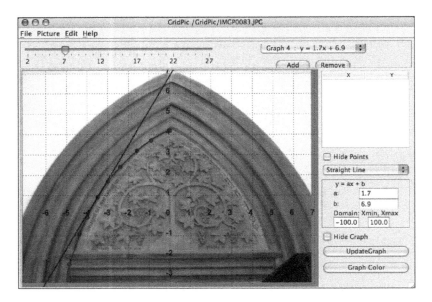

Figure 12.6 **Drawing a tangent to the curve at the point (−3.5, 0.9)**

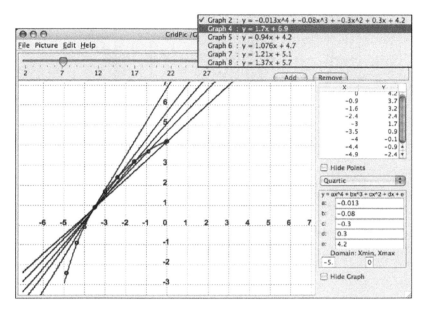

Figure 12.7 **Drawing secants approaching the tangent to the curve at the point (−3.5, 0.9)**

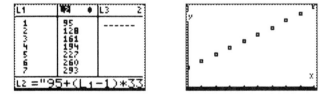

Figure 12.8 **Constant rate of change of height of a stack of plastic cups, per cup**

Students need many early experiences with constant (e.g., stacking plastic cups, see Figure 12.8) and non-constant rates of changes (e.g., shortest-path problems, see Figures 12.9 and 12.10) and being asked to distinguish between these, using information from all three representations.

The key question being asked is *how does the rate at which the y values are changing compare with the rate at which the x values are changing?* Distinguishing between change (i.e., Δx or Δy) and rate of change (i.e., $\frac{\Delta y}{\Delta x}$) causes problems for students. They have difficulty understanding the difference between a *difference* and a *ratio of differences.*

In the following exchange, for example, the students are trying to answer a question about the rate at which the total distance travelled by a runner is changing as the runner

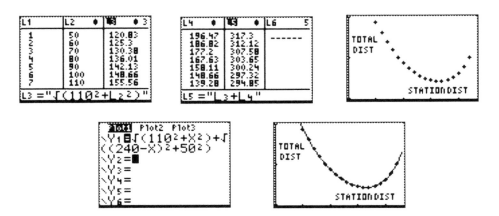

Figure 12.9 **Varying rate of change of total distance run with station distance**

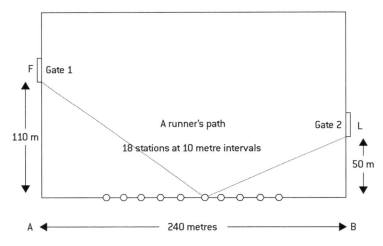

Figure 12.10 **Path and distance travelled by runner**

traverses a field via one of 18 equally spaced drink stations along one side of the field (see Figure 12.10). The runner must travel via only one drink station after entering through one gate and leaving via a second gate.

Thomas: Does the total distance change at the same rate as you travel via station 1, or 2 or 3 or 4?

Andrea: Yes.

Thomas: No it doesn't. No it doesn't.

Andrea: Total distance so it does.

Thomas: It says, 'does the rate change?' Look, it means if you go from 1 to 2 ...

Andrea: No.

Thomas: ... is it the same as going from 2 to 3?

Andrea: No, does the total distance at the same rate?

Maddie: [leaning towards the others] It is the amount, so it is increasing and decreasing.

Andrea: It means if you are running at 5 km/h, does it stay the same? No it doesn't stay the same.

Thomas: Does the total distance run change? I need to concentrate on this.

Andrea: [using scale plan and pointing to different drink stations] Does your total run, whether you go to that, to that, to that and to that change, if you are running at, if everyone is running at 5 km an hour, will it change? Yes, it will because the totals are different.

Source: Extract from RITEMATHS project (2006).

REVIEW AND REFLECT : After reading about the above situation, state what you know about each person's understanding of rate of change. What questions could you ask that would give you the best information about the true understanding of rate of change in this situation for each individual?

***

Devise a practical task for Year 9 students that will allow them to explore the differences between constant and varying rates of change. (See *Navigating through algebra* for examples.)

Key steps in the introduction of differential calculus

A suggested sequence for the introduction of differential calculus is as follows:

1. *rates of change* examples looking at positive, negative and zero constant rate examples and variable rate examples in a variety of contexts, including contexts in which time is not a variable

2. *constant* rate of change → linear graph → rate of change = gradient
3. *variable* rate of change → non-linear graph (e.g., Motion (distance–time) graphs and other contexts)
4. *average rate of change* calculated numerically and graphically, to show it is the gradient of secant
5. *instantaneous rate of change* as successive numerical approximations using a spreadsheet or lists in a graphics or CAS calculator and graphically to show it is the gradient of the limiting secant and use of local straightness to show it is the gradient of the tangent
6. *calculating instantaneous rate of change* by drawing the tangent by hand, and calculating the gradient or using a function grapher (see Figure 12.11).
7. *drawing of gradient functions.* This can be a sketch using the 'ruler as tangent' approach deducing from global and local properties (i.e., where the gradient is positive the gradient function will be above the X-axis, where the gradient is negative it will be below the X-axis, and where the gradient is zero the gradient function will intercept the X-axis) (see Figure 12.12).

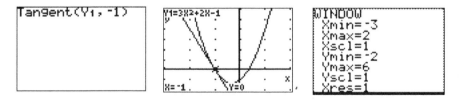

Figure 12.11 **Using Draw Tangent and a judicious choice of window**

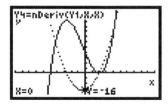

Figure 12.12 **Use of numerical derivative to sketch gradient function of $f(x) = (x-2)^2 (x+5)$**

REVIEW AND REFLECT :

- Make up a set of cards with various functions with which you, but not necessarily secondary school students, would be familiar—for example:

$$f(x)=2(x+4)^3 -3, f(x)=-\tfrac{1}{2}x^5, f(x)=\tfrac{2}{3}x+5.2x^4, f(x)=3\cos 2x.$$

- Keep these functions hidden.
- Everyone enter $Y_2 = [Y_0(X+h)-Y_0(X)]/h$ into the function window of a graphics calculator.
- Choose and enter a value for h, such as 0.001.
- In pairs, each person selects a card and enters this function into Y_0 of the function window of their partner's graphics calculator without the partner seeing.
- Individually, graph Y_2 and Y_0 in a suitable viewing window of your calculator without viewing Y_0 in your function window. Using your knowledge of functions, identify the algebraic function of the gradient function of the unknown function.
- When you think you have identified the function graphically, use the table to check. If you are correct, use different coloured pens (e.g., red for Y_0 and green for Y_2) to draw your function and its gradient function for display on the wall.
- As a group, discuss what knowledge and understandings secondary students would be drawing on when they were conjecturing relationships between functions and their gradient functions when completing a similar task, but with a function such as $y = x^3$.

Obstacles to learning calculus

'Beginning calculus students are not prepared to understand calculus' (Carlson et al. 2015, p. 223). Carlson et al. (2015) reported that many students in their study were unable to answer proportional reasoning questions and function word problems. Ferrini-Mundy and Gaudard (1992) suggest that most students only achieve, at best, procedural competence in the symbolic manipulations of functions, their derivatives and integrals rather than a deep understanding of the concepts underlying the procedures. Indeed, Tall (1985) asserts that an approach to teaching calculus in which the topic is broken up into small chunks and presented in a sequential logical series of lessons is not the most appropriate sequence, as it restricts understanding to procedural competence and later, Tall (1997) claims that the emphasis on the logic of the processes in the traditional approach to calculus hampers conceptual insight. The sequence may only appear logical to the teacher who can see the whole picture, whereas the isolated pieces presented may provide no structure for students and little access to the whole picture (Tall, 1991). Students may not even be aware that there is a whole picture. Later, Tall (2011) re-iterated that students' difficulties in the learning of calculus are a result of their lack of conceptual understanding.

Long ago, Orton (1985) and Tall (1985) expressed the concern that conceptual understanding should be promoted. Orton (1985) advocates that 'blind manipulation of a notation or the mechanical application of rules' (p. 13) was not good enough and that an intuitive introduction with a gradual development of concepts was required. More recently, Hashemi, Abu, Kashefi and Rahimi (2014) assert that 'focusing in symbolic aspect rather than graphical aspect, and the weakness of making connection between graphical and symbolic aspects of the concepts' (p. 363) were the main reasons students found difficulty with calculus.

Difficulties with limits

The limit concept is a sophisticated idea that is difficult to understand even at the tertiary level (Blaisdell, 2012), and problems at this level are attributed to intuitive views of limits coming from secondary schooling (Przenioslo, 2004). There is also confusion caused by the everyday use of the word 'limit' (Monaghan, 1991). Students may have heard or seen such phrases as height limits on traffic passing under bridges, legal blood alcohol limits for drivers or speed limits for traffic. These are all boundary values that must not be exceeded. This view of limit can constrain students' thoughts about functions and 'prevent them from understanding that functions can, for example, oscillate over and under the limit value and still tend to that limit' (Juter, 2005, p. 78).

The limit concept can be dealt with explicitly by considering expressions such as $\lim_{h \to 0} \frac{(x+h)^2 - x^2}{h}$. Informally, the limit can be considered at an intuitive level by considering what occurs as $h \to 0$ when h is varied dynamically. For $h \neq 0$ the rational expression simplifies to $2x + h$, and as h 'tends to zero', this expression visibly becomes $2x$ as is seen in Figure 12.13 when the symbolic facilities of a CAS calculator are used.

F1▾ Tools	F2▾ A13ebra	F3▾ Calc	F4▾ Other	F5 Pr3mIO	F6▾ Clean Up
■ Define y = $\dfrac{(x+h)^2 - x^2}{h}$					
					Done
■ y|h = 1.					2·x + 1
■ y|h = .1					2.·x + .1
y|h=0.1|					
MAIN	RAD EXACT		FUNC		3/30

F1▾ Tools	F2▾ A13ebra	F3▾ Calc	F4▾ Other	F5 Pr3mIO	F6▾ Clean Up
■ y|h = .1					2.·x + .1
■ y|h = .01					2.·x + .01
■ y|h = .001					2.·x + .001
■ y|h = 1.ε-4					2.·x + .0001
y|h=.0001|					
MAIN	RAD EXACT		FUNC		6/30

F1▾ Tools	F2▾ A13ebra	F3▾ Calc	F4▾ Other	F5 Pr3mIO	F6▾ Clean Up
■ y|h = ⁻1					2·x − 1
■ y|h = ⁻.1					2.·x − .1
■ y|h = ⁻.01					2.·x − .01
■ y|h = ⁻.001					2.·x − .001
y|h=⁻.001|					
MAIN	RAD EXACT		FUNC		4/30

Figure 12.13 **Finding the limit informally using the symbolic facilities of a CAS calculator**

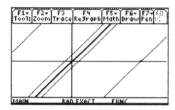

Figure 12.14 **Finding the limit informally using the graphing facility of a CAS calculator**

This is confirmed graphically using the graphing facility with the graphs approaching the line y = 2x from above and below (see Figure 12.14).

Martinovic and Karadag (2012) claim that the concept of limit is dynamic where the 'limit of the rate of change of a function over a diminishing interval, leads to its derivative' (p. 1). They suggest that dynamic, interactive learning environments are still not widely used by educators and the deficiency of representing inherently dynamic concepts (e.g., the limit process) in traditional ways (e.g., without technology, with an emphasis on formulas) makes their learning unnecessarily challenging. They report on the manner in which dynamic, interactive learning environments may be used to support students' development of conceptual understanding of the limit concept.

Even the mathematical language used presents problems, as terms such as 'tends to' or 'approaches' suggest that limits are not attainable (Juter, 2005). Confusion between *the limiting process* and *the attaining of a limit* can lead to students simultaneously holding quite different conceptions in theoretical, as opposed to problem, situations (Juter, 2005; Williams, 2001). There are thus incompatibilities in seeing limits as a process and as an object. Many of the difficulties students experience with other concepts such as continuity, differentiability and integration can be related to their difficulties with limits, according to Tall (1992, 1996) and Williams (2001).

Rate

The traditional approach to the introduction of the concept of derivative assumes a sound understanding of rate and illustrates the derivative as the gradient of the tangent to the curve at a point, then moves quickly to emphasise symbolic manipulation. The symbolic representations of a function are manipulated to establish a symbolic expression for instantaneous rate by taking the limit of the average rate. Some students become competent in this manipulation and can accurately produce the symbolic representation of the derivative (delos Santos & Thomas, 2005), but may not appreciate its meaning and connection to other mathematical concepts studied in earlier years.

Many texts use speed as a prototypical example on which to build an understanding of rate where this particular rate is emphasised with a detailed discussion of displacement, velocity, average velocity and instantaneous velocity. The slope of a linear function may be described as a purely abstract definition as the change in the dependent variable resulting from a unit change in the independent variable. This definition may be connected to the graphical representation emphasising the unit rate by the diagram seen in the left of Figure 12.15. Variable rate may be introduced through reference to the symbolic and graphic representations of the general function $y = f(x)$, again emphasising a unit rate approach (see the right of Figure 12.15). Exercises following this introduction usually relate to contexts in which distance and time are the rate-related variables. 'It appears to be assumed that since speed is experientially real for students, it is also well understood as an abstract concept' (Herbert, 2010, p. 18). Therefore, speed appears to be taken

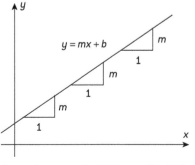
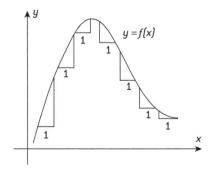

Source: Anton et al. (2005, pp. 153–154).

Figure 12.15 **Rate diagrams**

to be a familiar concept on which to build an understanding of rate and hence, derivative. However, understanding of speed may not transfer to an understanding of rate in non-motion contexts (Herbert & Pierce, 2011).

Calculus students' difficulties with rate manifest in many forms. One of the most significant of these is the confusion between the rate and the extensive quantities that constitute it (Thompson, 1994; Rowland & Jovanoski, 2004), for example, understanding of speed as a distance. Other difficulties with understanding of rate include confusion relating to symbols and their use as variables (White & Mitchelmore, 1996); lacking awareness of the relationship between slope, rate and the first derivative (Porzio, 1997); misunderstandings related to average and instantaneous rate (Hassan & Mitchelmore, 2006; Orton, 1984) and related-rates problems in speed (Billings & Kladerman, 2000); and confusion between average and instantaneous rate (Hassan & Mitchelmore, 2006) and geometric contexts (Martin, 2000). Hassan and Mitchelmore (2006) report on their study of 14 Australian senior secondary students' understanding of average and instantaneous rate. The students were interviewed twice, before and after a teaching intervention designed to address the difficulties with rate expressed in the first interview in which they found most students seemed to have a fair qualitative understanding of rate, but few demonstrated an understanding of either average or instantaneous rate at the start of the first interview. The interview tasks were embedded in the contexts of motion, population growth and cooling rate. None of these rate contexts addresses a rate where time is not a variable.

Students need to bring a robust understanding of rate to their study of differentiation, including qualitative understanding of constant and variable rate, average rate and instantaneous rate with awareness of the relationships between extensive variables to enable students to apply this robust understanding of rate to a variety of contexts.

Misconceptions with tangents

The description of the derivative as the gradient of the graph at a point is another cause of misconceptions. This is because the notion of tangency in many students' concept images (Tall & Vinner, 1981) is related to the special case of tangent to a circle. The notion of a tangent touching at a single point and not crossing the curve is held in opposition to the notion of a secant cutting in two points. Giraldo et al. (2002) point out that 'this leads to a narrowing of the concept image of a tangent that is not consistent with the notion of tangent in infinitesimal calculus' (p. 38).

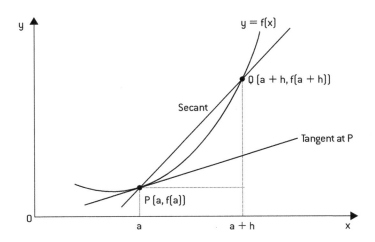

Figure 12.16 **Common textbook diagram for gradient at a point**

Another misconception concerned with tangent is related to the 'disappearing chord' focus that students often take when the common textbook (see Figure 12.16) and teacher explanation of the gradient at a point on a curve is used (Ryan, 1991; Tall, 1985). In Figure 12.16, the key idea to be accessed is what happens in the limiting process as Q moves closer to P along the curve, namely the secant (i.e., the *extended chord*) through P and Q becomes the tangent at the point P. Many students focus on the chord PQ (especially if this is being used instead of, or as, the secant as in Swedosh et al., 2006), which actually tends to zero length as the points become closer together. Other students see this as a static diagram and may not be able to visualise the dynamic image of Q moving closer and closer to P.

Confusions with notation

There is quite an array of new notation that is associated with the study of calculus at the secondary level that students are expected to be able to give sense to and to use meaningfully, namely, $f(x+h), f'(x), y', D_x f(x), \frac{dy}{dx}, \frac{d}{dx}(f(x)), \delta x, \Delta x, \int f(x)\,dx$ to name but a few. Some textbook authors such as Barnes (1991) restrict notation for the derivative, for example, initially *to f'(x) and y'*, deliberately leaving the introduction of the Leibniz notation of $\frac{dy}{dx}$ until later in the development of calculus. The extent to which teachers have the freedom to do this is dependent on the requirements of the intended curriculum and how closely these are followed, and the extent to which they use technological tools that use such notation as $\frac{dy}{dx}$. There are many sources of confusion here, so we will deal with only a couple. Students looking at $\lim_{\delta x \to 0} \frac{\delta y}{\delta x} = \frac{dy}{dx}$ can think that δy tends to dy and δx tends to dx so they think of dy and dx as very small non-zero numbers. However, this conflicts with their being told that $\frac{dy}{dx}$ is a single symbol and cannot be treated as a fraction or that $\frac{d}{dx}$ is an operator. Further conflicts arise when they meet the chain rule, such as $\frac{dt}{dx} = \frac{dy}{du} \cdot \frac{du}{dx}$, and are told that they cannot cancel the dus as du does not have a separate meaning, but in anti-differentiation they must write dx in $\int f(x)dx$ as dx means 'with respect to x'.

   As alluded to above, the advent of calculators that are capable of carrying out numerical and symbolic differentiation and integration has meant that particular notations are being privileged and there is the increased cognitive load of learning new notation, the required inputs and what these mean, and in some cases, the interpretation of output in non-standard mathematical forms. Consider the task, $f(x) = 3x^2 + 2x - 1$ *find* $f'(-2.7)$. Once the function has been entered in the function menu as $Y_1 = 3x^2 + 2x - 1$, a graphics calculator (see Figure 12.17), for example, can be used to find this using the numerical derivative function, nDeriv. This requires the inputs of the function or expression, the independent variable and the x-value; or use of $\frac{dy}{dx}$, which requires selection of the function using up or down arrows once the graph screen is displayed and the setting of $x =$ to the required value.

Approaches to teaching calculus

Awareness of the state of students' understanding is important in the preparation of instructional materials. For example, Carlson, Ochrtman and Thompson (2007) claim that the revisions they made to their calculus modules, on the basis of their analysis of data on calculus students' reasoning, are resulting in great improvement in calculus students' co-variational reasoning abilities.

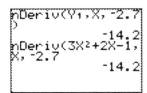
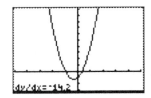

Figure 12.17 **Calculator notation for finding gradient at a point**

The value of procedural competence without conceptual understanding is called into question when the new generation of calculators and other advances in technology are considered. Hassan and Mitchelmore's (2006) teaching intervention, emphasising Tall and Watson's (2001) zooming-in approach, appears to facilitate the learning of instantaneous rate independently of average rate. Tall and Watson (2001) suggest that an appropriate introductory activity is to base 'the idea of "rate of change" on "local straightness" of the graph, actually *seeing* the gradient of the graph change as one moves one's eye along the graph from left to right' (p. 1). This activity provides a sequential process for sketching the gradient function of a graph. In addition, a calculator may facilitate the exploration of the rate of a non-linear function by the magnification of its graph to examine 'locally straight' sections (Tall, 1994). This may provide a link for students between constant rate and variable rate. In Tall's (1994) locally straight example, the calculator is used to provide the graph with the affordance of zooming in, thus allowing the student to concentrate on the notion of rate without the time-consuming effort and distraction of plotting a magnified graph. The use of CAS enables a shift in the teaching and learning of mathematics from procedural competence to conceptual understanding.

Tall (1985) suggests students' introduction to a mathematical concept should be qualitative and global followed by more formal description of concepts. He proposes the use of a graph to demonstrate the manner in which the dependent variable changes when the independent variable changes at a constant rate, in order to develop an intuitive feel for the derivative. This suggests the desirability of emphasising rate during a study of linear functions and again whenever a new function is introduced. In Tall and Watson's (2001) study of the manner in which students build up meaning to sketch the gradient graph of a given function, one teacher privileged 'a visual–enactive approach', whereby she followed the shape of graphs in the air with her hand, encouraging her students to follow her lead—thus building a physical sense of the changing gradient. Visual and symbolic ideas were deliberately linked by the teacher. The students of this teacher outperformed the other students in the study who were taught by other teachers using a more traditional approach to graph sketching and development of gradient of these graphs.

The obstacles discussed earlier come together in the traditional 'first principles' approach often adopted in upper secondary school. Further, White and Mitchelmore (2002) claim that 'a high school introduction to calculus which focuses on symbolic definitions and manipulations results in an abstract-apart concept of derivative, and students have no sense as to what calculus is about' (p. 250). The term 'abstract apart' is used to convey the idea of 'concepts formed in isolation from the contexts in which they occur' (White & Mitchelmore, 2002, pp. 239–241), as opposed to abstract-general concepts that the learner is able to recognise in a variety of contexts and to abstract common properties from similarities in a variety of base contexts. Alternatively, an intuitive understanding of instantaneous rates of change could be developed through familiar situations, ensuring that at least some non-motion contexts are used. This could be followed by the use of a graphical and numerical approach to the measurement of rates of change *with the aid of* technology such as graphics calculators, function-graphing software, animation software such as JavaMathWorlds and Geometer's Sketchpad (Herbert & Pierce, 2012), GeoGebra (Takači, Stankov & Milanovic, 2015) or lite applets available on the internet. These could then be justified, if need be, with first-principles derivations once the concepts have been established informally.

An investigative approach, such as the one outlined in the excellent series *Investigating change* by Barnes (1991), makes calculus accessible to a wider range of students. It develops a strong understanding of basic concepts while avoiding known conceptual obstacles. Both Barnes (2000b) and Williams (2000) give examples of students studying calculus at Year 11 and 12 level via a class collaboration approach (Williams, 2000, p. 658) experiencing 'magical moments' of excitement as they discover mathematical insights for themselves.

Recent advances in technology mean that technological approaches to teaching calculus can no longer be ignored. As computer software and handheld calculators can perform most, if not all, of the skills and manipulative procedures that have dominated calculus areas of study in secondary school in the past, there appear to be three options for how these might be used: as a tool to perform all the procedures, freeing students to explore real-world and mathematical applications; as an integral part of a learning environment to deepen understanding of underlying concepts; or as a mixture of the first two options.

Differential calculus

Differentiation

There are several frameworks about the development of understanding of differentiation (delos Santos & Thomas, 2003; Kendal & Stacey, 2003; Zandieh, 2000). Kendal and Stacey

(2003) have taken a systematic approach to trying to identify 'basic building blocks of competence in calculus' (p. 39) with the construction of the Differentiation Competency Framework guided by consideration of the cognitive demands required in learning about the concept of derivative (not merely the rules for differentiation). The focus of the framework is on students' early understanding of the concept of derivative in different representations—namely, numerical (N), graphical (G) and symbolic (S) especially in CAS supported environments. As a first step to producing the framework, a concept map of differentiation (see Figure 12.18) was produced, linking these three representations as well as physical representations of rate of change, which link most closely to the numerical representation as shown. Solid arrows show translations between representations. The heavy dotted lines within a representation circle separate situations in which the limit has been taken (e.g., gradient of a tangent at a point) from those in which it has not (e.g., gradient of a secant). Differentiating from first principles is shown as involving all three representations.

Since solving problems involving differentiation requires being able to work with all its common representations (N, G and S), and translating among these representations as identified in Figure 12.18, the Differentiation Competency Framework concentrates on these three external representations. Table 12.1 shows the set of 18 competencies provided by the framework that reflect basic understanding of the concept of derivative.

Table 12.1 Kendal and Stacey's (2003) Differentiation Competency Framework

	Representations associated with differentiation		
	Input derivative		
Output derivative	Numerical (N)	Graphical (G)	Symbolic (S)
Numerical (n)	FNn	FGn	FSn
	*I*nn	*I*Gn	*I*Sn
Graphical (g)	FNg	FGg	FSg
	*I*Ng	*I*Gg	*I*Sg
Symbolic (s)	FNs	FGs	FSs
	*I*ns	*I*Gs	*I*ss

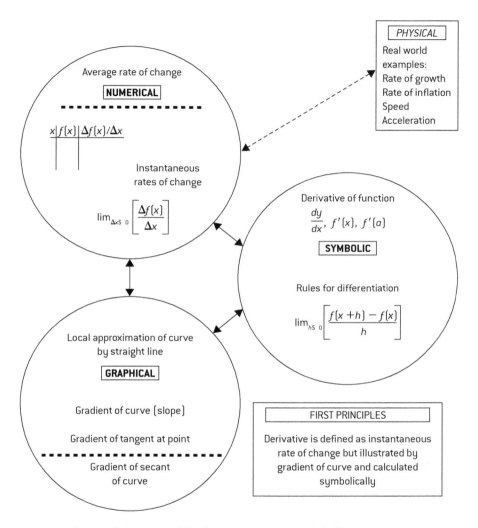

Figure 12.18 **Kendal and Stacey's (2003) concept map of differentiation in numerical, graphical and symbolic representations**

The 18 competencies shown in their framework could be used to structure teachers' topic development and monitor the learning that results. Kendal and Stacey (2003) have produced a test of the competencies. Illustrative examples (see the box below) show the utility of the framework for designing questions for assessing students' understanding of differentiation.

Examples of Differentiation Competency test items

ISn—Interpretation with translation between Symbolic and numeri-
cal representations: *circle the letter corresponding to your answer:* The derivative of the
function $g(t)$ is given by the rule $g'(t) = t^3 - 5t$. To find the rate of change of $g(t)$
at $t = 4$, you should:

A. Differentiate $g'(t)$ and then substitute $t = 4.2$.
B. Substitute t $= 4$ into $g'(t)$.
C. Find where $g'(t) = 0$.
D. Find the value of $g'(0)$.
E. None of the above.

FGg—Formulation within a Graphical representation: use a graph of $y = x^2$
$+ x - 10$ to find the gradient of the curve at $x = 3$.

INs—Interpretation with translation from a Numerical representation
to a symbolic representation: an eagle follows a flight path where its height
depends on the time since it flew out of its nest. The rule for finding the height
of the bird (H in metres) above its nest is a function of $H(t)$ of t, the flight time
(in seconds). Five seconds after takeoff, the 4 Kilogram eagle was observed to be
100 meter above its nest and climbing at the rate of 3 metres per second. What is
the value of $H'(5)$?

Source: Kendal and Stacey (2003).

Delos Santos and Thomas (2003) have developed a framework that maps students'
dimensions of knowing derivative across their representational preferences (symbolic,
graphical, numeric or tabular). The dimensions of knowing are procedure-oriented,
process oriented, object-oriented, concept-oriented and versatile (see Table 12.2).

Using this framework, delos Santos and Thomas (2003, 2005) analysed the changes in
thinking and understanding of target students from Form 7 (18 years of age) from four NZ
schools. Students were involved in an eight-week study of derivative using TI–83 graphics
calculators.

Table 12.2 Delos Santos and Thomas's (2003) Representational Framework of Knowing Derivative

Dimensions	Representations		
	Symbolic	Graphical	Numeric/Tabular
Procedure-oriented	Manipulate symbols according to rules.	Calculate from graphical forms.	Use procedures to obtain numerical results from tables.
Process-oriented	Interpret the meaning of symbols as a differentiation process.	Have a pointwise approach to derivative of graphs. Can understand second derivative as rate of change of gradient.	Understand and apply rate of change, differentiation and gradient processes in a tabular setting.
Object-oriented	Operate on the symbols for derivatives as objects. Interpret nth derivatives as functions.	Interpret derivative graphs as representing functions. Operate on graphs as an entity.	Interpret a table of values as representing a discrete approximation to a continuous function.
Concept-oriented	Relate the differentiation procedures, and processes applicable in one representation to each other and to relevant concepts.		
	Make procedural and conceptual connections between corresponding differentiation procedures, processes and objects in different representations.		
	Identify and operate on conceptual objects such as derivative and function presented in different representational forms.		
Versatile	Have sufficiently well-formed differentiation procedure, process, object and concept oriented knowledge to be able to identify and use appropriate objects, processes and procedures in their various representational manifestations. Choose appropriate representational system perspectives to solve a differentiation problem. Move seamlessly and fluently between the chosen perspectives as required.		

Zandieh (2000) developed a framework for mapping the development of two students' understanding of derivative, from observation of textbook descriptions of derivative and the manner in which it was discussed by mathematics education researchers, mathematicians, mathematics graduates and students in calculus classrooms (see Figure 12.19). It is

	Contexts				
	Graphical	Verbal	Paradigmatic Physical	Symbolic	Other
Process-object layer	Slope	Rate	Velocity	Difference Quotient	
Ratio					
Limit					
Function					

Source: Zandieh and Knapp (2006, p. 4).

Figure 12.19 **Zandieh's (2000) framework**

organised around the ideas of multiple representations and process–object pairs involving ratio, limit and function:

> The derivative framework has two main components: multiple representations or contexts and layers of process–object pairs. The concept of derivative can be represented (a) graphically as the slope of the tangent line to a curve at a point or as the slope of the line a curve seems to approach under magnification; (b) verbally as the instantaneous rate of change; (c) physically as speed or velocity; and (d) symbolically as the limit of the difference quotient. Many other physical examples are possible, and there are variations possible in the graphical, verbal, and symbolic descriptions. (Zandieh & Knapp, 2006, p. 4)

Roorda, Vos and Goedhart (2007), reporting on the suitability of Zandieh's (2000) theoretical framework, describe the manner in which they used the framework to design tasks for a task-based interview with two students. They found that a 'great advantage of the matrix [framework] is the possibility to describe the solution strategy followed by a student by indicating patterns in different cells' (Roorda et al. 2007, p. 7).

REVIEW AND REFLECT : Concept maps prepared by students are considered by many (e.g., delos Santos & Thomas, 2005; Williams, 1998) 'to be an externalisation of conceptual schemas' (delos Santos & Thomas, 2005, p. 378) and as such a useful tool to tap into students' current conceptual schemas and changes in these over time.

Ask two senior secondary students to draw a concept map of their understanding of derivative. Emphasise that it is important for object and procedural links in the concept maps to be labelled (e.g., a link from, say, 'derivative' to 'rate of change' should be clearly labelled as meaning 'derivative IS A rate of change' or 'derivative IS USED TO FIND a rate of change').

Use one of the frameworks mentioned above to compare and contrast the students' concept maps.

What insights have the concept maps given into the students' apparent understanding of derivative?

Developing rules of differentiation

Once the notion of a derivative has been established, the rules for differentiation can be developed using an investigative approach involving the graphical representation and gradient functions or pattern searching with the algebraic representation. Both of these could be done by hand or with technology. In Figure 12.20, the rule for the derivative of the sum or difference of two functions is developed through a graphical approach using the difference quotient and a graphics calculator. In Figure 12.21, an algebraic pattern-searching approach is used to find the product rule, which is verified using CAS.

Anti-differentiation

Undoing or reversing the process of differentiation is called anti-differentiation. This process becomes necessary in situations in which we know the rate at which something is changing but not the function itself. One way of doing this is by 'guess and check'. You ask *what function could be differentiated to give this result?* If you were given $\frac{dy}{dt}=t$, a possible candidate would be $f(t)=\frac{1}{2}t^2$ as we know $y=t^2, \frac{dy}{dt}=2t$. However, it could also be that $f(t)=\frac{1}{2}t^2+3$ or $f(t)=\frac{1}{2}t^2-1$ Geometrically, the set of anti-derivatives of $2t$ represents a family of parabolas given by $y=f(t)=\frac{1}{2}t^2+c, c\in\mathfrak{R}$. The curves have the same shape and can be obtained from each other shifting up or down (i.e., translation). To find which particular curve was being targeted, we would need some more information such as an initial condition.

Example—Adding and subtracting functions $\frac{d}{dx}[f(x) \pm g(x)]$

(a)   Using your graphics calculator, graph $y = x^2, x \in [-5, 5]$ and its gradient function then $y = x, x \in [-5, 5]$ and its gradient function as shown. Set h to 0.001.

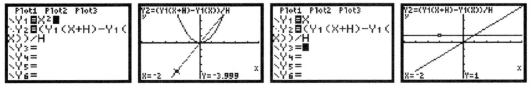

What are the derivatives of $f(x) = x^2$ and $g(x) = x$?

(b)   Using your graphics calculator, graph $y = x^2 + x, x \in [-5, 5]$ and its gradient function then $y = x^2 - x$, $x \in [-5, 5]$ and its gradient function as shown. Set h to 0.001.

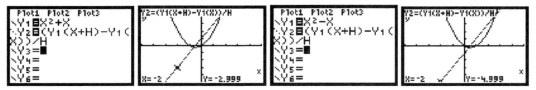

Compare the graphs of the gradient functions of $y = x^2, x \in [-5, 5], y = x^2 + x, x \in [-5, 5]$ and $y = x^2 - x$, $x \in [-5, 5]$. What do you notice?

What do you think is the derivative of $f(x) + g(x) = x^2 + x$?

What do you think is the derivative of $f(x) - g(x) = x^2 - x$?

(c)   Explain in words the connection between the derivatives in (a) and (b).

(d)   Repeat (a) to (b) for $f_1(x) = x^3$ and $g_1(x) = 2x$ using an appropriate viewing window.

(e)   Describe in words how you think you could find the derivative of the sum or difference of any two functions. Write this as a rule.

Use this to predict the derivatives for $y = x^2 - 6x$ and $y = x^4 + 2x$. Check your prediction with your calculator. Modify your rule for finding the derivative of the sum or difference of two functions if need be.

Figure 12.20 **Developing a rule for the derivative of the sum or difference of two functions**

Another method to find anti-derivatives is to use *direction (slope) fields*. This can be done using the differential equation graphing facility of calculators such as the TI–Nspire, as in Figure 12.22. First, the slope field is drawn with no initial condition set. Short line segments are drawn at points all over the plane. The gradient of each is equal to the value of $\frac{dy}{dt}$ at its midpoint. A diagram like this could be used to draw graphs of functions of the form $y = f(t) = \frac{1}{2}t^2 + c, c \in \Re$, which satisfy $\frac{dy}{dt} = t$. This can be done by hand or by setting initial conditions as has been done in Figure 12.22 where c is set to $-1, 0, 1, 2, 3$. Other activities related to investigating anti-derivatives can be found in Barnes (1991).

Example—Product of two functions $\dfrac{d}{dx}[f(x)g(x)]$

(a)   Using a CAS calculator, find derivatives of $f(x) = x + 1$ and $g(x) = x + 2$ and $h(x) = (x + 1)(x + 2)$.

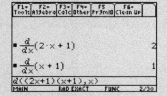

Can you see a relationship between $\dfrac{d}{dx}h(x)$ and $f(x)$ and $g(x)$?

Complete the following table using a CAS calculator.

$f(x)$	$g(x)$	$\dfrac{d}{dx}f(x)$	$\dfrac{d}{dx}g(x)$	$\dfrac{d}{dx}h(x) = \dfrac{d}{dx}[f(x)g(x)]$
$x + 1$	$x + 2$	1	1	$2x + 3$
$x + 2$	$x + 3$			
$x + 3$	$x + 4$			
$x + 1$	$x - 2$			

Make a conjecture based on what you see in the table about the relationship between the derivative of a product of functions and its factors.

Use your conjecture to predict the derivative of $h(x) = (2x + 1)(x + 1)$. Use the CAS calculator to find the derivatives as before to check.

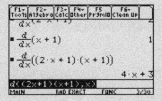

If your conjecture is correct, use it to complete the table below; otherwise, complete the table using a CAS calculator as before writing the derivative for $h(x)$ in terms of its factors as shown in the first row.

$f(x)$	$g(x)$	$\dfrac{d}{dx}f(x)$	$\dfrac{d}{dx}g(x)$	$\dfrac{d}{dx}h(x) = \dfrac{d}{dx}[f(x)g(x)]$
$2x + 1$	$x + 1$	2	1	$4x + 3 = 2x + 1 + 2(x + 1)$
$x + 2$	$x + 3$			
$x + 3$	$x + 4$			
$x + 1$	$x - 2$			

Look for a relationship between $f(x)$, $g(x)$, their derivatives and the derivative of the product $h(x)$. Describe in words how you think you could find the derivative of the product of any two linear functions. Write this as a rule for when $f(x) = ax + b$ and $g(x) = cx + d$. Using a CAS calculator find derivatives of $f(x) = ax + b$ and $g(x) = cx + d$ and $h(x) = (ax + b)(cx + d)$ to check your result.

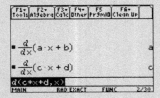 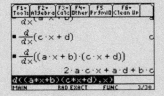

Now continue your exploration for products where the factors are non-linear to see if you can develop a rule for derivatives of products of any functions. Write this rule in your own words. Predict the derivative for $h(x) = (5x^2 + 2)(3x - 9)$ and check by using your calculator. Modify your rule if necessary. The rule can be expressed in a general form using $\frac{d}{dx}[f(x)g(x)]$. See if you can use your rule to work out the derivative. Finally, use your CAS calculator to check your general rule.

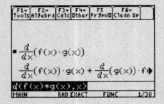 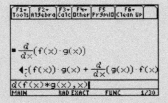

Figure 12.21 **Developing a rule for the derivative of the product of two functions**

The process of anti-differentiation is also known as integration and then the anti-derivative is called an *indefinite integral*. Using function notation $\int f(x)\,dx$ denotes the anti-derivative or integral of $f(x)$ with respect to x. We can obtain rules for indefinite integrals (anti-derivatives) by reversing rules of differentiation, for example, if $\frac{d}{dx}[f(x) \pm g(x)] = \frac{d}{dx}f(x) \pm \frac{d}{dx}g(x)$ then the reverse of this must be $\int[f(x) \pm g(x)]dx = \int f(x)dx \pm \int f(x)dx$.

Integration

The calculus had its origins in two problems, both of which were studied by mathematicians from ancient times. The first, which we have already met, was finding the unique tangent (if it exists) that can be drawn at a given point on a curve, and this problem led to the differential calculus. This problem was not solved as a general method for all curves until the seventeenth century. The second problem involved finding the area bounded by

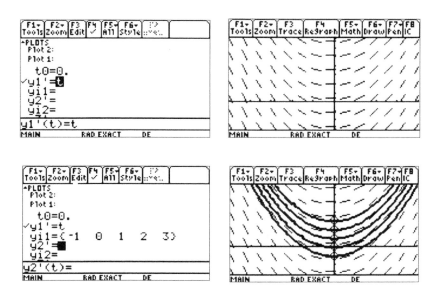

Figure 12.22 **Use of slope diagrams to find anti-derivatives**

a given curve. The solution of this problem led to integral calculus. Archimedes solved the problem for particular curves such as the area of a parabolic segment using the method of exhaustion, so called because polygons are used to eventually 'exhaust' (use up) the area concerned. This method was the basis of more generalised approaches until Newton and Leibniz showed how calculus can be used to find the area bounded by the curve.

In integral calculus, the area of a region bounded by a curve can be found using the areas of polygons as an approximation. An informal consideration of limits could also be used as the introduction to numerical integration (Herbert, 2013). Consider dividing a circle up into sectors (see Figure 12.23), find the area of a triangles bounded by two radii and a chord. Recording the combined areas of the triangles as the number of sectors is increased allows students to see this area approaching the familiar formula for the area of a circle (see Figure 12.24), thus linking numerical integration with students' prior knowledge about area and circles.

Similarly, Zakaria and Salleh (2015) propose the use of technology to support students learning of integration. They claim that CAS has 'the capability to manipulate algebraic symbolic forms of an equation'; improves students' understanding of concepts through the visual supports; saves a 'lot of calculation time'; and 'facilitates the balance, the sequence, and the priorities of conceptual and procedural knowledge in the mathematics

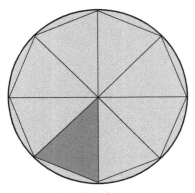

Source: Herbert (2013).

Figure 12.23 **Introduction to numerical integration**

No. of Triangles	Angle at centre	Total area
8	π/4	2.82843r²
16	π/8	3.06147r²
32	π/16	3.12145r²
64		
128		

Figure 12.24 **Demonstration of limiting process leading to formula**

curriculum' (Zakaria & Salleh, 2015, p. 145). The example below in Figure 12.25 illustrates finding the area bounded by a curve using upper and lower rectangles.

This example and procedure could then be linked to the notion of the definite integral defined as: $\int_a^b f(x)dx = \lim_{\Delta x \to 0} \sum_a^b f(x)\Delta x$ provided the limit exists. The integral sign is an elongated S (from the German 'somme' for sum). The notation was introduced by Leibniz. This definition of the definite integral is called the *Riemann Integral* after the German mathematician Georg Riemann.

There have been reported student difficulties with this notion of using successive approximations with more and more rectangles. Schneider (1992), for example, reports that some students think 'as long as the rectangles have a thickness, they do not fill up the surface under the curve, and when they become reduced to lines, their areas are equal to zero and cannot be added' (pp. 32–33). Tall (1996) suggests that a figure like that in Figure 12.26 be used instead.

I have just leased an agistment block down by the river for my horses during the drought. I have been able to determine that the side of the block along the river can be modelled by the function $f(x) = (x - 1)^3 + 3$, $x \in [0,2]$ where x is in kilometres. The side of the block opposite the river is 2 kilometres long and the other two sides are 2 and 4 kilometres in length.

In order to find an approximate area for my block, I can use my graphics calculator to divide the block into a series of ten rectangles each 200 metres wide (0.2 km) and sum the areas. To do this I will use this function entered into my calculator as Y_5 and use the sequence command to enter x values of the leading vertical side of my rectangles into List 1, followed by the lengths of the rectangles (given by $Y_5(L_1)$) into List 2. I can then use this to draw a histogram. *Note:* I need to set the X scale in the viewing window to the width of my rectangles.

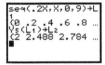 

The area of each rectangle is the width (0.2 km) multiplied by the length, $Y_5(L_1)$. By using SUM I can add up the areas of all the rectangles to determine the total area enclosed by the rectangles. As the set of rectangles are all lower than the curve, I will call this sum S_L and expect this area of 5.8 km² to be less than the actual area of land, which I will call A.

To improve my estimate I can repeat my summation process using narrower rectangles (e.g. 100 m, 50 m, 25 m... wide).

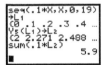  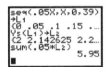 

These rectangles are all below the curve I have used to model the river bank. I could have used rectangles above the curve. This time my estimates, which I will call S_U as they are the sums of the upper rectangles, will be too high but will become closer to the actual area as the width of the rectangles is reduced. Thus, $S_L < A < S_U$.

 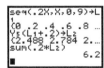 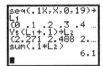 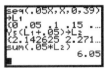

By continuing this process we could show $S_L \to 6$ and $S_U \to 6$.

Figure 12.25 **Using summation to find area bounded by a curve**

The Fundamental Theorem of Calculus says that, given an area $A(x)$ from a fixed point a to a variable point x, $A'(x) = f(x)$. The additional area under the curve from x to $x + h$ is $A(x + h) - A(x)$. In this figure, there is now only one strip, which tends towards a rectangle and it is clear visually that as $h \to 0$ $\dfrac{A(x+h) - A(x)}{h} \to f(x)$.

This can also be shown symbolically connecting the notion of numerical integration with differentiation from first principles, as seen in Figure 12.27.

However, students still face the usual obstacles associated with limits that an approximation becomes an equality in the limiting case. Historically, integral calculus and the process of integration were developed using this summation approach, not as the undoing of differentiation—that is, anti-differentiation. It was not until differential calculus was developed in the seventeenth century that the relationship between these two branches of calculus became apparent. Rules for integration can be developed using investigative

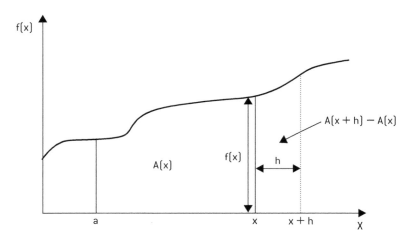

Figure 12.26 **Area bounded by a curve**

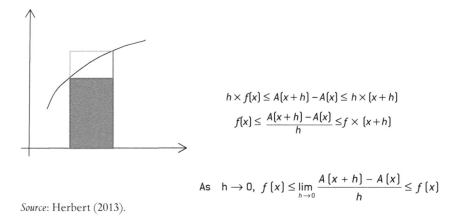

$$h \times f(x) \le A(x+h) - A(x) \le h \times (x+h)$$

$$f(x) \le \frac{A(x+h) - A(x)}{h} \le f \times (x+h)$$

As $h \to 0$, $f(x) \le \lim_{h \to 0} \frac{A(x+h) - A(x)}{h} \le f(x)$

Source: Herbert (2013).

Figure 12.27 **Development of the Fundamental Theorem of Calculus**

graphical and algebraic methods with technology support similar to developing rules for differentiation. However, the paradigm shift in the teaching and learning of this subject cannot be implemented abruptly without a proper investigation of students' readiness in using this technology (Herbert, 2013, p. 145).

REVIEW AND REFLECT : Examine textbooks from two different textbook series. Compare the extent to which the authors attempt to address the difficulties that students are known to have with introductory calculus concepts, and then adopt a multiple-representation approach.

Conclusion

Despite predictions in the 1980s that calculus would wither and die in secondary schools (Tall, 1987), calculus and the underpinning concepts of function still hold a central place in most mathematical subjects in the upper secondary curriculum. With ever-increasing access to technological tools that are rapidly advancing in their mathematical capabilities for teaching, learning and doing mathematics, calculus appears to be becoming stronger as it becomes more accessible to more students because emphases have changed.

Recommended reading

Barnes, M. (1991). *Investigating change: An introduction to calculus for Australian schools.* Melbourne: Curriculum Corporation.

Hohenwarter, M., Hohenwarter, J., Kreis, Y. & Lavicza, Z. (2008). Teaching and learning calculus with free dynamic mathematics software GeoGebra. In *11th International Congress on Mathematical Education.* Monterrey, Nuevo Leon, Mexico.

Jaafar, R. (2016). Writing-to-learn activities to provoke deeper learning in calculus. *PRIMUS,* 26(1), 67–82. doi: 10.1080/10511970.2015.1053642

Kaput, J. & Schorr, R. Y. (2002). Changing representational infrastructures changes most everything: The case of SimCalc, algebra & calculus. In M. K. Heid & G. Blume

(Eds), *Research on the impact of technology on the teaching and learning of mathematics* (pp. 47–75). Mahwah, NJ: Lawrence Erlbaum.

Kendal, M. & Stacey, K. (2003). Tracing learning of three representations with the differentiation competency framework. *Mathematics Education Research Journal*, 15(1), 22–41.

Kouropatov, A. & Dreyfus, T. (2014). Learning the integral concept by constructing knowledge about accumulation. *ZDM*, 46(4), 533–548.

Shell Centre for Mathematical Education. (1985). *The language of functions and graphs.* Nottingham, UK: Joint Matriculation Board and Shell Centre for Mathematical Education.

Tall, D. (1996). Functions and calculus. In A. J. Bishop, K. Clements, C. Keitel, J. Kilpatrick & C. Laborde (Eds), *International handbook of mathematics education* (Vol. 1, pp. 289–325). Dordrecht, The Netherlands: Kluwer.

White, P. (1995). An introduction to teaching calculus. In L. Grimson & J. Pegg (Eds), *Teaching secondary school mathematics: Theory into practice* (pp. 165–85). Sydney: Harcourt Brace.

PART 4

Equity and diversity in
mathematics education

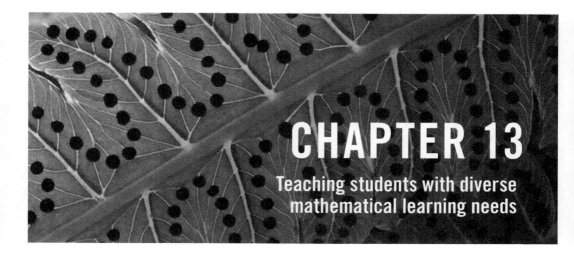

CHAPTER 13
Teaching students with diverse mathematical learning needs

Mathematics achievement is a significant factor affecting success in schooling (Rothman, 2002). Success, confidence and creativity for all students are specified in the *Melbourne Declaration* (MCEETYA, 2008). However, recent studies show that gaps between the highest and lowest mathematics achievers among Australian secondary students persist (Thomson et al., 2013a; Thomson et al., 2012; Vale et al., 2013) and the proportions of students undertaking the intermediate and most demanding senior secondary mathematics subjects is declining (Mack & Wilson, 2014; Wienk, 2015). The chapters in the previous sections have focused on teachers' knowledge of curriculum knowledge and their PCK regarding how students learn mathematics in general and for particular mathematics concepts and skills. The chapters in this section focus on identifying reasons for the wide gap in achievement and using knowledge of students and school context to improve the mathematics learning of all mathematics students.

In this chapter, we consider what is probably the most vexatious issue and the most difficult task for teachers of mathematics: teaching students with diverse mathematical knowledge, skills and attitudes. Teachers can expect students' mathematics achievement to span seven years of schooling in their junior secondary classrooms (Siemon et al., 2001) and we know that students at both ends of the achievement spectrum (low achievers and high achievers) are at risk of under-achieving (Diezmann et al., 2012). Teachers most readily recognise the classroom management problems that arise, and the subsequent impact on the learning of students, when not everyone is engaged. Of equal concern is the sense we have of failing our students when we don't cater for their mathematical learning needs.

Often mathematics teachers talk about teaching a 'mixed-ability' class. However, using the term 'ability' to describe differences among students is contentious since it implies differences in potential to learn. Our focus in this chapter is on the belief—held by excellent teachers of mathematics—that all students can learn mathematics (AAMT, 2006). We will discuss the various approaches and programs used by schools and teachers for students with diverse mathematics learning needs, as well as those students with mathematics learning difficulties, talented students of mathematics and differently abled students.

Responding to the diverse needs of students

Teachers' attitudes to mathematics and mathematics teaching and learning (discussed in Chapter 1) come to the foreground when teachers tackle the complexity of teaching a class of students with not only diverse social and cultural backgrounds but also diverse prior learning experiences and achievement. The belief that 'some people have a maths mind and some don't' evident among some practising primary and secondary teachers of mathematics has implications for improving the success of lower achievers (Beswick, 2007–8). A. Watson (2006a) agrees: 'in some primary schools learners were seen to be deficient but curable, and they did return to mainstream classes. In secondary schools, learners are more likely to be seen as deficient and incurable' (p. 65).

Consequently, teachers neglect students who struggle with mathematics because, in our community, it is acceptable 'not to be good at' mathematics (Carroll, 2006). Teachers sometimes use the notion of different learning styles—especially the multiple intelligences literature (Gardner, 1993)—to justify this belief, rather than conducting appropriate formative assessment to get to know their students and their learning needs and develop an appropriate and targeted learning program that builds on students' strengths or preferences while also addressing their weaknesses.

Knowing your students is one of the AITSL (2014) standards of teaching, which involves not only awareness of students' social and cultural contexts, their interests and aspirations, but also their attitudes to mathematics, their prior learning and knowledge of mathematics and their approach to mathematical thinking, problem solving and reasoning. Finding out about students' attitudes was discussed in Chapter 1 and formative assessment, including student self-assessment, was discussed in Chapter 6. The process of using formative assessment to set achievable learning goals (or intentions) and enable students to make connections with prior knowledge is now an expected approach for planning high-quality teaching of mathematics. Typically, when planning for diverse classrooms, secondary teachers choose tasks according to students' level of prior knowledge, prepare extensions for high-achieving students, use open-ended tasks with inquiry or provide examples of solutions and support students to follow procedures (Clarke et al., 2014). Teachers also sort students according to achievement within the classroom or between classes.

Streaming

Most schools choose to respond to the diversity of students' mathematical achievement by implementing a streaming policy—that is, sorting students into mathematics class groups according to mathematics achievement (Forgasz, 2010, 2012). (Different terms are used to describe such policies in other countries—e.g., 'setting' in the UK and 'tracking' in the USA.) While streaming is more commonly used in Years 9 and 10, many schools sort students from their first year in secondary school, often based on the results of a single written mathematics test.

Teachers claim that streaming enables them to focus content and teaching strategies according to students' prior knowledge (Clarke et al., 2014; Forgasz, 2010; Turner, 2007). However, research shows that rather than narrowing the achievement gap between the lowest and highest achievers, the gap widens across the years of secondary schooling when streaming occurs (Tate & Rousseau, 2002; Vale et al., 2013). The negative impact of streaming was most evident when a low socio-economic secondary school, which was achieving higher than expected annual growth in student mathematics achievement using heterogeneous classes, introduced a streaming program. Annual growth in mathematical learning fell below expected levels (Vale et al., 2011).

Streaming fails to narrow the gap because teachers have different expectations—and therefore goals—for students in the high- and low-achieving groups, and consequently they offer different and more intellectually demanding learning experiences to the higher stream students while tending to set routine computational tasks for students in the lower stream. Teachers too often re-teach the whole process or procedure or do the exercise for them when students experience difficulty. This fails to help the student identify the exact point of difficulty and overcome the problem.

Despite claims of flexibility, streaming programs rarely provide opportunities for students to move to higher achieving groups once the groups have been set (Clarke & Clarke, 2008; Zevenbergen, 2003a). This is especially concerning when a single 'snapshot' of student achievement using one test instrument is used to sort students and when teachers' recommendations for assigning students to class groups exclude students from disadvantaged backgrounds from participating in higher achieving mathematically challenging class groups (Jorgensen et al., 2014). Poor achievement outcomes for the lowest achieving class groups also occur because less qualified and less experienced teachers of mathematics are allocated to teach these groups (Clarke & Clarke, 2008: Mawson, 2013; Turner, 2007), with less opportunity for these teachers to collaborate when planning (Vale et al., 2011).

Streaming also impacts on students' mathematical self-efficacy, as they accept assessments of themselves as 'bright', 'mediocre' or 'dumb' and this affects their engagement with mathematical tasks and ongoing participation in mathematics learning. For example, a pre-service teacher who investigated his practicum school's approach to streaming observed one junior secondary student exclaim, 'This is too hard. I thought we were supposed to be in the dumb group' (Mawson, 2013).

As streaming can deny low-achieving students access to a rich and challenging mathematics curriculum relevant to their broader interests in the junior secondary years, they gain only limited experience of the mathematical content needed in a range of higher education and vocational education programs beyond secondary schooling. Hence, streaming of students is reinforcing, rather than removing, achievement differences between top- and bottom-streamed students (A. Watson, 2006a; Jorgensen et al., 2014; Vale et al., 2013; Zevenbergen, 2003a). It might be assumed that streaming benefits high-achieving students, and provides a pathway for mathematics learning in the post-compulsory years. However, declining participation in intermediate and challenging mathematics subjects in Year 12 (Mack & Wilson, 2014; Wienk, 2015) suggests that the content and teaching approaches used with the highest achieving groups are not contributing to their continued participation and success in mathematics. In the following sections, we consider alternate approaches when responding to diverse learning needs.

REVIEW AND REFLECT : Research the policy and practice of streaming in a secondary school by scrutinising relevant school policy and planning documents (including the school website), observing lessons for students in different groups in the same year level, and interviewing the teachers.

Policy and planning

What is the school's rationale for streaming?

How does it work? What information does the school use to assign students to groups? What is the policy and practice about movement between groups?

What is the curriculum for the different groups and what resources do they use?

Lesson observation

Compare the content, teaching and learning activities, and student engagement and learning, in the different classes.

Teacher interviews

What are the teachers' attitudes to streaming and expectations of their students?

Evaluation

How effective do you think the school's program will be in achieving Goal 2 of the *Melbourne Declaration* (MCEETYA, 2008)?

Student-centred approaches in heterogeneous classrooms

Alternatives to streaming use student-centred approaches in heterogeneous classrooms. A heterogeneous classroom is more commonly known as a 'mixed' or 'mixed-ability' classroom, comprising students in the same age cohort and year level. According to Black (2007), a student-centred approach:

- Is based on a challenging curriculum connected to students' lives.
- Caters for individual differences in interest, achievement and learning styles.
- Develops students' ability to take control over their own learning.
- Uses authentic tasks that require complex thought and allow time for exploration.

- Emphasises building meaning and understanding rather than completing tasks.
- Involves cooperation, communication and negotiation.
- Connects learning to the community. (p. 1)

Three models for student-centred teaching include inquiry- and problem-based learning in which students have control over their learning and there are high levels of cooperation among learners; authentic curriculum in which learning is connected to students' interests and needs using rich and authentic tasks; and differentiated learning or targeted teaching in which teachers tailor their instruction to students' learning needs using 'just right' tasks (Reilly & Parsons, 2011) and students take responsibility for their learning.

Inquiry- and problem-based learning

A number of researchers have found that using inquiry-based learning, in which students work in small groups to explore mathematical contexts, model and solve authentic and mathematical problems and present their solutions and reasoning to the whole class, engages and benefits learners at all levels of achievement in heterogeneous classrooms (Boaler, 2008b; Boaler & Staples, 2008; Brown, 2011; Clarke et al., 2014; Inoue, 2011; Sullivan et al., 2006; A. Watson, 2006a). As discussed in Chapters 2 and 3, rich and challenging tasks invite students to use prior knowledge, problem-solving skills and mathematical reasoning to explore mathematical concepts and solve mathematics and authentic problems. Such tasks may be solved using various methods and solutions expressed in different ways and at different levels of efficiency or abstraction (see Figure 13.1). Open-ended tasks have multiple solutions; some open-ended tasks may enable generalisable solutions.

To ensure that these tasks promote learning for all students, learning goals (or intentions) need to be specified and enable students at the diverse levels of prior knowledge to develop their understanding, fluency, problem solving and/or reasoning. When planning these lessons, teachers need to complete the problem themselves to identify alternate methods and solutions. With this knowledge, teachers can then anticipate students' solutions and potential difficulties to plan enabling and challenging prompts (Clarke

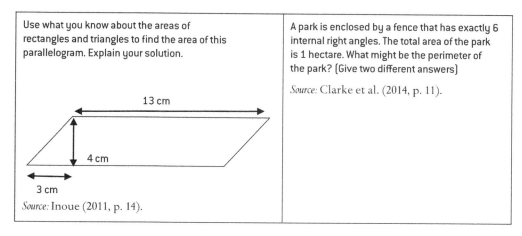

Use what you know about the areas of rectangles and triangles to find the area of this parallelogram. Explain your solution.

13 cm

4 cm

3 cm

Source: Inoue (2011, p. 14).

A park is enclosed by a fence that has exactly 6 internal right angles. The total area of the park is 1 hectare. What might be the perimeter of the park? (Give two different answers)

Source: Clarke et al. (2014, p. 11).

Figure 13.1 **A rich mathematical task and an open-ended mathematical task**

et al., 2014; Inoue, 2011; Sullivan et al., 2006), and orchestrate presentation and whole-class discussion of student solutions (Inoue, 2011; Stein et al., 2008). Lessons are structured using the following steps:

- *Launching the task.* The teacher poses posing the problem and articulates the learning intention. This provides the opportunity for students to clarificy the problem and have individual think time to enable all students to contribute to small group work (Brown, 2011; Clarke et al., 2014).
- *Exploring and finding solution(s) in small groups.* During this period of the lesson, teachers monitor group work to observe the range of solutions and strategies used, provide enabling or challenging prompts, and resist telling (Clarke et al., 2014; Inoue, 2011; Stein et al., 2008; Sullivan et al., 2006).
- *Presentation and whole-class discussion of various solutions.* Students should feel comfortable reporting to the whole class, and it is recommended that presentation of group solutions be ordered from least efficient to most efficient, so that all students can feel valued and students may learn from others' solutions that make sense to them (Inoue, 2011; Stein et al., 2008). Students are encouraged to ask questions to make sense of various solutions and connect their thinking and solutions with others.

- *Summarising learning*. The teacher supports students to draw conclusions related to the learning intention and write reflections on their learning (Inoue, 2011; Stein et al., 2008; Reilly et al., 2009).

The success of inquiry approaches within heterogeneous classrooms depends of developing a community of inquiry in which students have respect for each other, listen and contribute to the learning of each other and take responsibility for their own learning (Boaler, 2008b; Brown, 2011; Reilly et al., 2009). Teacher questioning and listening is important for modelling and supporting these practices in all phases of the lesson.

Reilly et al. (2009) use reciprocal teaching, a literacy approach, to support students' work on the task. It involves four stages: predicting, that is, to think about what the question is asking and what the solution might look like; clarifying the problem by listing unknown terms, what is known and information needed to work on the problem; solving; and summarising, involving recording the solution and methods used and justifying the solution in readiness for presentation to the class.

Sullivan et al. (2006) found that successful teachers plan and use 'enabling prompts' in their classrooms. These prompts, prepared in advance by the teacher:

- Reduce the required number of steps (e.g., make the problem simpler or provide a drawing).
- Reduce the required number of variables (e.g., make the problem simpler).
- Simplify the modes of representing results (e.g., provide a recording format).
- Reduce the written elements in recording (e.g., provide an alternative medium for recording).
- Make the problem more concrete (e.g., provide materials or drawings of representations).
- Reduce the size of the numbers involved.
- Simplify the language.
- Reduce the physical demands of manipulatives. (Sullivan et al., 2006, pp. 502–503)

Teachers also prepare 'challenging prompts' that encourage students to find further solutions, generalise solutions, justify solutions and/or create similar problems for others to solve (Clarke et al., 2014; Sullivan et al., 2006). These prompts are intended to extend and deepen students' thinking rather than requiring them to do additional tasks that leave them feeling punished for finishing early.

REVIEW AND REFLECT : Plan an inquiry lesson for a heterogeneous class comprising students with diverse needs.

Choose an inquiry task and solve the problem.

Identify the learning goal(s) (intention) and document anticipated student solutions.

Plan enactment of the lesson.

Explain how the lesson will engage all students in mathematical activity to improve skills, challenge their intellect, build knowledge, generate understanding and achieve success in mathematics.

Targeted teaching

In heterogeneous classrooms, when teachers use targeted teaching, students negotiate 'just right' tasks for common content encouraging them to take responsibility for their learning (Reilly & Parsons, 2011). 'Just right' tasks enable students to work within their Zone of Proximal Development (ZPD): 'all students within the one classroom work on a similar learning outcome, e.g. area of composite shapes, but at a level that maximises their opportunity to learn … Students access the achievable tasks by identifying work that is just right for them' (Parsons & Reilly, 2012, p. 504).

Formative assessment, including student self-assessment at the beginning of a unit, enables students to set their own learning goals (see Figure 13.2 for a sample self-assessment task). Teachers plan tasks at three different entry levels for each lesson and students select the level of the task they will work on in negotiation with the teacher at the beginning of the lesson following a common warm-up/fluency task. Figure 13.3 is an example of a tiered task. Selecting tasks at the beginning of each lesson ensures students are assigned to flexible working groups. Negotiating the selection of tasks ensures that students do not select tasks that are too hard or too easy. At the end of the lesson, students write a reflection on their learning.

Math in the classroom book | (Technology solution)
©Jodie Parsons & Yvonne Reilly 2012

PARTY PLANET

Things I understand	I don't know this yet	I am begining to understand this.	I understand this.	I've got it! I could teach someone else.	Show the teacher evidence
I can identify equivalent fractions.					
I can place percentages, fractions and decimals on the same number line.					
I can add & subtract fractions with the same denominator.					
I can add & subtract fractions with a different denominator.					
I can multiply fractions.					
I can divide fractions.					

Number of cakes I started with............ Number of cakes I finished with........... Improvement...........

Source: Parsons and Reilly (2012, p. 5).

Figure 13.2 'Just right' self-assessment task

REVIEW AND REFLECT : View the following AITSL videos about differentiating learning using 'just right' tasks:

- 'Teaching to the point of need' (Yvonne Reilly, Sunshine SC) www.aitsl.edu. au/australian-professional-standards-for-teachers/illustrations-of-practice/ detail?id=IOP00415
- 'Differentiation in maths' (Jodie Parsons, Sunshine SC) www.aitsl.edu.au/ australian-professional-standards-for-teachers/illustrations-of-practice/ detail?id=IOP00402

Compare and contrast the illustration of the practice of targeting teaching. What do you notice about how these teachers enact targeted teaching using 'just right' tasks?

Choose a task from a resource or textbook and design three different versions of the task to fit different levels of prior knowledge of the concept or skill.

Resource sheet 4.9: Party activity cards

Draw the following areas on your party bag

1/2 of the bag is lollies and 1/2 of the bag is toys

1/2 of 1/2 is Chocolate beans
1/4 of 1/2 is football cards
1/4 of 1/2 are jelly snakes
1/3 of 1/2 are marbles

Party Bag

What fraction of the whole bag is each of your items?
How much space do you have left for bubble gum and coloured pencils?

Draw the following areas on your party bag

1/2 of the bag is lollies and 1/4 of the bag is toys

1/3 of 1/2 is Chocolate beans
1/4 of 1/2 is football cards
1/4 of 1/2 are jelly snakes

Party Bag

What fraction of the bag is not for lollies or toys?
What fraction of the whole bag is each of your items?
How much space do you have left for bubble gum and coloured pencils?

Draw the following areas on your party bag

50% of the bag is lollies, 1/3 of the bags is toys and the rest is cake
1/2 of 1/3 is football cards and 25% of 1/3 is marbles
0.25 of 50% is Chocolate beans and 1/2 of 50% is jelly snakes
25% of the cake is sprinkles and 1/4 of the cake is cream

Party Bag

What fraction of your whole bag is each item?
How much space do you have left for bubble gum and pencils?
If the cake is made of sponge, cream and sprinkles, what % of the cake is sponge?

20 PARTY PLANET

Math in the classroom book I [Technology solution]
©Jodie Parsons & Yvonne Reilly 2012

Source: Parsons and Reilly (2012, p. 5).

Figure 13.3 **'Just right' tasks**

Students with learning difficulties in mathematics

Defining and identifying students with learning difficulties

Teachers and schools use various terms to describe students who have difficulty learning mathematics. These terms include 'students with learning difficulties', 'students with learning disabilities', 'low-achieving students', 'students at risk of failure', 'students at educational risk' and 'students with special needs' (Diezmann et al., 2012; Gunn & Wyatt-Smith, 2011). Gunn and Wyatt-Smith (2011) make a distinction between students with learning difficulties and the much smaller proportion of students (about 3 per cent) with learning disabilities in mathematics, sometimes called *dyscalculia* (Williams, 2013). Students with learning difficulties in mathematics are generally understood to be those students who require extra assistance with mathematics due to their lack of, or failure to use, mathematical knowledge and skills. In some schools, this term is used to describe students who are performing below, or well below, the average level for students of their age (van Kraayenoord & Elkins, 2004). A Victorian study at the turn of the century revealed that up to 9 per cent of Years 8 and 9 students were assessed as performing at or below the standard expected for Year 4 students, with a further 10 per cent of students achieving below the standard expected for Year 6 students (Siemon et al., 2001).

Students with learning difficulties typically have under-developed knowledge, or gaps or misunderstandings in mathematical concepts and skills, poor motivation (including lack of persistence), problems recalling information and facts, difficulties in recalling and using problem-solving strategies, limited vocabulary and low levels of metacognition. Their mathematics work is likely to indicate:

- reliance on inefficient counting strategies for computation, such as counting on by ones using fingers or tally marks (a strategy used by students in the early primary years prior to learning basic facts)
- lack of partitioning (or decomposition) strategies for computation—that is, students have not extended facts from basic facts (e.g., have not developed knowledge of pairs of numbers that total 100)
- poor understanding of the structure of the number system—that is, place value of large numbers and decimals and representation of number using linear models

- that their knowledge of multiplication is limited to a repeated addition concept, and does not include the use of the commutative and distributive law for deriving number facts and mental computation
- that their knowledge of division is limited to sharing by ones and does not include understanding quotition
- poor understanding of fractions (e.g., equal partitioning when representing fractions) (Anthony & Walshaw, 2002; Department of Education and Training, New South Wales, 2002; McIntosh, 2002; Mulligan et al., 2006; Perry & Howard, 2001; Siemon et al., 2001; Vale, 2002b, 2004).

These difficulties with number concepts and skills contribute to the problems that students experience when applying number to measurement and developing algebra skills. Further, while inefficient computational strategies can often result in the correct answer, students have expended so much effort in completing the trivial computational tasks that they are unable to progress to more advanced concepts or complete multi-step or complex problems (Department of Education and Training, New South Wales, 2002). Recent research on the learning difficulties of children in the early years has underlined the importance of structural understanding, including spatial structure (Mulligan et al., 2006). A lack of spatial and patterning skills may have significance for upper primary and junior secondary students experiencing difficulties in measurement, geometry and functions.

While some children begin school with very limited experiences of mathematical concepts and learning in their early childhood, and thus start schooling a long way behind other students, many of the causes of learning difficulties in mathematics can be attributed to insufficient or inappropriate instruction (Carroll, 2006; A. Watson, 2006a). We know that the achievement gap between the lowest and highest achievers widens during secondary schooling, and that inappropriate teaching and curriculum contribute to this widening gap. The following set of learning obstacles for students applies in both primary and secondary classrooms:

- The curriculum proceeds too quickly and students are not ready to assimilate new concepts and procedures.

- The teacher's language when explaining concepts or asking questions does not match the students' level of comprehension.
- Abstract concepts are introduced too early without the support of concrete materials or visual aids; real-life examples or visual aids may have been removed too soon.
- Mathematical investigations are not followed up with carefully structured tasks for consolidating concepts and skills.
- Students with reading difficulties are given pure arithmetic or algebra exercises only, and denied experience of worded and non-routine problems.
- There is an emphasis on teaching of procedures and tricks that are rapidly forgotten because they do not enable meaningful learning.
- Insufficient attention is given to revisiting and reviewing of concepts and processes at regular intervals and making connections with prior learning.
- There is insufficient corrective feedback (Chan & Dally, 2000; Siemon & Virgona, 2002).

Assessment tools

A variety of assessment tools have been developed to assist teachers to identify the particular misconceptions and needs of students who have difficulties with mathematics. These include pen-and-paper diagnostic tests, clinical interviews and rich assessment tasks. Pen-and-paper diagnostic tests such as the *Progressive Achievement Test in Mathematics* (ACER, 1997) or online tests such as *On Demand Testing* (Victorian Curriculum and Assessment Authority, 2015) assess content and procedural knowledge and assist teachers to construct a student profile. Clinical interview instruments, such as the *Counting on Intensive Assessment Interview* (Department of Education and Training, New South Wales, 2002), *Initial Clinical Assessment Procedure Mathematics Tasks—Upper Primary* (Hunting, Doig, Pearn & Nugent, 1995), *Fractions and Decimals Online Interview* (Department of Education and Training, Victoria, 2014a), *Assessment for Common Misunderstandings* (Department of Education and Training, Victoria, 2014b) and the *Middle Years Numeracy Intervention Interview* (Vale, n.d.) allow teachers to probe students' thinking by inviting them to express their ideas verbally in one-to-one or conferencing interactions. An example of an item and the criteria for recording the student's response are provided in Figure 13.4. Clinical interviews are used to find out how students think

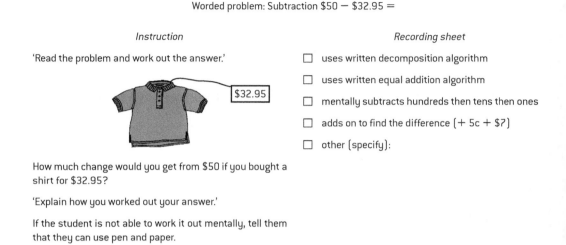

Worded problem: Subtraction $50 − $32.95 =

Instruction

'Read the problem and work out the answer.'

$32.95

How much change would you get from $50 if you bought a shirt for $32.95?

'Explain how you worked out your answer.'

If the student is not able to work it out mentally, tell them that they can use pen and paper.

Recording sheet

☐ uses written decomposition algorithm

☐ uses written equal addition algorithm

☐ mentally subtracts hundreds then tens then ones

☐ adds on to find the difference (+ 5c + $7)

☐ other (specify):

Figure 13.4 **Sample clinical interview item**

about and solve a problem and whether they use inefficient strategies, rather than only focusing on right or wrong answers and procedural errors.

When conducting clinical interviews, teachers should resist giving hints, ask students to explain their thinking and invite them to work through the task step-by-step to identify point of difficulty. When analysing students' mathematical work, teachers should find out whether these errors are:

- random or repetitive, and in what circumstances they occur
- general or conceptual errors, or errors relating to particular cases
- based on weak recording skills, or misuse of equipment
- based on inappropriate generalisation, or using rules adapted from elsewhere
- based on unhelpful images of the concept
- based on faulty memory (A. Watson, 2006a, p. 65).

A. Watson (2006a) reminds us to be careful about the judgements we make when assessing students. As we noted above, poor instruction is often the cause of students' learning difficulties, so we must not 'blame the victim' and assume that they have 'low ability'.

With the information gathered from formative assessment, it is possible to plan learning activities to improve students' mathematics learning and achievement.

REVIEW AND REFLECT : Three Year 7 students were shown a flashcard with the fraction $\frac{2}{3}$ written on it and asked to 'draw a diagram to show what this card means' and then to 'explain your diagram in words'. Copies of drawings produced by three students are shown in Figure 13.5.

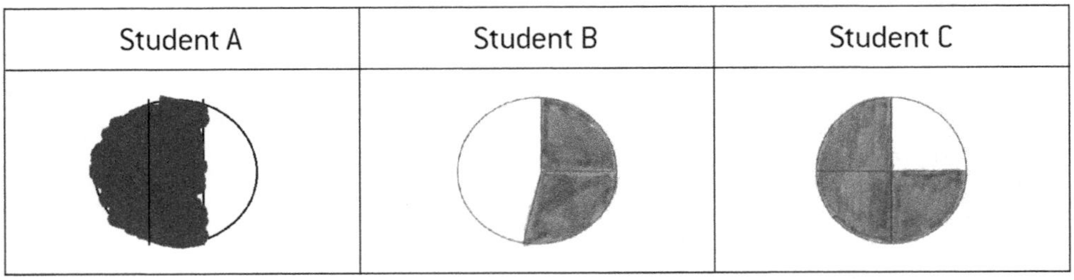

Student A	Student B	Student C

Figure 13.5 **Three students' drawings of** $\frac{2}{3}$

How do you think they would have explained their diagram?

In what way are these representations similar and different?

What do their representations suggest about how each of these students thinks about and understands fractions?

What other questions would you like to ask these students to find out what they know about fractions and how to represent them?

Teaching students with learning difficulties

Students with learning difficulties have traditionally been excluded from learning mathematics or had limited access to the domain of mathematics (Diezmann et al., 2004; Dole, 2003; A. Watson, 2006a). Diezmann et al. (2004) argue that there has been a shift in approach from the medical model of diagnosis and remediation towards making the curriculum accessible to all by designing effective pedagogies. Two main approaches are advocated: explicit teaching and inquiry approaches (Gunn & Wyatt-Smith, 2011). These approaches are used in intervention programs, withdrawal programs or within a streaming program.

Inquiry approaches

A. Watson's (2006a) approach with low-achieving students is to prompt discussion of mathematical investigations to promote exemplification, symbolisation and structure. She provides the following examples of questions that she uses in the classroom:

'What is the same or different about …?' (Encouraging learners to give attention to pattern and classification)

'Describe what happens in general.' (Nudging learners through generalisation towards abstraction)

'Can you give me an example from your own experience …?' (Prompting exemplification)

'Can you show me one which does not work?' (Prompting counter-exemplification)

'Show me …' or 'Tell me …' (Eliciting information about images and other aspects of their understanding)

'Can you show me this using a diagram/letters/numbers/graphs?' (Prompting flexible use of representations)

'If this is an answer, what might the question be?' (Shifting the focus on to structures rather than answers) (A. Watson, 2006, p. 109)

Ocean and Miller-Reilly (1997) propose a connected knowing model for teaching students with learning difficulties. Their model focuses on developing confidence, encouraging students to discuss their ideas, and providing first-hand experience through the use of materials. They recommend that teachers:

Discover and affirm what the student already knows, respect the students' existing ideas … ask students for explanations even when they are right, so that question-asking does not become synonymous with doubt, assume the students have reasons for their opinions and listen to them, and ask for details. (Ocean & Miller-Reilly, 1997, p. 19)

Intervention programs

Targeted intervention early in schooling and for a relatively short period has proved to be effective for primary school children with mathematics learning difficulties. However, there is no agreement on the relative merits of 'in-class' or withdrawal programs for students in the middle years (Carroll, 2006). Some schools rely on online individualised intervention programs such as *Mathletics*. Intervention programs typically use an explicit teaching approach, which involves a focus on identifying and explaining terms, symbols and diagrams in problems, modelling procedures for solutions and guided practice of routine problems. Carefully designing and ordering exercises are important for building fluency and understanding as students notice similarities and differences to generalise and abstract concepts and procedures (Watson & Mason, 2006).

QuickSmart (Graham & Pegg, 2010; Pegg, Graham & Bellert, 2005) is a withdrawal program for students in the middle years, conducted with pairs of students with similar learning difficulties. The program runs for five 30-minute sessions over 25 weeks. The focus of the program is on learning basic facts and extended basic facts with whole numbers for the four operations, and the objective is to improve the speed and accuracy of automatic recall of these facts. Each 30-minute session has four components: revision of the previous session; guided practice activities featuring overt self-talk and the modelling of strategies to develop and extend basic facts; discussion and practice of strategies for memory and retrieval; and specially targeted games or worksheet timed activities.

Building Accuracy and Skills in Core Skills (BASICS) (See Byers, 2009) is an intervention program comprising three levels of instruction: direct instruction of basic rules, concepts and skills; developing problem-solving skills and strategies; and hands-on inquiry-based learning. The model allocates sequentially more time to each level and content is related to the curriculum at the particular year level.

Middle-years students in withdrawal programs have been found to enjoy skill-building activities, and have improved and developed mastery of basic skills (Graham & Pegg, 2010; Perry & Howard, 2002). However, withdrawal programs often do not include applications for real-world contexts that enhance mathematical literacy (Dole, 2003). Dole (2003) argues that application problems provide the motivation for students to learn basic skills, and students with learning difficulties need support to develop and communicate mathematically using multiple representations. However, a focus on routine application problems excludes students from making sense of mathematical concepts and engaging with big ideas (Jorgensen, 2014).

Stephens (2000) also criticises withdrawal programs as an intervention method, arguing that such programs remove the classroom teacher from the picture and hinder their ability to monitor students with learning difficulties and provide adequate follow-up and support during mainstream classes. Another problem occurs when staff allocated the task of teaching students in these withdrawal programs are not appropriately trained or do not like teaching students with learning difficulties (Walshaw & Siber, 2005). It is surprising how often pre-service teachers are assigned the role of providing additional support or conducting intervention programs during practicum experience when it would be far more appropriate for the experienced teacher of mathematics to take on this role.

REVIEW AND REFLECT : Investigate an intervention program—for example:

- *Counting On* (Department of Education and Training, New South Wales)
- *QuickSmart* (SiMMER, University of New England)
- another published intervention program
- an electronic tutorial program
- an intervention program used by your school.

For whom is this program designed? How are students selected for the program? How is the intervention program related to the mainstream mathematics class and program?

Note the objectives, structure, content, materials and teaching approaches used in the program. Comment on the strengths and weaknesses of the program.

Compile an annotated portfolio of resources that you could use with students with learning difficulties. The annotations could include reflection on your experience of trialling these resources.

Students with special learning needs

Students with special learning needs provide a particular challenge for classroom teachers. These students may have physical impairments, such as hearing and visual impairments, emotional and social disabilities (e.g., ADHD) and other intellectual impairments and special qualities, such as cerebral palsy or a type of autism. Government policy mandates inclusion in regular schools but in reality, many of these students are excluded from engaging mathematical learning experiences (Healy & Powell, 2013; Slee, 2011). Teachers should not jump to the conclusion that students with special learning needs have mathematics learning disabilities or difficulties; indeed, some of these students may be mathematically gifted or talented (Lambert, 2015; Norris & Dixon, 2011). Rather, teachers should seek to understand the way in which school and classroom culture and discourse excludes these students. Teachers can address disadvantage by:

> Analyzing how learning scenarios and teaching practices can be more finely tuned to the needs of particular groups of learners, empowering them to demonstrate abilities beyond what is generally expected by dominant discourses. (Healy & Powell, 2013, p. 69)

Healy and Powell (2013) refer to Vygotsky's (1997) research on how differently abled students use sensory functions, the eye, speech and touch as 'instruments' to see and think and develop and the way in which learning materials can be designed and used so that differently abled students may substitute sensory functions for learning. For examples of the way in which tools, language and instruments can be adapted, substituted and used to support visually and hearing impaired students, see Fernandez and Healy (2012) and Pagliaro (2006). Lambert (2015) discusses the way in which procedurally focused and inquiry- or discussion-based pedagogical practices impact differently on the mathematical engagement, self-efficacy and achievement of two students with learning disabilities.

Students with mathematical talent

Students who are capable of high-level performance in mathematics are described as 'gifted', 'mathematically talented students', 'highly able students', 'promising students' and students with 'a mathematical cast of mind' (Diezmann et al., 2004). The Australian Senate Inquiry into the Education of Gifted and Talented Children reported that often teachers think the gifted or talented students are the high achievers, and they fail to identify gifted and talented students among the under-achievers, divergent thinkers, visual–spatial learners and children who mask their ability (Collins, 2001). An interest in mathematics along with creative thinking skills, including advanced inductive and deductive mathematical reasoning and problem-solving skills, are defining characteristics (Bicknell, 2009; Sharma, 2013). It is important to realise that gifted or talented girls and boys have diverse cultural characteristics: they may live in a family of low socio-economic status (SES), be an Indigenous person, have a physical or intellectual disability or live in a geographically isolated place.

Attempts to define 'giftedness' need to go beyond notions of general intellectual ability or specific academic aptitude. Mathematically talented students demonstrate creative and divergent thinking skills, demonstrated through responses that display fluency (a large number of responses), flexibility (a variety of representations and the ability to change between these readily), originality (unusual or uncommon responses) and elaboration (embellishment or expansion of ideas). Additionally, they may display the following cognitive behaviours:

- risk-taking—willingness to try different or difficult things
- curiosity—ability to seek alternatives and study in depth
- complexity—capacity to explore and discover
- imagination—power to visualise or conceive symbolically.

In mathematics, these attributes are normally manifested through the approach to and strategies used on non-routine problems or modelling tasks (Neiderer & Irwin, 2001; Sriraman, 2003). Talented mathematics students take longer to orientate themselves to the problem and understand the problem situation than other students. They develop a plan that is more general. They use mathematical reasoning rather than the application of routine algorithms, draw on many strategies and, depending on the characteristics of the problem, will begin with simpler cases to control the variability, or systematically explore a large number of possibilities for open-ended problems. These students seek to form generalisations—that is, they look for similarities, structures and relationships by abstracting from the content of the problem. Studies have also shown that students with mathematical talent have strong spatial-visualisation skills (Diezmann et al., 2004).

Assessment tools

Unfortunately, there is no readily available assessment tool that teachers can use to identify gifted or talented mathematics students. Routine mathematical problems and standard mathematics tests (such as the *NAPLAN* (ACARA, 2013b), or *Progressive Achievement Test in Mathematics* (ACER, 1997)) do not enable students to display the creative and reasoning skills listed above, and teachers' assessments of students' mathematical talent have also been found to be unreliable (Neiderer, Irwin & Reilly, 2003). Assessment should include challenging problem-solving tasks and analysis of students' spatial ability (Diezmann et al., 2004).

The Australian Mathematics Competition (Australian Mathematics Trust, n.d.) provides an opportunity for students from Years 3 to 12 to engage with a range of mathematical problems, including problems that are challenging for the most able students. Results from this competition indicate the suitability of curriculum content for students with different mathematical abilities. Leder (2006) analysed the results for the items that were common to papers for more than one secondary year level in the Australian Mathematics Competition. She was able to identify different types of problems. For one group of items, performance steadily improved over the secondary year levels, with most of the top 2 per cent of students already able to complete this problem in Year 7. A second group of items could be correctly solved by about half the top 2 per cent of students irrespective of the year level. An example of these two kinds of problems and a description of the level of thinking needed to solve these problems are shown in Table 13.1.

Studies of the highest performing students in mathematics have consistently reported a gender difference in favour of males. One of the significant findings of Leder's (2006) study was that, in the Australian Mathematics Competition, boys outnumber girls three to one in the top 2 per cent of students. Explanations for gender differences in achievement are discussed in Chapter 15. Teachers need to ensure that they do not apply gender-stereotyped

Table 13.1 Items from the Australian Mathematics Competition

Category of item	Example item	Level of thinking required
Solved correctly by top 2 per cent of Year 7 students; performance by all students improved with higher grade level	Seven consecutive integers are listed. The sum of the smallest three is 33. What is the sum of the largest three? [A] 39; [B] 37; [C] 42; [D] 45	Sequential multi-step problem.
Solved by most of top 2 per cent of students irrespective of grade level; little improvement in performance by all students with higher grade level	A 3 x 3 square is divided into nine 1 x 1 unit squares. Different integers from 1 to 9 are written into these unit squares. Consider the pairs of numbers in squares sharing a common edge. What is the largest number of pairs where one number is a factor of the other number? [A] 7; [B] 8; [C] 9; [D] 10; [E] 12.	A lot of integration and synthesis of information required.

Source: Leder (2006).

attributes for 'high-achieving girls' and 'high-achieving boys' when assessing students to identify students with mathematical talent.

REVIEW AND REFLECT : Complete the two problems in Table 13.1. Record all your thinking and working. Share your approach, strategies and solution with your peers. Compare your approaches with respect to the four phases of problem solving:

- orientation (understanding the problem)
- organisation (setting goals and planning)
- execution (carrying out the plan)
- verification (evaluating and reflecting on your solution).

Identify any instances of creative and divergent thinking (fluency, flexibility, etc.). To what extent were you able to generalise during the process of finding a solution to these problems?

Teaching mathematically talented students

> [Gifted] children have special needs in the educational system; for many their needs
> are not met; and many suffer underachievement, boredom, frustration, and psy-
> chological distress as a result ... The common belief that the gifted do not need
> special help because they will succeed anyway is contradicted by many studies of
> underachievement and demotivation among gifted children. (Collins, 2001, p. xiii)

In our community, there are negative attitudes about people who are gifted mathematically and, unfortunately, little attention is given to their needs (Collins, 2001; Diezmann
et al., 2004). Important for the development of mathematical talent among gifted students
is the provision of immediate and long-term extrinsic rewards, together with enjoyment
when doing mathematics and support from teachers and the school (Csikszentmihalyi,
Rathune & Whalen, 1997). Parents also play an important role as motivator, mathematics
content adviser, resource provider, monitor and learning counsellor to their gifted and
talented children (Bicknell, 2009).

Teachers should not assume that gifted and talented students will enjoy all mathematics tasks and work productively in mathematics lessons. These students still need
instruction, but it needs to move quickly without unwanted repetition. Nor should
teachers suppose that gifted and talented students will want to collaborate with peers,
because these students usually see no benefit in working with others on exercises or
trivial and routine application problems. They are more likely to collaborate and use
higher-order thinking skills when working on challenging problems when placed in
extended curriculum programs with similarly talented students (Kronberg & Plunkett,
2015). However, higher-order thinking does not come automatically, even among
gifted students, and the use of mathematical argument to explain and justify a solution
is often preceded by the use of more pragmatic reasoning, such as trial and error, and
systematic reasoning, such as the organisation of information, data or strategy (Lee,
2005). Interaction with other students in a small-group setting is crucial for advancing
to these sophisticated levels of mathematical thinking, and discussion with other students provides an opportunity to understand why verification and proof are needed in
mathematics. Teachers are responsible for helping all students develop the social skills

required for collaboration, and this is especially relevant in the light of Barnes's (2000a) observation that some talented mathematics students do not have the language skills to take advantage of collaborative learning settings.

In spite of the lack of attention to mathematically talented students, structured programs of various kinds have been designed and used to develop autonomy, self-reliance and social skills. These programs, organised by schools, mathematical organisations or education systems, include selective entry public schools, accelerated class groups of mathematically talented students within a comprehensive school, enrichment programs typically involving part-time withdrawal from mainstream classes for gifted students, and a differentiated curriculum to cater for students with diverse learning needs in a heterogeneous mathematics class. We consider accelerated and enrichment programs below in the context of provision for talented students, and differentiated curriculum later in the chapter when discussing responses to diversity.

Accelerated programs

Accelerated programs enable students to complete the required curriculum in a shorter time span—for example, they might complete the work normally covered in three years of the mathematics curriculum in only two years. Typically, accelerated groups of students are part of a year-level streaming program, which selects the highest achieving students for acceleration. Diezmann et al. (2012) acknowledge that selection for these programs may not involve identification of giftedness. Normally, students in an accelerated class program work on a different curriculum, one or more years ahead of the students at the same age. Hence, students in the 'top' group, or an accelerated program, may simply be experiencing a mainstream curriculum of the next year level, rather than having their mathematical understanding and thinking challenged with higher-order tasks (Diezmann et al., 2012; Forgasz, 2010).

Successful accelerated programs involve more advanced content, delivered at a faster rate, and activities that require higher-order, abstract thinking and the connection of mathematical ideas in ways that would not normally be expected of students at that particular age level (Anthony, Rawlins & Winsley, 2002; Diezmann et al., 2012; Kronberg & Plunkett, 2008, 2015). Challenging tasks involve complexity, and creative and critical thinking; they may also be open ended and require group interaction and real-world deadlines for real-world problems and audiences. Suitable tasks involving mathematical reasoning, conjecture and verification or proof can be found in a variety of resources (e.g., *Maths300*).

Enrichment programs

Enrichment programs often involve one-off activities, a withdrawal program for a defined period of time, or elective subjects in Years 7–9 conducted by the school or mathematics associations at the local, state or national level (Dharmadasa et al., 2014; Diezmann et al., 2012; Sharma, 2013). They include competitions, after-school and holiday programs, and mathematics camps or activities, often conducted in cooperation with university mathematics departments. In some cases, withdrawal programs for individuals or small groups of students are delivered online, especially to cater for gifted and talented students in isolated locations (see Clarke & Bana, 2001).

Some withdrawal enrichment programs have been criticised as time wasting and trivial, designed to occupy the highest-achieving students while the rest of the class catches up (Collins, 2001). Enriching mathematical activities involving problem solving and application are appropriate tasks for talented students, but these tasks should be part of the curriculum for all mathematics students, not just the gifted and talented. The Australian Senate Inquiry into the Education of Gifted and Talented Students concluded that ad hoc enrichment activities were insufficient, and that gifted and talented students needed a differentiated curriculum (Collins, 2001). Provided that they are emotionally and socially ready, these students also benefit from accelerated programs.

REVIEW AND REFLECT : Investigate an enrichment or accelerated program for gifted and talented mathematics students—for example, the Australian Mathematics Competition, the Young Australian Mathematical Challenge, the International Mathematics Olympiad, the Mathematics Talent Quest, or other competitions or programs for talented students. (Search the AAMT website for state and national programs catering for high-achieving, talented and gifted students, www.aamt.edu.au; see also Australian Mathematics Trust, n.d.).

For whom is this program designed? How are students selected for the program? How is it related to the mainstream mathematics program?

Note the objectives, structure, content, materials and teaching approaches used in the program. Comment on the strengths and weaknesses of the program for gifted and talented students.

Compile an annotated portfolio of resources that you could use with gifted and talented students of mathematics. The annotations could include reflection on your experience of trialling these resources.

Conclusion

Studies of classroom organisation do not show that one form is best for all. Rather, employing a range of organisational settings is most effective, since different arrangements will suit particular students at different stages of their learning or for specific mathematical tasks. The teacher's role is important in each of these organisational settings. As Anthony and Walshaw (2007) explain, quality teaching provides a space for:

> The individual, partnerships, small groups and whole class arrangements … [All] students need some time alone to think and work quietly away from the demands of a group … Teachers who make a difference to all learners work at establishing a web of relationships within the classroom community. They do this to encourage active participation, taking into account the different purposes of, and roles within, the particular social arrangements they establish for their students. Organisational structures are established with a view towards the potential of those arrangements in developing students' mathematical competencies and identities and in providing other positive outcomes for students in particular contexts. But more significantly … the effective teacher constantly monitors, reflects upon, and makes necessary changes to, those arrangements on the basis of their inclusiveness and effectiveness for the classroom community. (p. 533)

Collaborative practices in the planning, enactment and evaluation of teaching are essential to meet the diversity of student learners and to enable continuity of high-quality learning experiences as students transition between year levels, programs and streamed classes or classroom groupings (Diezmann et al., 2012; Kronberg & Plunkett, 2015; Vale et al., 2011).

In the next two chapters, we discuss the diverse interests and needs of students in mathematics classrooms from the point of view of SES, geographic location, indigeneity, ethnicity and language, and finally gender.

Recommended reading

Anthony, G. & Walshaw, M. (2007). *Effective pedagogy in mathematics/pāngarau: Best evidence synthesis iteration [BES].* Wellington: New Zealand Ministry of Education.

Boaler, J. (2008b). Promoting 'relational equity' and high mathematics achievement through an innovative mixed ability approach. *British Educational Research Journal,* 34(2), 167–194.

Byers, T. (2009). The BASICS intervention program for at-risk students. *The Australian Mathematics Teacher,* 65(1), 6–11.

Dharmadasa, K., Nakos, A., Bament, J., Edwards, A. & Reeves, H. (2014). Beyond *NAPLAN* testing … Nurturing mathematical talent. *The Australian Mathematics Teacher,* 70(2), 22–27.

Stein, M., Engle, R., Smith, M. & Hughes, E. (2008). Orchestrating productive mathematical discussions: Five practices for helping teachers move beyond show and tell. *Mathematical Thinking and Learning,* 10(4), 313–340.

Watson, A. (2006a). *Raising achievement in secondary education.* Maidenhead, Berkshire, UK: Open University Press.

CHAPTER 14

Equity and social justice in mathematics teaching and learning

The *Melbourne Declaration* (MCEETYA, 2008) includes the following two goals:

- Goal 1: Australian schooling promotes equity and excellence
- Goal 2: All young Australians become: successful learners; confident and creative individuals; active and informed citizens.

The rhetoric of Western countries is that mathematics learning is to prepare students for active citizenship, but often the reality is that mathematics education maintains the social order (Jorgensen, 2014; Skovmose & Valero, 2002). Studies show that cultural heritage and social backgrounds of students continue to be related to advantages (and disadvantages) in schooling, and hence educational success (Thomson, De Bortoli et al., & Buckley 2013; Thomson et al., 2012). While mathematics teachers recognise the diverse social and cultural experiences and identities of students in their classroom, teaching practices, such as streaming discussed in Chapter 13, contribute to continued lower self-efficacy and achievement of disadvantaged students. In this chapter, we discuss social and cultural issues and teachers' knowledge of context for teaching mathematics and the ways these issues are related to students' mathematics achievement and attitudes. Finally, we explore socially just teaching approaches that enable success in mathematics for all students and reduce the inequities in our society. Gender justice is discussed in the next chapter.

Sociocultural factors and students' mathematics learning

In this section, we will discuss the relationship between various socio-cultural factors and mathematics participation and achievement. Figures 14.1, 14.2, 14.3 and 14.4 are taken from the report of the 2012 PISA study of mathematical literacy (Thomson et al., 2013a) regarding results for Australian students according to SES, geographic location, Indigenous

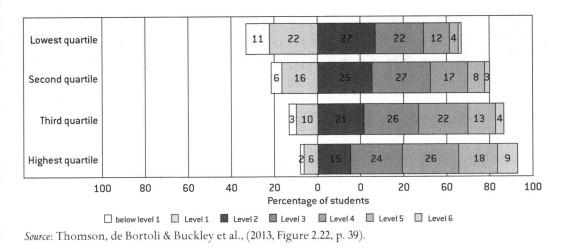

Source: Thomson, de Bortoli & Buckley et al., (2013, Figure 2.22, p. 39).

Figure 14.1 **Percentage of students for each level of mathematics literacy by socio-economic background from PISA 2012**

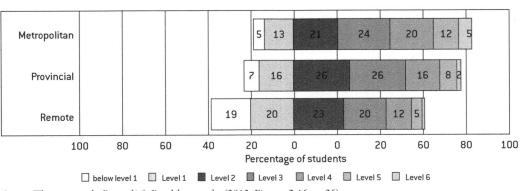

Source: Thomson, de Bortoli & Buckley et al., (2013, Figure 2.16, p. 36).

Figure 14.2 **Percentage of students for each level of mathematics literacy by geographic location from PISA 2012**

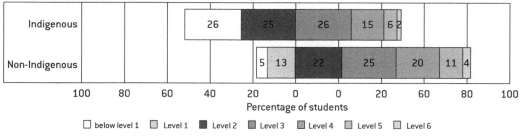

Source: Thomson, de Bortoli & Buckley et al., (2013, Figure 2.22, p. 37).

Figure 14.3 **Percentage of students for each level of mathematics literacy by Indigenous background from PISA 2012**

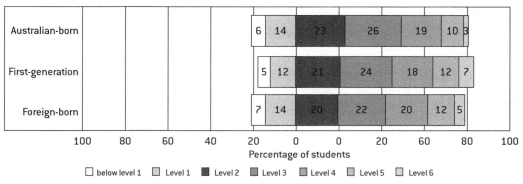

Source: Thomson, de Bortoli & Buckley et al., (2013, Figure 2.24, p. 40).

Figure 14.4 **Percentage of students for each level of mathematics literacy by immigrant background from PISA 2012**

background and immigrant background. The PISA study assesses the mathematical literacy of 15-year-old students from 65 countries including Australia. Mathematical literacy is defined as the ability to apply mathematical 'knowledge and skills to real life problem and situations' (Thomson, de Bortoli et al., & Buckley 2013, p. viii). The items on the PISA instrument include multiple-choice questions and short-answer questions. The content of these items concern change and relationships (number, algebra and functions), space and shape (geometry), quantity (number and measurement), and uncertainty and data (probability and statistics). They require students to model situations mathematically apply concepts, procedures and reasoning, and interpret solutions.

> **REVIEW AND REFLECT :** Compare and contrast the mathematical literacy reported in each of these figures and discuss these research findings with colleagues.
>
> Which cohort(s) of students are the most advantaged (likely to achieve Level 6)? Which are the most disadvantaged (likely to achieve below Level 1)?
>
> Why do you think that mathematical literacy is related to these socio-cultural factors?

Socio-economic status (SES)

National and international studies of Australian students at different levels of secondary schooling show that there is a significant relationship between socio-economic background and mathematics achievement (Ainley, Kos & Nicolas, 2008; Forgasz, et al., Leder & Halliday, 2013; Thomson, de Bortoli & Buckley et al., 2013; Thomson et al., 2012). As the findings presented in the tables above suggest, SES interacts with other factors, geographic location and cultural background (Indigenous or immigrant) to compound inequalities (Forgasz et al., 2013; Jorgensen, 2012a; Jorgensen & Lowrie, 2013). Studies of growth in mathematics achievement over time show that teachers are able to implement teaching approaches that close the gap during the school year, especially during Terms 2 and 3. However, the gains made are not sustained across Term 4 and the summer break, with secondary students of low SES not improving on previous results and in some regions falling further behind by the end of Term 1 in the following year (Vale et al., 2013).

Given these results, it is not surprising that participation and achievement in post-compulsory mathematics in Australia is also related to socio-economic background. Analysing Year 12 mathematics results from Victoria, Teese (2000) found that:

In the urban regions [of Melbourne] where working-class and migrant families are highly concentrated, every third girl can expect to receive fail grades in the least demanding mathematics subject. Among boys—whose attraction to mathematics is even more fatal—failure strikes more than 40 per cent. The better students gravitate to the mathematics subjects that lead to university. But here too failure awaits them, one in three. (p. 2)

Social capital and power is the main explanation for the relationship between SES and achievement (Jorgensen, 2014; Jorgensen et al., 2014; Teese, 2013). This notion of social capital relates to economic resources within the family to support learning (such as internet access from home), the cultural knowledge of education within the family, such as parents' level of education, including level of mathematics learning, and other community and cultural connections of the family that support academic success. However, this perspective of social capital takes a deficit position and blames the 'victim', rather than looking at how power is structured and sustained to deny access and empowerment of learning for the least advantaged (Jorgensen et al., 2014). Differences in social capital are sustained through the private/public schooling system and the failure of systems to provide adequate funding (e.g., the Gonski funding model) and shortages of experienced and qualified teachers of mathematics in schools in disadvantaged communities. Teese (2013) argues that changes in curriculum and assessment at the senior secondary level over the past 50 years show that when changes in curriculum occur, it benefits the most advantaged communities. In addition, researchers have argued that low expectations of teachers, the use of routine mathematics tasks that focus on procedures, and low-level questioning techniques act to 'dumb down' the curriculum and deny access to mathematical knowledge making (Atweh & Brady, 2009; Cooper & Dunne, 2000; Jorgensen, 2012a; Jorgensen et al., 2014, McGaw, 2004).

Cooper and Dunne (2000) analysed the responses of students to real-world worded problems. They found that, on average, students from a lower socio-economic background scored lower on these problems than other students. They observed that low SES students often used their personal experiences of these 'real' problems when providing solutions, whereas middle-class students recognised that the context of the problem was irrelevant to the solution (i.e., a 'wrapper' problem, see Chapter 3), and were able to interpret the problem and provide the correct solution. Peled and Suzan (2011) noticed that students' cultural values also contributed to their solutions of 'wrapper' problems.

REVIEW AND REFLECT : Solve these two problems (Peled & Suzan, 2011, pp. 83–84).

- Two friends, Anne and John, bought a $5 lottery ticket together. Anne paid $3 and John paid $2. Their ticket won $40. How should they share the winning amount?
- During the Country Fair Tammy and Abby put up a lemonade stand. Tammy bought disposable cups for $10 and Abby bought concentrated lemon-juice for $14 and mineral water bottles for $36. They sold lemonade for a total of $480. How should they split the money?

Anticipate possible Years 7 and 8 student solutions, then analyse the following student solutions of these problems (Peled & Suzan, 2011, pp. 83–84):

Student solutions Problem A:

Aviv: *They should split it evenly. 40:2 = 20 so each gets $20.*

Liam: *Since Anne paid one dollar more than John, she should get 20^1/_2$ and John should get 19^1/_2$.*

Sula: *They should get according to what they paid. 40:5 = 8, 8 x 3 = 24, 8 x 2 = 16, so Anne should get $24 and John should get $16.*

Student solutions Problem B:

Latika: *They should split the amount according to the ratio of invest-ment. Total investment = $60, Total earnings = $480. Investment ratio = 60:480 = 1:8. Tammy invested $10 so she should receive $8 x 10 = 80; Abby invested $50 so she should receive $8 x 50.*

Van: *They should repay the costs and split the profit evenly. The profit was $480 – $60 = $420. An equal share of the profit is $210. Tammy should receive $210 + $10 = $220. Abbey should receive $210 + $50 = $260.*

Which of these solutions are correct? Why or why not?

Identify the mathematical knowledge and cultural values that these students used to solve these problems? How do their socio-cultural values influence the mathematical skills they use to solve these problems?

How would you modify these 'wrapper' problems (see Chapter 3) so that they ensure that students use proportional thinking?

Indigenous students

Participation rates of Indigenous secondary school students vary according to geographic location and SES across Australia. 'An Indigenous child is more likely to be locked up in prison than they are to finish high school' (Grant, 2015). While many Indigenous people live in highly populated locations, the reality for Indigenous students living in remote communities is that they must leave home if they want a secondary education. While the findings from the PISA study (Thomson, de Bortoli, & Buckey et al., 2013, Figure 14.3) show that high proportions of Indigenous student achieve at the lowest achievement levels, Indigenous student do achieve at the higher levels. Unfortunately, many Indigenous students experience school differently from non-Indigenous students. They may be stereotyped and marginalised in classrooms by teachers, by teaching practices, by the curriculum and by the culture of schooling (see the box below).

Che: a case study

Che, a Koori student, was a member of a multi-ethnic Year 9 class, and the only Indigenous student in an urban government school. The class was studying geometry and the students used their laptop computers and the Geometer's Sketchpad during five lessons in the period that I observed the class. Che showed a lot of interest in the computer-based mathematics lessons. He collaborated with other students to learn how to use the animate function in the software and discover some mathematical ideas. Like other students in the class, Che was sometimes off-task. He used the computer to do things other than the set mathematics task, and sometimes moved around the room to talk to other students. On one of these occasions, he collaborated with other students to load other software (MicroWorlds) from one laptop to another. In the following extract of the transcript of the lesson about the sum of exterior angles in polygons, Che discussed the results of the guided investigation with two other students, Lawrie and Darren. They had been following instructions for this investigation from a worksheet. Che was the first student to complete the task. He had completed the task for homework, and Lawrie asked him what he wrote for the conjecture:

Che: I done all that. I done all the way to here.

Lawrie: What do you do here, what did you write?

Che: Um, I wrote, um, I found out that all the angles equal up to 360 degrees.

Lawrie: Not matter what shape as long as its perimeter ...

Che: I found it for all pol ... polygons or something like that equals up to ...

Darren: The hexagon equals up to ...

Che: It's not a hexagon. Do control later on. No, no you don't. You go to calculator. Where's your calculator?

Lawrie: I already calculated it [points to the result on the screen].

Che: Yeah, well there you go. You done it all. Now you just write there [points to the screen] that all the angles equal up to 360 degrees. That's your conjuncture [*sic*]. [Waves his hands as if to say 'or whatever it is'.]

Lawrie: Where's the text box?

Darren: [At the same time] Conjuncture.

Che: You go [he points to the tool bar] text box.

In a subsequent lesson, students were constructing a series of geometric shapes. Some students constructed by eye and did not measure lengths or angles. Those who did measure angles tended to erase incorrect line segments and try again. No students were observed using the parallel line tool in the construct menu. Che was the only student who used the drag facility to make changes to his shapes.

> **REVIEW AND REFLECT :** Discuss the above case with colleagues.
>
> What surprises you about this case? How is this case similar to, or different from, your knowledge or expectations of Indigenous students?
>
> Interactions between the student and the teacher have not been included in this case. What do you think would be the likely content of the interactions between Che and the teacher? What feedback would you give Che?
>
> What advice would you give a pre-service teacher colleague who was teaching Che?

Ethnicity and language context

With the exception of Indigenous students, mathematical achievement is not related to cultural background or language spoken at home (Thomson et al., 2012, Figure 14.4; Thomson, de Bortoli & Buckley et al., 2013). Differences within and between cultures with regard to SES and study habits explain more of the variation in achievement than does non-English language background. Nevertheless, we know that language plays an important role in mathematics learning, and so we can expect that, when teachers and learners do not share the same language or culture or system of reasoning, mathematical learning will be impeded (Averill, 2012; Edmonds-Wathen, 2014; McMurchy-Pilkington, Trinick & Meaney, 2013). In Chapter 2, we indicated that language and communication are important aspects of constructivist and socio-cultural theories of learning, and in Chapter 3, we showed how language is important for establishing meaning in mathematics and making connections between concepts, symbols and real situations.

Even though language background is not related to achievement, the vocabulary and grammar in problems and textbooks, and classroom metaphors and gestures that can create culturally bound concepts, can be difficult for students whose first language is not English, especially for Indigenous students and students who are recent migrants to Australia (Edmonds-Wathen, 2014; Wilson & Barkatsas, 2014). MacGregor and Moore (1991) explain that the language of counting, measuring and comparing, the meaning of articles and prepositions, and the verb 'to be' are particularly difficult. Further, some problem-solving tasks and investigations place a high level of demand on comprehension and writing skills. Students need to be able to formulate conjectures, explain and justify their methods of solution, and pose mathematical problems. These skills are especially demanding for students whose first language is not English.

REVIEW AND REFLECT : Consider the following problem.

Emma wants to buy a car for $28,000. She plans to pay a deposit and make equal monthly payments over four years. If she pays a deposit of $3000 how much will her monthly payments be?

Compare your interpretation of this problem with that of a colleague. What difficulties do you think students would have with the language in this problem? Rewrite this problem so that it could more easily be interpreted.

Compare and discuss the vocabulary and structure of the language used in the explanations, examples and problems of two different textbooks for the same mathematical topic.

Geographic location

National and international testing has drawn attention to the disadvantage of students in schools in provincial and remote locations. It is important to realise that students attending these schools come from a range of socio-economic and ethnic backgrounds, including diverse Indigenous backgrounds, and that these factors of diversity are not common across geographic locations (Atweh et al., 2012). Further, transient and fluctuating populations, immigration, rural economic circumstances, and seasonal conditions and natural disasters (droughts, floods and bushfires) are issues for schools in rural Australia (Pegg, 2009).

Education policy, programs and resources also have an impact on teaching in rural Australia. There is a shortage of fully qualified mathematics teachers, and new graduates tend to return to the coast once initial contracts are completed (Handal, Watson, Petocz & Maher, 2013; Pegg, Lyons et al., 2005). Teachers of mathematics need to be flexible and adaptable and be prepared to work alongside and mentor out-of-field teachers, and seek support within the community and with professional colleagues in the region as mathematics professional learning inside and outside the school is not readily provided (Hobbs, 2013).

Socio-cultural norms and teachers' practices

The socio-cultural norms of the classroom refer to the language, styles of communication and classroom rules used by teachers and learners that convey meanings about mathematics, what it is and how it is practised. In Chapter 3, we discussed the different approaches that teachers use to control or direct the discussion of mathematics. A number of researchers have observed the kinds of questions and styles of communication that include or exclude students (e.g., Jorgensen et al., 2014). For example, students whose first language is not English may be reluctant to respond to questions in a public forum or contribute to discussion in small-group tasks. Further, a teacher may not accept a method of solution that is commonly used and taught in another country. The assessment of students' work is another element of teachers' practice that may unconsciously be influenced by the socio-cultural background of students (Morgan & Watson, 2002), such as the case of Che above.

REVIEW AND REFLECT : Observe a teaching episode and write a record of the teacher's questions and students' reactions and response. Alternatively, videotape (or audiotape) one of your teaching episodes. Use the following prompts to reflect on the socio-cultural norms and inclusion of students in this lesson:

- What were the rules (explicit or implicit) for engagement in this lesson?
- Who participated and who did not participate?
- What language knowledge did students need in order to participate?
- What types of questions were used?
- How did the teacher control the discussion?
- How do you think the students felt about their inclusion or exclusion?
- What would you change about this teaching episode and how would you manage the discussion? Why?

Equity, ethics and social justice

Teachers of the most disadvantaged students are often acutely aware of the difficulties they and their students face. A mathematics teacher in a school with many students from disadvantaged backgrounds described the way in which these socio-cultural factors affected her students' mathematics learning:

> I always use the analogy that we're all running the same race in the end, but our kids are jumping hurdles. Some kids are running flat races. If you're at [a private school in an affluent suburb], you've got a pretty easy hundred metre run. Our kids tend to fall because they're jumping over stuff. So to me the school has to make up for that, so [social justice is] about taking away those hurdles. (Mathematics teacher, western suburbs school, Melbourne, personal communication)

This teacher believed it was both the school's and the teachers' responsibility to 'pre-empt what the barriers are going to be and teach kids how to go round them, go over them, or knock 'em down'.

> REVIEW AND REFLECT : What does equity and social justice mean to you? Discuss with your colleagues in a small group and write a definition upon which you can all agree.

Equity is one of the goals of the *Melbourne Declaration* (MCEETYA, 2008). But what does it mean? Researchers argue that equitable practice is more than redistributing resources and providing equal opportunity, equal treatment, and recognition (or inclusion). It involves establishing respectful relationships, paying attention to the different needs of students and teaching to these needs, and engaging with and giving voice to the community when making decisions (Atweh, Vale & Walshaw, 2012; Fraser, 2013; Vale, Atweh, Averill & Skoudoumbis, 2016). It involves establishing trust and acting ethically by taking responsibility for fairness and justice (Atweh & Brady, 2009). A secondary mathematics teacher explained it this way:

> If you are fair to everyone and if you are just to everyone then they respect you … [The students] should have a feeling that they are in a very just society and that in

the classroom they should find justice everywhere. (Mathematics teacher, western suburbs school, Melbourne, personal communication)

Social justice or response-able teaching is about empowering students. Teaching mathematics well gives students access to knowledge of mathematics, and the power and success in society that this elicits. That is, learners are empowered by their mathematical literacy so that they can function effectively and critically as citizens of a democratic society, and be agents in their own use and learning of mathematics (Gutstein & Peterson, 2006; Jorgensen, 2014; Skovmose & Valero, 2002; Steen, n.d.).

Teachers have developed a number of different appoaches for taking into account the social and cultural factors influencing the performance of disadvantaged students, and to resolve the conflict between their experiences of mathematics outside and inside the classroom. Bishop (1994) tracked changes in teaching approaches concerning cultural background to provide a framework to describe and discuss approaches as they evolved overtime: *assimilation, accommodation* and *amalgamation. Assimilation*, the initial approach with migrant students, provided English as second language teaching (now known as EAL—English as an additional language) and some cultural contexts were included, typically of mathematical historical significance. *Accommodation* offered further acknowledgement of cultural background and the inclusion of contexts and tasks that were relevant and meaningful to students. These teaching practices make use of students' 'funds of knowledge' (Moll, Amanti, Neff & Gonzales, 1992) that we now refer to as 'inclusive curriculum'. Ethnomathematics also fits with this approach but goes further so that the mathematics curriculum builds on the mathematical knowledge of traditional and non-Western cultures (D'Ambrosio, 1998). The most recent approach, *amalgamation*, called 'democratic or critical curriculum' is jointly organised by teachers and the community embracing bi-lingual language and bi-cultural teaching. This approach respects and builds on language, cultural knowledge and ways of thinking. More recently, Fraser (2013) has traced the history of feminism to show three similar periods of action for gender justice: *redistribution, recognition* and *participation*.

The Productive Pedagogies framework (Hayes et al., 2000), proposes six main attributes of an equitable mathematics classroom:

- *Equal access.* This concept includes access to the classroom, resources and materials, and also to the discourse of the mathematics lesson—that is, the language and norms of mathematical practice and thinking—such that no one feels left out or marginalised in the classroom.

- *Connected learning.* The teaching and learning program builds on prior knowledge and experiences of students. The program is negotiated with them, and the contexts of mathematical applications and investigations are socially, culturally and politically relevant and empowering for them.

- *Collaborative methods.* The practices of the classroom recognise the importance of discussion and social interaction for the learning of mathematics. Hence, collaborative tasks and practices are valued, and students are encouraged to share their knowledge and skills and to explain their thinking.

- *Supportive environments.* Teachers construct an environment based on the belief that everyone can learn mathematics and establish classroom norms so that students feel safe, free from abuse and respected. They make explicit their expectations for mathematical thinking and practices, and they model and scaffold mathematical thinking in the classroom.

- *Intellectual quality.* Teachers in equitable classrooms have high expectations of their students and engage them in meaningful mathematical thinking. Through the mathematical skills and concepts that they learn, students are empowered to use mathematics in ways that enable them to participate effectively in school, work and the community.

- *Respect for difference.* Materials, problems and tasks reflect the gender, cultural and social diversity of the students in the classroom. The materials are free of gender and cultural bias. They are relevant and respectful of the students' interests, and teachers understand that the real contexts of mathematical problems may be experienced differently by girls and boys and by students from varying social and cultural backgrounds.

Social justice pedagogies: response-able teaching

The social justice approaches outlined above are elaborated and explained using examples drawn from research studies of particular approaches: caring teaching, culturally response-able pedagogy teaching, place-based pedagogy, culturally response-able pedagogy for

Indigenous students and socially response-able pedagogy. Gender response-able approaches are discussed in the next chapter.

Caring teaching practices

Averill and Clarke (2013) observed teachers using caring and respectful practices to provide safe and supportive environments and to enhance intellectual quality. These practices included being well prepared for teaching; directing questions to students who are willing to respond and using revoicing; providing positive feedback; enabling students to elaborate and add to previous responses; helping by prompting and asking, not telling; and giving time for students to think or identify their own error. The approaches are similar to those advocated by Sullivan et al. (2006). In these classrooms, teachers focused on mathematics learning, showed respect to students, were aware of the meaning of physical movement and gesture to communicate thinking, and developed a sense of community and personal responsibility (Averill, 2012).

Culturally response-able curriculum

Knowing one's students is an expectation for quality in mathematics teaching (AAMT, 2006; AITSL, 2014). In a multicultural classroom, teachers need to know about students' prior mathematics knowledge and skills, and their language knowledge. Getting to know students also enables teachers to make connections to contexts that have meaning for them. Equally important is some understanding of the values and attitudes that the students have about learning, and learning mathematics in particular. If they have had some schooling in other countries, they may have different expectations about how a mathematics lesson should be conducted. Students may also be accustomed to presenting solutions to problems in a different way. These differences need to be celebrated rather than minimised in a mathematics classroom, as they illustrate the richness of mathematical knowledge and the flexibility involved in making sense of mathematics.

Terry Tao

Australia's first Field Medal (the Nobel Prize in mathematics) was awarded to Terence 'Terry' Chi-Shen Tao in 2006. He was born in Adelaide and attended Flinders University. He is now a professor of mathematics at UCLA working on harmonic analysis, differential equations, additive combinatorics and other topics.

It is important not to stereotype students of particular ethnicities. See Figures 14.1 to 14.4 and the box above. Knowing one's students enables teachers to remain sensitive to their needs and to engage them more effectively in the mathematics lesson. Teachers are thus able to select problems and contexts that are more engaging for students, as well as select content and tasks that address their mathematical learning needs.

Further, many cultures have a very rich history of mathematics. Investigating the origins and historical development of mathematical ideas and algorithms is one way of including and valuing the mathematics of various cultures. It also enables teachers to challenge the Eurocentric historical record of mathematics (Shan & Bailey, 1994). For example, the Chinese had documented Pythagorean triples long before Pythagoras, a Persian, Al-Khowarizmi, developed two methods for solving equations, *al-jabr* and *al-musqabalah*, and the earliest known proof appears in an ancient Indian text.

When teaching students from a particular cultural group, studying the mathematics of their culture can give them a sense of ownership and provide a bridge to Western mathematics. However, a school-based curriculum that includes examples from traditional culture runs the risk of not only trivialising the mathematics but, more importantly, embarrassing the students from ethnic minorities because they are made to feel primitive, rather than engaging their interest and valuing their culture (Jorgensen, 2014; Lerman, 2006). Therefore, this approach needs to be used judiciously in an Indigenous classroom.

MacGregor and Moore (1991) provide advice on designing teaching and assessment materials for students whose first language is not English. They advise that teachers should read the materials with students and make sure that terms and instructions are understood. Teachers should also train students in techniques for completing assignments, problem-solving tasks and investigations, including the writing styles and formats. Student work from previous years can provide invaluable examples for students. MacGregor and Moore emphasise the need to teach vocabulary (such as *parallelogram* or *domain*) and terms that inform processes to be followed (such as *evaluate, simplify* and *factorise*). Key mathematical terms should be written down and visible, not just spoken. MacGregor and Moore also recommend that teachers assist students to comprehend and interpret word problems and diagrams. MacGregor and Moore document strategies and activities for developing mathematical language, based on the practices of teachers of English as an additional language. Some examples include:

- labelling exercises using key terms on cards
- true/false exercises for mathematical statements written in words
- cloze exercises, where sentences are completed using a list of words (see Figure 14.5)
- problem reconstruction, where steps in the solution process are written separately on cards using symbols and words and students arrange both sets of cards to show the solution process
- mix-and-match cards (e.g., cards to match graphs with written descriptions of their features, or mathematical vocabulary and their meaning)
- cooperative learning problems, where clues or pieces of information about the problem are written on separate cards and distributed among a group of students who have to share their information by reading and explaining and not showing, to solve the problem together.

Fill the blanks by choosing the correct words from the list.

Word list: **dilated reflected rotated translated vertically horizontally**

The basic parabola $y = x^2$ is shown ... as a reference.

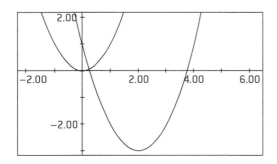

The graph of the basic parabola

has been ....

3 units ....

and 2 units ....

Source: MacGregor and Moore (1991, p. 89).

Figure 14.5 **Cloze task**

> **REVIEW AND REFLECT** : Use the strategies suggested by MacGregor and Moore (1991) above to design a set of language-support activities for a mathematics topic.

Place-based pedagogies

Deficit thinking and shallow understanding of community and culture by teachers contribute to low participation and success (Corbett, 2009; Jorgensen & Lowrie, 2013). Corbett (2009) argued that education in rural settings is disconnected with place and is about 'learning to leave'. He described curriculum and traditional approaches as 'urbanisation of the mind' and argued for recognition and valuing of difference through 'place-based' pedagogy. Place-based pedagogy is about connecting with local concerns, recognising and respecting local knowledge and values, and working collaboratively with the local community (Ell & Meissel, 2011; Wallace & Boylan, 2009). Getting to know your students and forming relationships with the community is paramount.

Many of the projects designed to support teachers in rural and remote schools have focused on using context to enhance students' numeracy skills (e.g., Ell & Meissel, 2011; Goos, Dole & Geiger, 2011; Graham & Pegg, 2010). See Chapter 5, Figure 5.4 for discussion of numeracy content and approaches. Others have used mathematical investigations and applied problems to develop problem-solving skills and deepen conceptual understanding. The *Working Mathematically in a Rural Context* website (Mitchelmore & McMaster, 2009) provides a number of authentic problems concerning agricultural activities in the Riverina, NSW. Goos, Geiger and Dole (2011) describe the development and enactment of a road construction project by a teacher in rural South Australia. Examples of other investigations and problems relevant to rural contexts are included in the chapters on number, measurement and statistics.

Culturally response-able pedagogy for Indigenous students

Mathematics educators who have worked closely with Indigenous people in Australia all make the point that it is important not to stereotype or essentialise Indigenous students—that is, don't think that all Indigenous students are the same (Buckskin, 2000; Jorgensen, 2012b; Matthews et al., 2005). In order to teach Indigenous students, teachers need to learn about Indigenous people, their culture and their ways of knowing, and to examine their own attitudes, beliefs and values about mathematics (Buckskin, 2000; Jorgensen, 2014; Owens, 2015; Perso, 2003). These principles apply to teachers of Indigenous students in urban and regional Australia as well as in remote communities.

Perso (2003) proposes that three aspects of practice need to be included and synchronised for the effective learning of Indigenous students: Indigenous people and their culture; the mathematical understandings of Indigenous students; and explicit teaching of mathematics. In remote Indigenous communities where English is not students' first language, teachers need to be sensitive to language needs and choose appropriate representations and gesture to support understanding (Miller, 2014). Owens (2015) goes further and argues that teachers need to engage with their Indigenous communities, understand their ways of knowing and involve them in decision making. Formal involvement of Indigenous people in the school as teachers or tutors or mentors, and Elders passing on their knowledge, has improved the educational outcomes of students (Buckskin, 2000; McCarthy, 2002). Indigenous leaders are demanding that Indigenous students are able to access challenging and empowering mathematics (Sarra, 2012). Jorgensen (2014) agrees and argues that Indigenous students must be involved in mathematical meaning making.

Cultural inclusion and participation is clearly important for enhancing the learning of Indigenous students. Buckskin (2000) explains that the principle of inclusion involves flexibility, participation of Indigenous people in educational management and delivery, teaching Indigenous languages, and increasing the cultural relevance of the mathematics curriculum. See Chapter 13 for discussion of flexible and targeted teaching approaches.

McCarthy (2002) describes a Year 10 numeracy project, *Where in the world is Spinifex Longifolius?* She was teaching at Tennant Creek, a small town in the Northern Territory, in a school in which half the students were Indigenous. After asking her students to tell her what they thought was missing from their mathematics lessons, she concluded, 'we needed something outdoors, involving technology, that would foster working in teams, something that supported and extended students' literacy skills development, that would make them aware of other cultures around them and that would keep them amused'

(McCarthy, 2002, p. 25). The unit that she designed required the active involvement of Indigenous women from the local Tennant Creek community, respect for Indigenous knowledge of medicine, and was based on the needs and interests of the students. The project is outlined in the box below.

Where in the world is Spinifex Longifolious?

The desired outcomes for the project were:

1. Students will identify the position of and best route to locations of local bush medicines.

    The project involved learning to navigate with a handheld global positioning system (GPS) and then using the information from the unit to produce both formal and informal (mud) maps.

2. Using appropriate hardware and software, students will produce an information package that will communicate the results to different audiences.

    This was suitably vague so as to give the students the opportunity to produce a package with which they were comfortable. Some of the options would be to produce a web-based presentation, a pictorial presentation, a highly literate or formal presentation and some students would produce an audio commentary.

3. The package will include charts, scale maps, diagrams, transcripts, photographs and video footage.

Source: McCarthy (2002, p. 25).

This numeracy project illustrates how the mathematics teacher can value Indigenous students' connection to place and Elders' knowledge and involvement. The *Make it Count* project (www.mic.aamt.edu.au), a national project designed to develop approaches for teaching Indigenous students, also provides resources for teaching numeracy developed through the project. In the Western Region of NSW, the NSW Department of Education and Communities has provided support for a project, *8 Aboriginal Ways of Learning* (www.8ways.wikispaces.com), to improve Indigenous engagement in schools and develop teachers' cultural competence in the classroom. Tyson Yunkaporta, the *8 Ways* project developer, encouraged school ownership of the project, and Aboriginal and non-Aboriginal teachers continue to develop the *8 Ways* framework and teaching and learning resources. The authors of the framework emphasise that this framework is about learning processes rather than learning content and that learning is promoted through:

- connecting through the stories that are shared
- picturing pathways to knowledge
- seeing, thinking, acting, making, and sharing without words
- keeping and sharing knowledge with art and objects
- working with lessons from land and nature
- putting different ideas together and creating new knowledge
- working from wholes to parts—watching then doing
- bringing new knowledge home to help the family/community. (*8 Aboriginal ways of learning: Developments* www.8ways.wikispaces.com/+Developments; Yunkaporta & McGinty, 2009, cited in Owens, 2015)

Each of these points is depicted in an image (see Figure 14.6) that schools use when planning and documenting their learning programs.

A critical element of this framework is the engagement with the Indigenous community to develop understanding of their ways of learning and to involve the Indigenous community in strategies the school is using to ensure successful learning for the Indigenous students. Owens (2015) reported on the changing pedagogical practices of secondary mathematics teachers in a provincial NSW school that was involved in these two projects and another project for school leaders, *Stronger Smarter Learning Communities* (www.strongersmarter.com.au). She reported that the mathematics teachers at the school used the image of river systems to communicate mathematical learning trajectories for parents and made more use of storytelling, links to land, visual representations and doing in their teaching. The *8 Ways* website provides further information on how to use this framework to design learning programs and teachers have contributed examples from their practice.

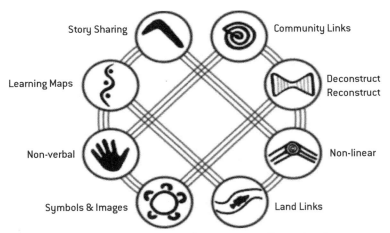

Story Sharing

Community Links

Learning Maps

Deconstruct
Reconstruct

Non-verbal

Non-linear

Symbols & Images

Land Links

Source: http://8ways.wikispaces.com/. ©2009 Allan Hall. Used with permission.

Figure 14.6 8 Aboriginal Ways of Learning

PLEASE NOTE: Ownership of Indigenous Knowledge work belongs to individuals and groups. Seeking permission from the original source is a requirement, out of courtesy. The *8 Way* framework is a NSW Department of Education initiative. Examples of work from communities and individuals guided by this framework should be respected at all times, so please follow protocol. If you are posting your own work on this website, please be mindful this is the public domain and not a NSW Department of Education website (this site is managed by the NSW Department of Education staff from the Wagga Network of Schools). Please drop an email or call 02 68413852 or 02 68701090.

REVIEW AND REFLECT : Observe the following video of a mathematics and history teacher discussing his reconciliation approach to teaching Indigenous students at his school.

AITSL Illustration of Practice: Promoting Reconciliation www.aitsl.edu.au/australian-professional-standards-for-teachers/illustrations-of-practice/detail?id=IOP00061.

Discuss his approach and compare it with the models of practice described in this section of the chapter. What approach will you use in your practice?

Socially response-able pedagogy: democratic or critical curriculum

In Chapters 2 and 13, we described Boaler's (1997a, 2002b, 2008b; Boaler & Staples, 2008) research on approaches to teaching and learning mathematics. The schools in these studies were located in low socio-economic and diverse cultural communities. In these studies, the students in the reform classrooms worked in groups to complete mathematical investigations of phenomena or topics that were connected to their experiences and environment.

In various countries around the world, teachers have adopted this method of teaching mathematics to engage students and enable them to develop and use mathematical knowledge to solve problems that they will encounter at school, at work and in the community (e.g., Frankenstein, 1997, 2001; Gutstein, 2003; Skovmose & Valero, 2002). Skovmose and Valero (2002) argue that teachers should use mathematical contexts of social significance to teach students about decision making.

Gutstein (2003) found that using mathematics to examine social issues in the community, including racism and discrimination, can be particularly empowering for marginalised students. He used projects in his middle-years classroom to gather data about issues of personal and social relevance to his students, such as teenage pregnancy and local housing programs. (See Chapter 5 for an example from Frankenstein, 2001 and see Gutstein & Peterson, 2006 for an excellent collection of teaching resources based on this work.) Some other examples are provided in Table 14.1. The first problem has personal relevance for students; the second had relevance for the community. The content of mathematical inquiry consistent with a democratic or critical curriculum approach contributes to the development of critical mathematical literacy for active citizenship.

An important aspect of the democratic approach is student involvement in making decisions about their learning, as illustrated in the *Where in the world is Spinifex Longifolious* example above. Typically, the teacher involves the students in making decisions about the topic for the project and the selection of a problem, or works through a process of investigation with the students to generate and pose a problem for solving. The teacher may negotiate with the students the means for conducting the project, and in this way, students will learn about the process of mathematical inquiry—including problem solving, research and mathematical reasoning. They will also discuss with students their

Table 14.1 Mathematics problems of personal or social relevance

Maths and booze

If you were planning to drive to and from a party and stay from 8.00 p.m. to 1.00 a.m., what number, capacity and strength of alcoholic drinks could you drink in that time and have a blood alcohol content below 0.05 any time after leaving the party? Explain. Show your working. Illustrate your blood alcohol content level over the time period in a graph.

Storm surge wall

How high should the storm surge wall need to be to protect the town from flooding if the cyclone hit the town at high tide?

Use the local data about high tides, the data from the day that the last cyclone hit the town and collect other data about the height above ground level of buildings in the town to model and solve this problem. Explain and justify your recommendation to the town council.

Source: Atweh and Ala'i (2012).

expectations, the assessment criteria and the ways in which to demonstrate their mathematical understanding.

Teachers who use this approach observe that their students show improved engagement with their mathematics learning, and improved achievement and attitudes to mathematics. Use of projects or extended problems can produce equitable outcomes (Boaler, 2002b, 2008b). Boaler, who studied teachers using this approach in the UK and the USA, found that successful teachers—that is, teachers who produce equitable outcomes—use particular practices in conjunction with this approach. These practices include:

- discussing the problems or project thoroughly with students when they are introduced so that the vocabulary and meaning of the problem or inquiry is understood
- making real-world contexts accessible to students—that is, recognising that girls and boys from different social, cultural and linguistic backgrounds encounter contexts differently, and taking this into account when explaining the problem
- providing more time working in groups on the problem than traditional lessons
- encouraging students to explain and justify their thinking
- supporting groups to present their solutions and encouraging students to question and learn from each other.

> **REVIEW AND REFLECT :** Work in a group to develop an open-ended investigation or design a problem about a social issue that is relevant to a group of students. Identify the resources and materials needed and design an assessment rubric.

Conclusion

To enable social justice in mathematics learning and teaching, education policy, mathematics curricula and teaching practice must recognise the needs of the most marginalised members of the community, and resources must flow to groups of social disadvantage. Schools and teachers must engage with their communities and enable them to participate in curriculum and teaching decisions that affect their children and community. Teaching and learning approaches should enable learners to be empowered through mathematical literacy and to develop agency as mathematicians. This requires teachers to be reflective and reformist.

Recommended reading

Atweh, B., Graven, M., Secada, W. & Valero, P. (2011). *Quality and equity agendas in mathematics education.* Dordrecht, The Netherlands: Springer.

Averill, R. (2012). Caring teaching practices in multi-ethnic mathematics classrooms: Attending to health and well-being. *Mathematics Education Research Journal,* 24, 105–128.

Boaler, J. & Staples, M. (2008). Creating mathematical futures through an equitable teaching approach: The case of Railside School. *Teachers College Record,* 110(3), 608–645.

Gutstein, E. & Peterson, B. (Eds). (2006). *Rethinking mathematics: Teaching social justice by the numbers.* Milwaukee, WI: Re-thinking Schools Ltd.

Jorgensen, R. (2014). Social theories of learning: A need for a new paradigm in mathematics education. In J. Anderson, M. Cavanagh & A. Prescott (Eds), *Proceedings of the*

37th annual conference of the Mathematics Education Research Group of Australasia (pp. 311–318). Sydney: MERGA.

Owens, K. (2015). Changing the teaching of mathematics for improved Indigenous education in a rural Australian city. *Journal of Mathematics Teacher Education*, 18(1), 53–78.

Teese, R. (2013). *Academic success and social power: Examinations and inequality.* Carlton South: Australian Scholarly Publishing.

Resources and websites

8 Ways of Aboriginal Learning: http://**8ways**.wikispaces.com/

Make It Count: http://mic.aamt.edu.au/

Maths300: www.maths300.com

Working Mathematically in a Rural Context: www.wmrural.net/activities.php

CHAPTER 15
Gender equity and justice in mathematics teaching and learning

Gender inequality in society, including discriminatory attitudes and behaviour, has re-emerged as a significant social and education issue in the second decade of the 21st century in Australia and elsewhere around the world. You may be surprised to know that in the 1990s, Australia was among world leaders in equitable gender participation and outcomes in mathematics education. Attending to gender differences in participation and achievement, and innovation in teaching practice and curriculum in the 1980s and 1990s did close the gap, but careful reading of the research shows that these improvements did not occur across all sections of the community (Vale, 2010).

At the turn of the century, the Australian government shifted its policy focus from improving the education of girls to considering the needs of boys (Parliament of the Commonwealth of Australia, 2000). For many social researchers, this signalled 'the end of equality' (Summers, 2003) and the advent of more conservative social policies. The final report from this inquiry did identify teaching strategies appropriate for boys in some fields of study, but it also stated that 'it is important to remember that while improvements to education outcomes for some groups of girls are real they have eluded many other girls' (Parliament of the Commonwealth of Australia, 2002, p. 18). Since then, Australian governments have agreed that all students should have access to high-quality schooling irrespective of gender and sexuality (MCEETYA, 2008).

It is important to remember that the debate and research concerning gender equity has an international context, especially in the era of globalisation. Indeed, at the beginning of the 21st century, two-thirds of the world's population who do not have access to, or are excluded from, a basic education are girls (United Nations Educational, Scientific and Cultural Organization, 2003). In the developed nations, the gap is closing but equity has not yet been realised (Vale, Forgasz & Horne, 2004). Globally, of the 67 countries participating in a study of mathematical literacy for 15-year-olds in 2012, boys outperformed girls in 57 countries; and girls significantly outperformed boys in three countries, Thailand, Malaysia and New Zealand (Thomson, De Bortoli & Buckley et al., 2013).

In this chapter, we describe the gender gaps in access to, and outcomes of, mathematics education, and present some of the explanations for these gaps. We consider practices in mathematics classrooms and curricula that affect the ways boys and girls see themselves as learners of mathematics, with consequent implications for attitudes, achievement and participation. Current understandings of teaching and learning practice that promote gender justice in mathematics are then presented. We begin the chapter by defining gender equity and gender justice.

Meanings of gender equity and gender justice

Over time, there has been a shift in focus on gender issues in mathematics from a concern about gender equality to gender equity and social justice and ethical practice (Atweh, Vale & Walshaw, 2012). Fraser (2013) describes three notions of feminism that reflect these shifts of focus and action over time: redistribution, recognition and participation.

Equality

Fennema (1995) described three aspects of equality: equal opportunity, equal treatment and equal outcomes. Equal opportunity is about access. Both girls and boys need to be able to participate in mathematics. This involves removing external barriers to girls' participation in schooling, as well as barriers constructed by schools, such as streaming or setting policies and pathways, and timetable structures that restrict girls' access to mathematics learning. It also means providing equal resources for mathematics learning in terms of learning time, teachers' time, materials and equipment, such as computers or handheld digital devices. During the 1980s and 1990s, researchers were occupied with equal treatment. They began to look closely at classrooms and discovered that teachers did not interact with girls and boys for the same amount of time, nor with the same quality of interaction or feedback (Jungwirth, 1991; Leder, 1993; Vale, 2003), and found inequitable representations of women and girls in textbooks and teaching materials (McKimmie, 2002). Fennema (1995) argued that equal access and equal treatment were not sufficient for gender equity. The pursuit of equity also involves a commitment to 'closing the gap' in outcomes, where outcomes include mathematical achievement, participation, retention and attitudes.

Equity and social justice

More recently, educators and researchers have also argued that equal access and equal treatment are not sufficient to overcome gender gaps and gender injustices in schooling (Anthony & Walshaw, 2007; Atweh, Vale & Walshaw, 2012). Equity discourse is concerned with fairness and reducing and eliminating difference between females and males, whereas diversity discourse is concerned with recognising and respecting difference as justice. An ethical view of social justice acknowledges social activity and interactions between individuals, where social justice is a moral obligation for care, respect and fairness rather than an act of charity or goodwill (Atweh, 2011; Averill & Clark, 2012). However, while it may be possible to establish gender-just classrooms, this may not be sufficient to establish equity without challenging unjust social and education structures.

Evidence of gender differences and gender gaps

In this section, we present a summary of the findings from recent studies into gender differences in achievement, attitude and participation. Researchers are now mindful that gender gaps in mathematical outcomes may vary for students of different socio-economic backgrounds. We invite you to explore some of these issues for yourself.

Participation in senior secondary mathematics

Gender inequities in mathematics are perhaps most obvious in the difference in participation rates for males and females for some mathematics courses in secondary schools and in tertiary mathematics courses (Forgasz, 2006; Mack & Wilson, 2014 Wienk, 2015). Women's increasing participation in non-traditional tertiary courses, including mathematics, appears to have reached a plateau, leaving a large gender gap. In 2014, only one-third of students taking a mathematics unit in their tertiary course were female (Wienk, 2015).

Since the turn of the century, the proportion of Australian students studying a mathematics subject in Year 12 has remained fairly constant: 80 per cent in 2001 and 82 per cent in 2013 (Wienk, 2015). While participation rates for girls and boys in the general mathematics subject was roughly equivalent in 2013, gender differences in participation in intermediate mathematics favoured boys, 20.7 per cent compared to 17.6 per cent of girls, and in advanced mathematics, 12.7 per cent compared to 6.7 per cent of girls (Wienk, 2015, p. 12). The shift away from studying advanced or intermediate mathematics (35 per cent in 2011 to 29 per cent in 2013) has been more pronounced for girls (Forgasz, 2006; Mack & Wilson, 2014; Wienk, 2015). Mack and Wilson (2014), reporting on participation of intermediate and advanced Year 12 mathematics subject in NSW, found that 'boys' total participation in at least one maths and one science subject in 2001 was 19.7% and in 2013 18.7%; not a substantial shift. Girls' total participation in 2001 was 16.8% and by 2013 this dropped to 13.9%' (p. 40).

> **REVIEW AND REFLECT :** Use the internet to locate the enrolment data for Year 12 mathematics subjects in your state. How do the participation rates for males and females differ according to the level of difficulty of the mathematics subject? Discuss possible explanations for this trend with colleagues.

Mathematics disposition and attitudes to mathematics

Various measures of attitudes have played a key role in research in gender and mathematics. Attitudes have been investigated in order to explain gender differences in performance and achievements, but they are an outcome of mathematical learning as well. Studies by Fennema and Sherman (1977) pioneered research into gender differences in attitudes, and their research has been replicated in many studies since.

Fennema and Sherman (1977) investigated secondary students' self-reported attitudes towards mathematics, and here researchers have continued to find gender differences that favour males. These attitudes include confidence in mathematics, perceived usefulness of mathematics, perception of ability, interest, persistence and intention to continue further study (Forgasz, 1995; Watt, 2000). However, differences are diminishing, and gender differences for confidence in, and liking of, mathematics were not statistically different for Year 8 students in the 2011 TIMSS study. However, findings from the PISA study showed a significantly higher proportion of boys than girls believed they were competent at mathematics, and boys scored significantly higher than girls did on mathematics self-efficacy (Thomson, de Bortoli & Buckley, 2013).

Previous studies have found that males are more likely than females to attribute their success to ability and females more likely than males to attribute their failure to lack of ability or task difficulty (Leder, 1993).

Australian researchers have also explored attitudes regarding the use of technology in mathematics (Forgasz, 2002; Tan, 2012; Vale, 2002a, 2003; Vale & Leder, 2004). Boys were also more likely to believe that computers would aid their mathematics learning and to enjoy the use of computers in mathematics. Tan (2012) found significant gender differences in ways of knowing and learning mathematics (Belencky et al., 1986), with males scoring higher on Connected Knowing—Deep Approach and girls scoring higher on Surface Knowing—Surface Approach. She also established association between these ways of knowing and the use of graphics and CAS calculators (Goos, Galbraith, Renshaw & Geiger, 2000). Digital tool 'as Master' was associated with Surface Knowing and digital tool 'as Collaborator' with Connected Knowing. Previous studies have shown that the positive attitude that boys display towards the use of computers is more strongly correlated with the aspiration to do well with technology than to excel in mathematics.

Mathematics achievement

A trend towards equal outcomes in mathematics achievement can be seen in studies of gender difference in achievement in developed nations, including Australia, towards the end of the twentieth century (Lokan et al., 1996, 1997). However, the most recent international studies have confirmed that gender differences in achievement for Australian secondary school students have increased since the turn of the century. TIMSS measures mathematics achievement of Year 8 students for knowing, applying and reasoning, whereas PISA measures achievement in mathematical literacy including problem solving. Both TIMSS and PISA found gender differences in outcomes favouring males in each cognitive domain and overall (Thomson et al., 2012; Thomson, de Bortoli & Buckley, 2013). These differences were significant for 14-year-old students and the gender differences were higher than the OECD average for 14-year-old students. These studies also found that males continued to be more highly represented in among the highest achievers: 12 per cent of males compared to 7 per cent of females among 14-year-olds (Thomson, de Bortoli & Buckley et al., 2013) and 10 per cent of males compared to 7 per cent for females among Year 8 students (Thomson et al., 2012). Analysis of the *NAPLAN* scores across Australia for Year 7 and Year 9 students show slight gender differences in average achievement favouring males but significantly higher proportions of males than females in the highest achievement band (Leder & Taylor, 2010; Wienk, 2015). The pattern of males being more highly represented than females in the highest achievement levels is also found in senior secondary mathematics achievement for all three mathematics subjects offered in Victoria (Forgasz & Hill, 2013). Forgasz and Hill (2013) found the gender gap increased with level of difficulty of the subject and the highest achieving students were males attending metropolitan schools in high socio-economic communities.

Studies have also investigated gender differences for particular content. Early studies identified gender differences in spatial visualisation as contributing to gender differences in mathematics (Fennema & Tartre, 1985) but these findings have not been consistent. The first PISA study found boys did significantly better than girls did on the questions that required them to interpret information that was presented in diagrams (Lokan et al., 2001). More recently, the 2011 TIMSS study found that Year 8 boys performed better than girls did on items about chance and data and number (Thomson et al., 2012), and Leder and Taylor (2010) found that girls outperformed boys on geometry and spatial perception items in the Australian Mathematics Competition. However, a study of out-of-school digital game playing found boys more likely to play games using map content and girls more likely to play games involving problem-solving strategies (Lowrie, Jorgensen & Logan, 2013). The strategies used when playing these games revealed boys' use of spatial and navigational tools, and girls' superior interpretation of text and other representations. A high demand on graphical interpretation was also a characteristic of the problems in which Year 12 boys performed better than girls when using graphics calculators in a Western Australian study (Forster & Mueller, 2002).

The type of assessment also appeared to have a major influence on gender differences. In post-compulsory mathematics, males perform better than females on timed short-answer questions, while extended problem-solving assignment tasks favour females; this pattern of gender difference is consistently reported in the literature (Leder, 2001). Forster and Mueller (2001) proposed that changes to content, and in particular the emphasis given to the use of technology in mathematically demanding subjects, may have contributed to the drift away from participation in the calculus subject in Western Australia, especially by girls.

The question of gender difference in mathematics achievement is thus a complex one. It depends on the content of the assessment tasks, the nature of mathematics knowledge and the mathematical skills being assessed, and the conditions under which assessment is completed. The inconsistencies in gender differences within age groups and across cultural groups show that the gap in achievement is not explained by sex alone.

Gender stereotyping and exclusive practices

Mathematics as a male domain

One attitude that Fennema and Sherman (1977) initially explored was the idea that students gender stereotyped mathematics as masculine or a 'male domain'. At the turn of the century, Leder and Forgasz (2000) surveyed a large number of students in Years 7–10 and found that most students did not gender stereotype mathematics. Students considered mathematics to be important and interesting for both girls and boys, that their parents believed in the importance of mathematics for girls and boys, and that teachers spend the same time with girls and boys in classrooms. Girls were considered more likely to be good at mathematics, to enjoy it and to think it important to understand the work. More recently, in the 2011 TIMSS study and 2012 PISA study Australian boys valued mathematics more than girls and this difference was significant (Thomson et al., 2012; Thomson, de Bortoli & Buckley et al., 2013). It seems that the gender-stereotyped attitudes of critical 'others' in students' homes and school and the beliefs of the general community contribute to students' gendered perceptions of the usefulness of mathematics (Forgasz & Leder 2011; Forgasz, Leder & Tan, 2014).

REVIEW AND REFLECT : For each of the statements below, state whether you think males (M) or females (F) are more likely to display the particular belief or behaviour. Use ND if you think there will be no difference.

Item	M or F or ND
Think it is important to understand the work.	
Think maths will be important in their adult life.	
Are asked more questions by the maths teacher.	
Maths teacher thinks they will do well.	
Find maths difficult.	
Think maths is interesting.	
Parents think it is important to do well.	
Teachers spend more time with them.	
Are good at using computers for learning maths.	
Think it is important for their future jobs to be able to use computers for mathematics learning.	

Source: Selected items from Leder and Forgasz (2000) and Forgasz (2002).

Compare your responses with those of other colleagues and with the findings reported below.

The influence of significant others is revealed in recent research. For example, a survey conducted face to face and online found that members of the public viewed mathematics as a male domain. It also indicated that these beliefs, along with survey respondent's participation and achievement in mathematics, was influenced by their teacher (Leder, et al., 2014). Carmichael (2014) asked parents how well their child was progressing in mathematics and found that 'parents of boys tended to have more positive perceptions about their son's mathematics achievements than parents of girls' (p. 124). An international online survey involving people from 81 countries found significant differences by country on gendered perceptions of mathematics capability, parents' gendered perceptions of mathematics and teachers' gendered perceptions. In most countries, but not all, these differences favoured males. Differences were highest in China, and Australia was one of six countries that did not record stereotyped perceptions (Leder et al., 2014). An intervention program conducted in the USA for parents aimed at enhancing their children's motivation to pursue STEM careers increased course taking for high-achieving daughters and low-achieving sons, but not low-achieving daughters (Rozek, Hyde & Svodoba, 2015). We might expect that gender stereotyping of mathematics as a male domain would impact on girls' decisions to continue studying mathematics, but studies also show that gender stereotyping also negatively impacts on female students' achievement and motivation to improve their performance (Fogliati & Bussey, 2013). The decline in motivation to improve is compounded when female students receive negative feedback.

Mathematical discourse: classroom environment and resources

Boaler (1997b, 2002b) has found that particular teaching approaches have different effects on the attitudes and performance of girls and boys. A number of studies have examined the norms and behaviours of teachers and students in mathematics classrooms and found differences among girls and boys (Barnes, 2000a; Chapman, 2001; Vale, 2002a, 2003). Each of these studies showed that the culture of mathematics and mathematics classrooms in the main advantages boys and particular groups of boys, and disadvantages girls and particular groups of girls.

In one classroom in which students were using laptop computers regularly in mathematics, the teacher interpreted one group of boys' interest and competence with computers as achievement in mathematics, while other boys and girls felt 'overpowered' by them (Vale, 2003). These boys competed for attention in general classroom discussions led by the teacher, denigrated girls' and other boys' mathematics and computer achievements, took over other students' computers to solve the problem or do it for them, and harassed students verbally and physically. In another classroom, in which students were using computers in small groups, Barnes (2000a) found that the attitude and behaviour of the dominant boys obstructed the learning of others, including the girls, and limited their capacity to learn in a small group on collaborative tasks. In the same classroom, another group of boys, who had the 'power' of technical and mathematical competence, could not take advantage of the cooperative learning environment because of their poor communication skills. Chapman (2001) shows that 'triadic dialogue'—that is, the question–answer–feedback (see Chapter 3 for an example) commonly used in a traditional classroom—advantages the dominant boys and disadvantages many other boys in the classroom. She advocates that teachers ought to adopt a 'language-sensitive approach' and use a range of literacy strategies and language representations in mathematics in order to include all boys and girls.

Walden and Walkerdine (1985) argue that girls are often placed in a 'no-win' situation because their success is taken to be achieved in the wrong way—that is, through rote learning, hard work, perseverance and carefulness rather than natural talent, flexibility and risk-taking. Girls with these 'masculine' attributes are judged, or perceive themselves to be judged, as not feminine. The following case study (see the box below) illustrates this point. It is taken from a study of a mathematics class that worked in a computer laboratory for one of their mathematics lessons each week (Vale, 2002a). The girls in this class were not homogenous and had differing attitudes to the use of computers in mathematics. Beckie was one of the girls who took risks and interacted with the computers in ways more usually associated with masculine culture, but she resisted a 'geek girl' identity and challenged the passive 'good girl' feminine identity.

Beckie: a case study

Beckie, the only Year 8 girl in the class who owned a laptop computer, sat with the boys who owned and brought their laptops to class. However, she did not bring her laptop to school but used a PC in the laboratory. She tutored other students about the software and mathematics, and collaborated with them to solve problems they encountered. For example, on one occasion, Beckie argued with Colin about the order of operations for solving the equation that was set for their slide show presentation. Instead of allowing them to work it out together, the teacher intercepted their argument and told them the answer to silence their noisy debate.

During off-task interactions with other students—and especially with boys—Beckie exchanged negative personal insults. The teacher monitored and managed Beckie's behaviour: 'Beckie, sit down and do some work'. She was one of two girls in the class to dominate the requests for assistance by the teacher. She would interrupt and call out on numerous occasions during the lessons that were observed. The teacher also held extended interactions with Beckie in which they argued about the mathematics or the software functions. For example, in the following episode, Beckie wanted to know whether she needed to record on a slide the next operation in the solution of the equation:

Beckie: Yeah, is that right?

Teacher: Yeah, that's right. Now you do that. Let's go back and highlight that little bar there and click on the underline.

Beckie: OK.

Teacher: Yeah. Is that the first step?

Beckie: Yep.

Teacher: If you want to put something down the bottom, you can write 'add five to both sides'.

Beckie: But that's what I did.

Teacher: No, you just write it. Add five to both sides.

Beckie was confused by the teacher's instruction. The teacher emphasised aesthetics in the presentation and description of the solution process, but Beckie was concerned with the method of solution and the number of steps to be included in the slide show.

Beckie appeared to enjoy the lessons and to be confident using computers. During off-task activity, she accessed and used other software. On three occasions during one lesson, Beckie praised her own work and sought praise from the teacher: 'My presentation's fantastic, sir'. On each of these occasions, there was no response from the teacher.

Later, during an interview, the teacher revealed that he regarded Beckie as a low achiever who required more of his attention.

REVIEW AND REFLECT : Why might the teacher have formed these views about Beckie? How else could the teacher have responded to Beckie's behaviour and needs?

Researchers have also analysed curriculum materials and found that males and females, and their interests and occupations, were not equally represented in materials and problems, and were typically represented in gender-stereotyped scenarios. A study of textbooks used in secondary schools in Victoria at the turn of the century found that references to males outnumbered references to females in two out of the three textbooks analysed (McKimmie, 2002). Studies from around the world continue to show bias.

REVIEW AND REFLECT : Analyse the content of a current mathematics textbook for gender bias. Collect and present data on:

- the number of males and females as protagonists in problems and examples
- the number of photographs and drawings of males and females
- the occupations of males and females in problems (are they stereotyped or non-stereotyped?)
- the race and ethnicity of males and females depicted in the textbook.

Gender justice in mathematics teaching

Over the decades, many researchers have sought explanations of gender differences in mathematics outcomes and identified approaches for gender equity. By tracing the historical development of gender awareness and theories to explain gender differences, we can show how current gender equity practice has evolved and what this practice involves. We have used the historical and theoretical schema of Kaiser and Rogers (1995), Jungwirth (2003) and Fraser (2013) to organise the main ideas regarding gender equity in Table 15.1. The dates in the table are a rough guide to the period in which these ideas and practices were being explored. The theories recorded in the table are the ideas being proposed to explain gender differences at the time, and the paradigms refer to the beliefs held by researchers and educators. The pedagogical practices summarise the related curriculum, materials or teaching approaches. Current thinking on best practice can be seen as a merging of recommended approaches in ways that redistribute resources, recognise diversity and provide for participation in making decisions about learning (Fraser, 2013). In the following sections, we discuss research studies and examples of curriculum and teaching practice to promote gender equity and social justice.

Table 15.1 Historical development of theories and practices for gender equity

Period	Theory	Paradigm	Pedagogical practice
1975–1980s	Deficit theory *Women have less talent, skills and interest for mathematics than men.*	Liberal progressive *Women are equal to men if given an educational environment to develop talent, skills and interests.*	Redistributing resources *Single-sex classrooms and programs; focus on enhancing particular knowledge and skills for women, e.g., spatial skills; equal treatment in co-educational settings.*
1980s–1990s	Difference theory *Women have different experiences, skills and interests with respect to mathematics from men.*	Radical feminist *Women's experience and knowledge of mathematics should be acknowledged and valued.*	Gender inclusive *Change the curriculum and classroom practice so that the things women and girls know about and are good at are included and women and girls can build on their strengths.*
1990s–2010s	Gender construction *Gender identity and the distribution of power is constructed through our social interactions; this may change in different environments.*	Post-modern *Gender is learned but not fixed; differences among women and among men should be recognised and respected.*	Learner-centred practice *Learners' interests and needs drive the teaching and learning program in a non-violent, caring and supportive environment.*
2010–present	Critical theory *Economic, social and political power is not equally distributed for women and men within and across nations.*	Emancipatory feminist *Equity and social justice to be achieved through redistribution of economic power, recognition of diversity and representation in decision making.*	Participatory practice *Learning needs and interests of all girls are recognised; girls participate in making decisions about their learning and resources are redistributed to address these decisions.*

Redistributing resources to promote equity

One of the initial strategies for providing access to high-quality mathematics education for women and girls was the promotion of single-sex schools and single-sex mathematics classes in co-educational schools. Forgasz et al. (2000) reviewed these studies of single-sex classrooms and reported that single-sex interventions, in themselves, did not achieve equity. Rather, teachers' beliefs and behaviours were more important. The views of parents about these programs were conflicting. Some parents were concerned that these programs were disadvantaging their boys, whereas other parents viewed them as an opportunity to improve the outcomes for their sons. The needs and interests of girls became lost in the policy settings at the turn of the century, as concerns about classroom management overrode needs for equal access for girls through single-sex classrooms or classes with an equitable distribution of girls and boys. Research suggests that teachers should use single-sex or equal-sex distribution when enacting group work activities in mathematics lessons.

With government attention on the importance of STEM for economic development, awareness-raising programs, replicating 'girls can do anything' and 'maths multiplies your choices' of the 1980s and 1990s, have re-emerged as strategies for interesting and engaging women and girls in mathematics. The 'choose maths' project (www.schools.amsi.org.au/choose-maths/) includes a career awareness campaign for women in mathematics using role models of women in mathematics to inspire girls to aspire to work in the STEM field. The project will provide awards to girls, teachers and schools for inspiring, motivating and preparing students to engage, enrol and achieve in mathematics.

In the 1980s and 1990s, educators focused intervention strategies on developing girls' spatial visualisation to support their mathematics learning, for example, *Space to Grow* (Wettenhall & O'Reilly, 1989). As discussed above, recent research shows spatial visualisation skills should not be assumed a weakness for all girls; rather, formative or diagnostic assessment should be used to identify the readiness for mathematical learning and learning needs of girls and boys. Research presented above suggests that formative assessment should focus on girls' readiness and support needs for using digital tools for mathematical inquiry and problem solving.

Recognising gender difference and diversity: inclusive curriculum

The deficit view of girls was criticised for assuming that there must be something wrong with girls. Instead, researchers argued that the curriculum and teachers' practices explained gender inequities, in that the content, methods of teaching and assessment practices favoured males (Boaler, 1997b, 2002a, 2008a; Burton, 1995). Further, feminist researchers observed that women and men learned in different ways (Belencky et al., 1986; Boaler, 1997b). Burton (1995) argued that inclusive mathematics would be humanising and value cultural, social and personal connections, as well as different methods of solution and forms of proof and different ways of thinking. In Chapter 2, we discussed two different pedagogical approaches investigated by Boaler (1997a). One classroom followed a traditional approach, the other an inquiry approach in which students worked on projects at their own pace. Boaler (1997a) also examined the responses of girls and boys to these approaches and, as a result, argued that the mathematics curriculum was the cause of girls' disaffection with mathematics. She found that both boys and girls preferred the mathematics program that enabled them to work at their own pace, but their reasoning was different. Boys emphasised speed and accuracy, and saw these as indicators of success. Conversely, girls valued experiences that allowed them to think, develop their own ideas and work as a group; they were concerned with achieving understanding. She argued that boys' preferences made them more able to adapt to the competitive environment of a traditional text-based mathematics classroom. The positive outcomes for girls' learning and disposition of using group work and inquiry approaches to learning have been confirmed in a range of other studies (Boaler, 2008b; Boaler & Staples, 2008; Brown, 2011).

Teachers became aware of gender bias in the content of mathematics curriculum, teaching materials and assessment methods, and early efforts to develop inclusive curriculum focused on developing units of work in which the content, language used and methods were consistent with the research on girls' preferences. Examples included *Investigating Change* (Barnes, 1991) and *Up, Up and Away* (Vale, 1987b). A key feature of these materials was an emphasis on group work, discussion and explanations of concepts, and relating mathematics to real situations. These materials were consistent with the development of mathematical reasoning, problem solving and modelling and constructivist approaches to teaching and learning. One example of inclusive materials that were developed in the 1980s is provided in the box below. This task continues to have relevance as the media regularly reports such incidents.

Baby and the heatwave

It was a February heatwave. Michael Jones was driving to the shops with his six-month-old son. He parked his car, grabbed the shopping list, looked at his son who was now asleep and thought 'I'll only be about 20 minutes, I won't wake him, I'll leave him in the car'. So he wound up all the windows, locked the doors and went off to do his shopping.

A little while later, on returning to the car, he saw someone smashing in the side window. He ran to the car. 'What do you think you are doing', he cried, 'trying to steal my son?'

'Steal him?' said the stranger. 'I'm trying to save his life.'

Why did the stranger think the baby's life was endangered? Was it? If so, would Michael Jones have been as unsafe in the car under the same conditions? To answer this question you will model a baby's body and adult's body to find out about the relationship between volume and surface area.

Source: Year 11 modelling, Curriculum Branch, Department of Education, Victoria.

Difference theory has been criticised for essentialising women and girls—that is, for assuming that their interests and preferred ways of learning are homogenous. The use of real applications in mathematics was thought to be universally attractive for girls, but this does not accord with the actual preferences of successful female research mathematicians (Day, 1997). These women were drawn to mathematics because of the creative and intuitive aspects of the discipline. Some were attracted to the abstract concepts of mathematics and others to the applicability of mathematics to problems with social benefit. Another study of women who completed doctorates in mathematics or mathematics education later in life found that their work experiences raised their mathematical curiosity (Harding et al., 2010). Women have contributed to the development of mathematical knowledge over the centuries though their contribution is rarely acknowledged in textbooks. Finally, in 2014, Maryam Mirzakhani, a professor of mathematics at Stanford, became the first woman to win the Fields Medal, the Nobel Prize in mathematics.

Source: Carey (2014).

Figure 15.1 **Maryam Mirzakhani, the first female Fields Medallist**

Maryam Mirzakhani was born and raised in Tehran, Iran. In primary school she aspired to be a writer. In high school, she discovered that she enjoyed problem solving and providing proofs: 'It is fun—it's like solving a puzzle or connecting the dots in a detective case,' she said. 'The more I spent time on maths, the more excited I got'.

She won gold medals at both the 1994 and 1995 International Math Olympiads and completed her undergraduate degree at Sharif University of Technology in 1999, and her doctorate at Harvard University.

She has worked in a number of mathematics fields including algebra, calculus and complex analysis, but her innovative approaches in the field of hyperbolic geometry earned her the Fields Medal. Her mathematical work contributed to the development of 3-D printers.

REVIEW AND REFLECT : Find out about famous female mathematicians, for example, Hypatia, Emilie Du Chatelet, Sophia Kovalevsky (Kennedy, 1983), Maryam Mirzakhani. Alternately, interview a female mathematician at your institution.

What attracted them to mathematics and what challenges did they face?

What are their special interests or research activity and what contribution have they made to mathematics knowledge?

Compare their interests and experiences with your female colleagues who are training to be secondary mathematics teachers.

Participating: gender response-able curriculum

More recent inclusive teaching practices have engaged students in making decisions about their learning by encouraging and enabling them to take responsibility for their learning in junior secondary and vocational pathway subjects and courses (Atweh & Ala'i, 2012; Reilly, Parsons & Bortolot, 2010; Tanko & Atweh, 2012). These developments build on pedagogical and curriculum approaches of negotiating mathematics learning with students that were used in alternate senior secondary curricula (e.g., Vale, 1987a), inquiry-based curricula such as the International Baccalaureate, and mathematics projects formerly used as assessment tasks in some states. In these classrooms, students participate in defining their learning objectives, choosing learning tasks and/or designing mathematics projects and inquiries, as well as selecting support materials or tools to aid their learning. Girls therefore are able to engage in mathematical inquiry related to their interests and aspirations. Proposed changes to assessment practices currently under development in the senior secondary curriculum in some jurisdictions will make it possible for girls to have more say in the problem-solving and modelling projects in these courses in the near future. If enacted successfully, it may improve girls' participation in mathematics as it has done in the past.

Gender equity and social justice in practice

The studies discussed in this chapter have shown that the beliefs of the teacher, the nature of classroom tasks, the nature of mathematical discourse and the practices of the immediate working group often place males in an advantageous or privileged position within the mathematics classroom and the discipline. Teachers need to be aware of the needs and preferences of all students and recognise between- and within-gender differences. Given that a range of needs and learning preferences exist among the boys and girls in the classroom, teachers need to find out about the learners, be flexible and use various approaches in their teaching. One single approach will not suit all:

> Setting up equitable arrangements for learners means different pedagogical treatment and paying attention to different needs resulting from different home environments, different mathematical identifications and different perspectives. (Anthony & Walshaw, 2007, p. 10)

Teachers need to be aware of the learning behaviours that they value and be explicit about the mathematical understandings and practices to be learned and demonstrated in the classroom. They also need to employ effective classroom management strategies to ensure that everyone in the classroom is supported and valued. Goodell and Parker (2001) list 12 practices for teachers and curriculum designers to follow to create a 'connected equitable mathematics classroom (CEMC)':

- All students have access to academically challenging mathematics curricula.
- Students are encouraged to develop confidence in their mathematics ability and positive attitudes to mathematics.
- Basic skills are developed that will enable students to be mathematically literate in the world outside school.
- The learning environment encourages students to develop their own voice and construct their own knowledge.
- Teachers have high expectations of all their students.
- Teachers connect mathematics with the real world.
- Teachers are able to recognise and act on inequities in their classroom.
- Teachers use a variety of teaching and assessment practices.
- The curriculum is designed within a social and cultural context, challenges stereotypes and values the contributions of women and minority groups.
- The curriculum includes real-world problems.
- The curriculum includes a focus on issues of social justice and world problems.
- The curriculum explicitly states equity goals. (pp. 419–421)

In practice, equity will come about through the adoption of the various practices that have been developed over time to address gender inequities. You will notice that these practices are consistent with the ideas and approaches discussed in other chapters concerning equity and social justice.

Recommended reading

Boaler, J. (1997b). Reclaiming school mathematics: The girls fight back. *Gender and Education,* 9(3), 285–306.

Goodell, J. & Parker, L. (2001). Creating a connected, equitable mathematics classroom: Facilitating gender equity. In B. Atweh, H. Forgasz & B. Nebres (Eds), *Sociocultural research on mathematics education: An international perspective* (pp. 411–431). Mahwah, NJ: Lawrence Erlbaum.

Tan, J. (2012). Students' ways of knowing and learning mathematics and their ways of interacting with advanced calculators. In J. Dindyal, L. P. Cheng & S. F. Ng (Eds), *Proceedings of the 35th annual conference of the Mathematics Education Research Group of Australasia* (pp. 704–711). Singapore: MERGA.

Vale, C. (2010). Gender mainstreaming: Maintaining attention on gender equality. In H. J. Forgasz, J. Rossi-Becker, K. Lee & O. B. Steinthorsdottir (Eds), *International perspective on gender in mathematics education* (pp. 111–143). Charlotte, NC: Information Age Publishing Inc.

PART 5

Professional and community engagement

CHAPTER 16
Working with parents and communities

Mathematics teachers who are doing their job well know that their responsibility extends beyond the mathematics classroom and the school to include interaction with other professionals, students' families and the community surrounding the school, and advocacy for mathematics and its learning in the wider community. This chapter looks more closely at the nature of educational partnerships between schools, parents and communities that support students' mathematics learning. We begin by examining parental and community attitudes towards mathematics, and ways in which mismatches between home and school cultures can create barriers to partnerships. Next, we outline a theoretical framework for comparing school-centred, family-centred and community-centred perspectives on mathematics education partnerships, and explore its implications for extending the often-limited ways in which schools and teachers view their interactions with parents. The final part of the chapter offers some guidelines for working productively with parents and communities.

Parental and community attitudes towards mathematics

It often seems to teachers that parental and community attitudes towards mathematics are at odds with contemporary views about mathematics learning and effective teaching practices (such as those presented in this book), or that many parents are simply uninterested in becoming involved in their children's schooling.

REVIEW AND REFLECT : Read the following pairs of quotations and discuss your immediate reaction with a peer or small group.

(A1) *School Principal:* Our parents here now still want these rows of algorithms and closed tasks and what we do here is catering for the differing abilities of our children … and we've had teachers very rudely spoken to by parents about 'who designed these'—it was a maths homework task to do with looking at snowfalls and look at the weather map you know, 'who designed this—stupid! We want REAL maths, we want real maths.' (Goos et al., 2004, p. 143)

(A2) *Parent:* I don't feel that [the teachers and administration are] always ready to listen to ideas that we might have … I think [it's] because they're trained, they've done their degrees and they know what they're doing about that kind of thing. Sometimes some of them feel that we're not qualified to offer that kind of advice. (Mills & Gale, 2004, p. 275)

(B1) *Teacher:* The students come from families that really don't care about school. Most of them are dropouts themselves, so school has no place in their lives. Many of the kids will say that their parents hated school and were no good at maths, so they believe it is their gene pool. (Zevenbergen, 2003b, p. 141)

(B2) *Parent:* I don't know why [other parents don't get involved]. I don't know whether it's their own experience at [secondary] school was pretty horrible when they were kids, but they do seem to be a lot less willing to be involved in the [secondary] school than they are with the primary school. (Mills & Gale, 2004, p. 272)

(C1) *Employer:* Young people do not seem to have the ability to calculate [mentally] things like we used to. They need to use a calculator to work out change when the cash register does not work. They do not know when the change they are giving is incorrect. (Zevenbergen, 2004, p. 108)

(C2) *Trainee draftsman:* The boss went ape at me the other day. He just bought this new computer software package and had it installed. I started playing with it and he told me I have to wait till the expert from the company came or I might crash the computer. As if! He has got no idea and the company is ripping him off as it is pretty basic. I had it going in no time. (Zevenbergen, 2004, p. 112)

According to the AAMT (2006) *Standards,* excellent teachers of mathematics are 'positive advocates for mathematics and its learning in the school and the wider community. They ensure effective interaction with families including provision of information about students' learning and progress.' How might such a teacher respond to the people who made the comments quoted above?

Cairney (2000) suggests that educators and parents (and, we might add, employers) need to go beyond the kind of deficit views of each other illustrated in the quotations provided above. He defines two types of deficit views in relationships between the school and the home. The first—*the family deficit view*—perceives the homes of children from diverse social and cultural backgrounds as providing limited learning environments and placing little value on education (quotation B1). The second—*the educational inadequacy view*—suggests that differential achievements are largely due to the failure of school to develop students' skills (quotations A1, C1). Cairney (2000) argues that neither of these two explanations is helpful because deficit views fail to recognise that much of the variability in student achievement reflects discrepancies, not deficiencies, between school resources, teaching approaches and the cultural practices of the home.

The roles of parents and communities in educational partnerships

When considering the nature of parents' involvement in their children's schooling, it is important for teachers to understand the variety of different family types and the major social changes affecting families and young people's home environments. Families are becoming more diverse in their composition, their ways of living and their cultural backgrounds. Although the 'nuclear family', consisting of two biological parents and their children, is still the most common family model, a significant proportion of Australian children under 18 live with only one of their natural parents, in step- or blended families, or in extended and same-sex parent families.

Funkhouser and Gonzales (1997) argue that developing effective partnerships with families requires a shift from policies and practices oriented to the typical (middle-class, monocultural, 'nuclear') family to those that are inclusive of greater social and cultural heterogeneity. The recognition of parental diversity can help overcome barriers to partnerships, such as the construction of 'good' and 'bad' parents (Crozier, 2000). Often, this conception is based on the SES of parents. Some people claim that low-income and single parents are uninterested in school and unsupportive of children's learning, which implies that the academic achievement of children is directly influenced by the SES of the family. While low-income or single-parent status can create additional constraints on parental involvement, researchers who analysed parent participation policies found that low SES parents and single mothers valued education as much as other parents did (Reay, 1998). Further, the researchers suggest that it is too simplistic to blame low-income, migrant or Indigenous parents for not helping their children with homework (Ewing, 2009; Meaney et al., 2008; Pena, 2000; Pérez Carreón et al., 2005).

Communities can become involved in the education of children and young people by offering a wide range of resources that are valuable to schools and the families they serve. These resources include people who volunteer their time in the school, organisations that offer enrichment opportunities, businesses that offer career-related information and workplace experiences, and agencies that provide various social services for students and families. However, in any of these types of collaborations, teaching and learning must be a central focus, and community involvement must not be activated only when the students are in trouble (Epstein, 2001).

Communities are powerful learning environments, creating potential for young people's development as they engage in social practices with others. This approach to learning suggests that teachers need to understand their students' communities, acknowledge the learning that takes place there and the way that learning takes place (Owens, 2015; Saxe, 2002; Warren et al., 2009). Drawing on communities' funds of knowledge, and Indigenous ways of knowing, can capitalise on the diverse cultures often found in mathematics classrooms, and overcome any mismatch between students' home environments and the culture of school (Owens, 2010, 2015; Peressini, 1997).

The Values in Mathematics Project, conducted by researchers at Monash University (Bishop, 2001; Fitzsimons, Seah, Bishop & Clarkson, 2001), supports these ideas. These researchers have examined teachers' awareness of what values they teach in their mathematics instruction, how this teaching takes place and, perhaps most importantly, what values students are learning from their mathematics teachers. All mathematics teaching practices—planning curriculum, choosing textbooks, assigning homework, and so on— embed socio-cultural values. As mathematics teaching is a form of cultural induction, teachers must be aware of cultural difference in the classroom.

Not all students' cultural communities share the same values, and this has implications for how students and their families might respond to unfamiliar teaching practices. Wong and Veloo (2001) highlighted the potential for mismatches in cultural values when they examined how national ideologies have been incorporated into school curricula in Brunei, Malaysia and Singapore. In these countries, respect for authority and for one's elders contributes to social cohesion, and teachers are traditionally held in high esteem. A strongly hierarchical social structure places the teacher in an authoritative position, so it is not surprising that whole-class teaching is common, with students sitting in neat rows paying careful attention to the teacher's explanations at the blackboard. Given these differences in cultural norms, teachers in Australian classrooms must be sensitive to values they teach in a class in which some students do not share the mainstream culture. Such students may feel uncomfortable—for example, in explaining their ideas to others. Thus, the classroom approach should focus on the notion of difference, rather than deficit, in learning.

REVIEW AND REFLECT : List any examples of parent and community participation in students' mathematics education you observed during your practice teaching, or experienced during your own schooling. What roles were parents and community members or organisations expected to play? Who initiated these interactions or relationships? What value did the students, parents and communities gain from these interactions? What barriers to partnerships were evident? Discuss and compare your experiences with a peer or in a small group. If you have few or no experiences of parent or community participation to share, consider why this may be so.

Why are partnerships important for mathematics education?

The arguments presented above may well apply to teaching in all school subjects, not just mathematics. However, there are good reasons why mathematics teachers in particular need to be aware of issues affecting parent and community participation in the mathematical education of young people.

Many adults have developed negative attitudes towards mathematics as a result of their experiences at school (Ewing, 2004), and this in turn can have a detrimental effect on their children's attitudes (Horne, 1998).

Due to recent changes in mathematics curricula, teaching methods and assessment, many parents are unfamiliar with current classroom practices (e.g., emphasising collaborative group work, use of manipulatives and technology), and may question or criticise the approaches taken by the teacher (Peressini, 1998).

Common perceptions of mathematics as consisting only of number and computation can lead to distorted views about which aspects of mathematics are important in the workplace, often expressed as criticism about the use of calculators eroding mental computation abilities (Jorgensen, 2014).

'School–community relationships need to be strong if Indigenous learners are to gain the most out of their schooling' (Meaney et al., 2008, p. 80). Building trusting relationships with Indigenous families and their community generates respect for Indigenous knowledge and ways of knowing and enables teachers to incorporate these ways of knowing in their teaching, and for the aspirations of Indigenous students and their families to be realised (Owen, 2015; Warren et al., 2009).

Families from different cultural backgrounds may have expectations regarding mathematical content, teaching and assessment methods that differ from those now common in Australian classrooms, and these parents may be accustomed to playing different roles in supporting their children's mathematics learning in the home setting (Cai, 2003).

Another important reason why mathematics teachers need to engage with parents and community members relates to the current emphasis on numeracy education in Australia. Numeracy has become a high priority for national and state or territory governments, and policies formulated to address this area typically capitalise on the need to build partnerships with homes and communities to support numeracy learning (e.g., MCEETYA, 2008). This position on partnerships is consistent with the definition of numeracy proposed by leading Australian mathematics educators: 'to be numerate is to use mathematics effectively to meet the general demands of life at *home,* in paid *work,* and for participation in *community and civic life'* (Department of Employment, Education, Training and Youth Affairs, 1997, p. 15, emphasis added). Such an approach to numeracy implies that it is the responsibility of all members of society—schools, families and communities—to ensure that young people gain not only mathematical knowledge and skills, but also a repertoire of problem-solving and decision-making strategies needed for intelligent citizenship in a rapidly changing world. As discussed in Chapters 1 and 5, numeracy is now a general capability included in the *Australian Curriculum* (ACARA, 2014).

However, while government policies aim to encourage schools to develop partnerships with families and communities in their local educational contexts (e.g., Department of Education and Training, Victoria, 2016), there are often discrepancies between the rhetoric of policy documents and the practice of family and community involvement in education. For example, a national research project investigating home–school–community partnerships that support primary school children's numeracy development found only limited evidence of parents and communities taking a leading role in shaping numeracy education partnerships (Goos, 2004a). It was also clear that most participants in the study held a narrow view of numeracy as school mathematics or 'number' learning, and this meant that the rich variety of numeracy learning opportunities in everyday (non-school) settings often remained invisible to teachers, parents and community members. Nevertheless, one significant and consistent finding from this project concerned the central role of teachers in either enabling or hindering the formation of partnerships. To develop good relationships with students' parents and communities, it is vital for teachers to have a clear understanding of the nature of 'partnerships' and how the participants view their roles.

REVIEW AND REFLECT : Search the website of your state or territory's education department for information and policies on *partnerships* or *networks* involving *parents* and *communities.* In discussion with peers, decide how well you think this information:

- addresses the importance of positive attitudes towards mathematics
- explains current mathematics teaching and assessment practices

- presents a broad view of mathematics that extends beyond number and calculation
- acknowledges the diverse cultural backgrounds and educational experiences of Australian families.

Stakeholder perspectives on partnerships

Epstein (1995) defines home–school–community partnerships as exemplifying a relationship between 'three major contexts in which students live and grow' (p. 702). In Figure 16.1, we have represented this relationship as a network with student learning at its centre. We will use this model to analyse different kinds of activities described in the literature on home, school and community connections, to investigate these activities in practice, and to consider how teachers and schools might use these ideas to improve relations with parents and communities.

Some activities take place at the school and represent a typical vision of parental involvement from the perspective of the school (parental attendance at school open days, parent–teacher conferences, and so on). Other activities take place in the home and can represent either parents' response to school initiatives (such as checking that homework is done) or their response to the demands of larger social practices and cultural values (e.g., involvement driven by parent aspirations for a child's education and well-being). Yet other activities represent community-centred connections between home and school, ranging from informal ties to formal partnerships (e.g., sponsorship of mathematics competitions, provision of work experience placements). Therefore, we can classify these links as being school-centred, home-centred or community-centred according to the different perspectives of the stakeholders (Goos, 2004a; Goos et al., 2004).

School-centred perspectives on partnerships

Epstein (1995) has defined six dimensions of home–school partnerships: parenting, communicating, volunteering, learning at home, decision making and collaborating with the community.

Parenting refers to the support provided to families to develop parenting skills that prepare children for school and to build positive home conditions that support learning. This

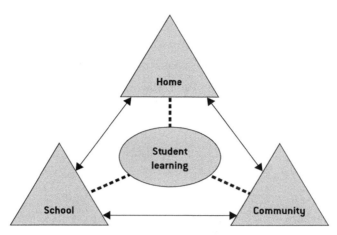

Figure 16.1 **Network model for home–school–community partnerships**

type of involvement is most often outside the classroom teacher's realm and is usually not a component of a mathematics teacher's parent-involvement strategies unless the focus is specifically on creating supportive environments for doing homework in mathematics.

Communicating involves establishing effective forms of interaction between school and home. This type of connection is perhaps the most common way in which teachers have traditionally involved families of students in education, and is especially needed in situations when parents feel uncomfortable in school, do not speak English well or come from different cultural backgrounds from teachers. Teachers can set up effective communication with parents by publishing mathematics newsletters, establishing networks of families to share information about their children's mathematics education, organising back-to-school nights at the start of the school year, offering formal and informal teacher–parent conferences and workshops, sending home mathematics portfolios, and telephoning or visiting families regularly to understand the cultural background and experiences of students and families (Barber et al., 2000; Miedel & Reynolds, 1999).

The most common way of communicating with students' parents and families is via telephone. Teachers often dread phoning home because it means telling parents about problems with their child's academic work, classroom behaviour or attendance. To initiate positive relationships with parents, it is better to take a proactive approach by calling parents *before* problems arise. One beginning teacher of mathematics described how she developed an index card system to keep track of these telephone calls (Brader-Araje, 2004). This systematic approach proved to be useful for several reasons. First, it provided evidence to address problems in meetings with parents and the school administration. Second, telephone calls often headed off potential problems because the teacher was able to discuss concerns before they escalated. Finally, the system ensured that parents received positive feedback about their child's successes, and demonstrated to them that the teacher genuinely cared about all of her students.

Volunteering expresses parents' and families' support for school programs by working with students on learning activities in classrooms, and participating in other activities outside the classroom or outside the school. While encouraging parents to become active in the mathematics classroom is a powerful way of helping them understand the changes in their children's mathematics education, this type of participation is probably more common in primary than secondary schools. Nevertheless, parents can be encouraged to volunteer their time and expertise in many other ways, such as through acting as guest speakers about their jobs and career opportunities in mathematics, excursion chaperones, and tutors or mentors to students.

Learning at home can involve parents in monitoring and assisting their children with homework and other mathematical activities, and this might be the most common way that parents expect to be involved in mathematics education. However, as mathematical content and pedagogy continue to develop and change, parents may find themselves on unfamiliar ground when they attempt to work with their children on school mathematics tasks. Organising and offering activities that are meant to be completed by both the parent and the student can make parents' efforts to be involved at home more productive (Ehnebuske, 1998).

Decision making refers to parents' participation in school decisions and advocacy activities through curriculum committees, school councils and Parents and Citizens' associations (Horne, 1998). Involvement in decision making is viewed by many to be the most empowering and productive type of parental involvement; however, it is also the most challenging type to organise and implement. This situation is particularly true for mathematics education because the mathematics community has made great efforts to enhance its professional status, and parental involvement in decision-making activities may be perceived as challenging the professional position of mathematics teachers and educators. Nevertheless, Horne (1998) argues, 'the involvement of parents in these roles can mean that the school becomes more responsive to the needs and culture of the local community' (p. 117). This is especially important for Indigenous communities, in which cultural negotiation can make explicit the hidden values and processes of the school while at the same time valuing the community's knowledge and the goals it holds for its children (Meaney & Fairhall, 2003; Owens, 2015).

Collaborating with the community reflects the increasing interest of many schools in making connections with local businesses, higher education institutions and community-based agencies. For example, schools might solicit financial or material support from the business community to provide computers, mathematics software, calculators, manipulatives and other materials for hands-on activities. Schools can also be involved in raising community awareness of the importance of all students developing numeracy abilities by participating in National Literacy and Numeracy Week or mathematics carnivals and Olympiads in which students showcase their achievements and demonstrate mathematics projects and activities. Schools can approach local universities to become involved in such events by providing intellectual and material support for teachers of mathematics, or by working directly with students on enrichment activities.

REVIEW AND REFLECT : Parents often ask how they can help their children with mathematics at home. It is not necessary for parents to be expert in the mathematical content being taught; instead, the best way they can help is by asking leading questions that build students' confidence and encourage mathematical thinking and communication. Design a handout to distribute to parents at the beginning of the school year, or at the first parent–teacher evening, with suggestions on how to help their child with mathematics learning at home. Include sample questions they could ask to help their child get started on a problem, represent and organise the information, deal with obstacles when they are 'stuck', and reflect on their solution. (See Mirra, 2004 for examples.)

Work with a group of colleagues to design a mathematics newsletter to send home to parents of classes you teach. You could include items such as a description of topics the classes have been studying, together with information about their historical development or relevance to everyday life or careers; examples of the types of mathematical activities and problems that students have worked on in different subjects and year levels; simple projects or data collection activities that students and their families could work on together; information about mathematics competitions or other extra-curricular activities; and articles that explain the philosophy of the classroom (e.g., use of calculators or group work).

Search the internet to find mathematics activities offered through National Literacy and Numeracy Week (www.literacyandnumeracy.gov.au). Design a week of school-based activities that develop family and community awareness of the importance of numeracy.

Family-centred perspectives on partnerships

We can also ask how families see their roles in connecting with schools and communities to support their children's mathematics education. Although there has been much less work undertaken from this family-centred perspective, we can identify the following six categories:

- creating supportive learning environments at home
- parental support for the child
- parents as role models for the value of education
- home practices that support numeracy development
- parent-directed activities that connect children to out-of-school opportunities for numeracy development
- parent–child discussions and interactions about school-related issues and activities (James, Jurich & Estes, 2001; Y. Katz, 2000).

Creating supportive learning environments at home includes the supervision and structure that parents give children to support their education—for example, by providing time and space for homework and limiting the time spent watching television and playing computer games.

Parental support for the child includes emotional and academic support, and the expression of parental aspirations and expectations regarding a child's current school performance. Research in this area shows that parents' educational aspirations are stable, high and certain over the pre-primary and primary years of schooling, but expectations can become lower, less stable and subject to considerably more uncertainty by the start of secondary school.

Parents as role models for the value of education refers to ways in which parents can model why school is important and share their own experiences that reinforce the value of education. Literature that describes this dimension of family involvement in children's education is rich in examples of the traditional (middle-class) mediation of educational values. It also shows that the vast majority of marginalised families fall into the 'uninvolved' category, and hence parents are represented as uncaring and as failing to provide a positive model for their children. Current research into practices of disadvantaged families illustrates that parents often engage in activities that are outside conventional understanding of involvement. For example, they may show the value of education to their children through the medium of hard work, thus teaching 'real-life' lessons that such work is both difficult and without adequate compensation and that without education, they may end up working in a similar type of job (Lopez, 2001).

Home practices that support numeracy development refer to such activities as parents doing problem-solving tasks and engaging in mathematical games. Studies in this area typically highlight the role that parents play in their children's early numeracy and literacy learning prior to school entry and in the primary school years, and less attention has been given to the numeracy learning opportunities available for secondary school-aged children in the everyday activities of families and communities (such as budgeting, shopping, scheduling, playing sport, travelling, measuring or building things).

Parent-directed activities that connect children to out-of-school opportunities for numeracy development may involve private tutoring, enrolment in different enrichment programs run by after-school and church organisations, museums and libraries, and community schools that teach migrant children about their home culture and language. Studies show that parents with higher levels of education (and predominantly mothers) are more likely to initiate these kinds of connections.

Parent–child discussions and interactions about school-related issues and activities involve parents in asking their children what mathematics they learned in school that day. Very often, this kind of discussion drives parental activism with regard to their involvement in school policies, representing and advocating for the interests and needs of their children.

REVIEW AND REFLECT : Various state education departments and parent organi-sations have developed information for parents to assist their students at home with numeracy.

Do a search on your digital device for 'Numeracy: families working it out together, the opportunities are everywhere', 'Helping your student in Year 7 with literacy and numeracy' and 'Helping your child with literacy and numeracy in the middle years: 6–8'.

Design a poster for families of lower secondary students or a poster that could be displayed in classrooms.

Find out about mathematics enrichment programs in your local area run by after-school organisations, museums and libraries, and community schools.

Community-centred perspectives on partnerships

Analysing numeracy learning in communities is very complex because of the multiple communities (social, cultural, religious and economic) in which a young person may participate. The literature in this area suggests the following ways in which communities may form educational partnerships with schools and families (Jordan et al., 2001; p. 489 Keith, 1999):

- community-driven school reform and curricular enrichment efforts
- business–school partnerships
- university–school partnerships
- community service learning programs
- after-school programs
- more extended programs that target family numeracy.

Community-driven school reform and curricular enrichment efforts use community resources to over-come the view of the school as the sole transmitter of knowledge. Horne (1998) illustrates this dimension of community–school connections in schools where Mathematics Task Centres operate, involving parents and community members as mathematics tutors.

Business–school partnerships may provide schools with resources, expertise and volunteers. Peressini (1998) argues that resources provided to schools should not be limited to financial help only—for example, local businesses can supply teachers with such resources as restaurant menus and grocery flyers to develop classroom mathematics tasks in life-like contexts. Organisations may also establish partnerships with schools so that students can spend a day at the particular business and observe how mathematics is applied in the real world.

University–school partnerships may provide expertise, resources and professional development to schools while schools participate in research studies or other collaborative projects. These partnerships can serve as a catalyst for mathematics educational reform where resources and expertise for change are lacking.

Community service learning programs link academic content with activities that allow students to contribute to the well-being of the community. Through service learning, the community enriches the students' education by providing real-world learning opportunities outside the classroom; simultaneously, the students and school contribute to the community as they perform needed service for individuals, organisations and wider community purposes.

After-school programs provide help with homework, or remedial and enrichment learning activities. Many community after-school programs also fulfil parents' needs for childcare and other social services.

More extended programs that target family numeracy are exemplified by the Family Mathematics Program, which originated in the USA but also flourished in Australia (Horne, 1998). These programs place parents and their children together in workshops with stimulating joint activities to learn and use at home. Trainers include other parents, school personnel and volunteers from community organisations. Studies show that most parents who have participated in Family Maths Programs engage in more learning activities at home with their children and that more student participants enjoy mathematics classes. Schools, too, have changed their approach to communication with parents after being involved in offering such programs. The fact that these programs are more common in primary than secondary schools suggests that teachers may see less need to involve parents in mathematics education as their children get older and begin to study more specialised mathematical topics.

REVIEW AND REFLECT : Find out what kind of involvement local businesses and community organisations have with your school's mathematics department. Are they providing resources or expertise, or work/service placements for mathematics students? What benefits do the teachers, students, businesses and community organisations see in these arrangements?

Search the internet and investigate the role of Homework Centres in providing after-school support for students in your state and community.

Developing partnerships with community

In order for all young Australians to become 'successful learners, confident and creative individuals and informed and active citizens' (MCEETYA, 2008, p. 7), the governments of Australia agreed on an action plan, involving eight interrelated strategies. One of these strategies was developing stronger partnerships:

Parents, carers and families are the first and most important influence in a child's life, instilling the attitudes and values that will support young people to participate in schooling and contribute to broader local and global communities. Partnerships between students, parents, carers and families, the broader community, business, schools and other education and training providers bring mutual benefits and maximise student engagement and achievement. Partnerships engender support for the development and wellbeing of young people and their families and can provide opportunities for young Australians to connect with their communities, participate in civic life and develop a sense of responsible citizenship. In particular, the development of partnerships between schools and Indigenous communities, based on cross-cultural respect, is the main way of achieving highly effective schooling for Indigenous students. (MCEETYA, 2009, p. 5)

This means that schools and teachers need to engage in conversations with parents and the community about the mathematics that students are doing at school, and their progress with mathematics learning, and encourage parents to initiate discussion and listen to their children talk about mathematics. Mathematics teachers also need to meet and talk with members of the broader community, business and industry to identify mathematics used in their fields. Such engagement can then lead to the development of meaning and relevant mathematics investigations and tasks (as discussed in Chapters 5 and 14) and work experience programs that foster and develop numeracy and an interest in mathematics.

Conclusion

When teachers and schools work with families and communities to enhance young people's mathematics learning, we would warn against inferring that the term *partnership* implies there should be similar contributions from, and roles for, all participants. This is especially important when considering the roles of parents and teachers in educational partnerships. While research has found plenty of evidence that parents genuinely care about their children's education, it is equally clear that not all parents want to be actively involved in all aspects of schooling, and many see their role primarily as a supportive one. Perhaps the most productive way forwards is to focus on what each participant—parent, teacher and community member—can bring to the partnership that will make best use of their diverse expertise, background and interests in supporting students' learning.

Recommended reading

Meaney, T. & Fairhall, U. (2003). Tensions and possibilities: Indigenous parents doing mathematics curriculum development. In L. Bragg, C. Campbell, G. Herbert & J. Mousley (Eds), *Mathematics education research: Innovation, networking, opportunity* (Proceedings of the 26th annual conference of the Mathematics Education Research Group of Australasia, pp. 507–514). Sydney: MERGA.

Mirra, A. (Ed.). (2004). *A family's guide: Fostering your child's success in school mathematics.* Reston, VA: NCTM.

Peressini, D. (1997). Parental involvement in the reform of mathematics education. *The Mathematics Teacher*, 90(6), 421–427.

Warren, E., Cooper, T. & Baturo, A. (2009). Bridging the educational gap: Indigenous and non-Indigenous beliefs, attitudes and practices in a remote Australian school. In J. Zajda & K. Freeman (Eds), *Race, ethnicity and gender in education* (pp. 161–184). New York, NY: Springer.

Wong, K.Y. & Veloo, P. (2001). Situated sociocultural mathematics education: Vignettes from southeast Asian practices. In B. Atweh, H. Forgasz & B. Nebres (Eds), *Sociocultural research on mathematics education* (pp. 113–134). Mahwah, NJ: Lawrence Erlbaum.

CHAPTER 17
Continuing professional learning

The idea of lifelong learning is highly relevant to teachers' professional lives. Graduation from a pre-service teacher education program is an important moment in your teaching career, but this does not mark the end of professional growth as a teacher. Qualified professionals are expected to take responsibility for their own *continuing* development as mathematics teachers. 'Engage in professional learning' is Standard 6 in the *Australian Professional Standards for Teachers* (AITSL, 2014), and applies to teachers at all stages and positions in their career. This requires a shift in the common view of professional development from a one-shot, short-term experience towards a commitment to long-term, incremental improvement. In this chapter, we discuss approaches to career-long professional learning and development that bring together critical self-reflection and collaborative interaction with others. These approaches are interpreted in the context of beginning teachers' professional socialisation and development of a professional identity. We also explore the professional standards framework developed by the AAMT to consider issues involved in planning for continuing professional learning.

Dimensions of professional practice

A significant concern in contemporary professional development programs is the need to foster teachers' reflection on their practice so they continue to learn about themselves as teachers and their students as learners. However, professional development activities can do more than promote the growth of individual teachers' knowledge about their practice by also encouraging collegiality among groups of teachers within and beyond their schools.

For example, in telling the story of a mathematics teacher and her struggle for professional growth throughout her career, Krainer (2001) comments on the transition of the group of teachers with whom she worked 'from an *assembly of lone fighters* to a *network of critical friends'* (p. 287, original emphasis). This change highlights what Krainer refers to as the four dimensions of teachers' professional practice: action, reflection, autonomy and networking. While each of these dimensions is important, he explains that it is necessary to achieve a balance between *action and reflection,* and between *autonomy and networking.* Krainer claims that the practice of most teachers and schools does not achieve this balance: there is a lot of action and autonomy (hence, the 'lone fighters'), and not much reflection and networking (as in 'critical friends'). Thus, promoting the latter practices represents a powerful strategy for the ongoing professional learning of teachers (Jaworski, 2008). Indeed, education systems now promote collaborative reflection and research of practice. In the following sections, we discuss the meanings of 'reflection' and 'networking', and consider how mathematics teachers can become reflective individuals within a networked professional community.

Becoming a reflective teacher of mathematics

Pre-service teachers are routinely urged to 'reflect' on their lessons, but what does this really mean? Artzt and Armour-Thomas (2002) describe reflection as thinking about teaching before, during and after enactment of a lesson. As the purpose of reflection is to evaluate the effectiveness of one's teaching in order to bring about improvement in student engagement and learning, it is important that reflection involves analysis rather than description, and uses information from a variety of sources instead of relying only on introspection. Sources of data for reflection include the teacher's own self-analysis of lessons, analysis of student work samples, comments made by students, students' self-assessments, and feedback from colleagues. Video-recordings of lessons is increasingly being used as a source of data for reflection on practice (Geiger, Muir & Lamb, 2015).

Killen (2013) identified three levels of reflective analysis: technical, practical and critical reflection:

- *Technical reflection:* teachers are concerned with the technical application of educational knowledge in the classroom to maintain order and achieve pre-determined outcomes.

- *Practical reflection:* teachers are concerned with goals, the connections between principles and practice, and the assumptions that underlie their practices and the value of their goals.
- *Critical reflection:* teachers become concerned with issues beyond the classroom, so that moral and social issues such as equity and emancipation can inform their reflections on classroom practices. (p. 113)

These levels of reflection show that as the depth of analysis increases, teachers pay attention to more elements of their practice, to notice student thinking and engagement, and knowledge of their students, to make sense of what is happening in their classrooms, and make adjustments to their practice to improve student learning.

Self-analysis of lessons

In Chapter 2, we presented a classroom scenario involving Damien, a pre-service teacher, and his Year 10 mathematics class. Damien's post-lesson debriefing notes—the reflections he recorded during an interview with his university supervisor—illustrate a framework for self-analysis of lessons. The prompt for self-analysis was a *reflection card* similar to that shown in Figure 17.1. The rows correspond to important lesson features for the students in the classroom: engagement and involvement; learning processes; progress made during the lesson; and the social context in which they learned. The columns refer to a selected set of lesson features of importance to a pre-service or beginning teacher: expectations and actions concerned with teaching approaches; and student actions that the teacher noticed during the lesson as a form of immediate feedback. The reflection card was designed for use in a research project in which a mentor, such as a university supervisor or supervising teacher, elicited the pre-service teacher's reflections in each of the cells (Goos, 1999). However, the reflection card can also be used independently for lesson self-analysis. This is not a strategy that needs to be applied to every lesson taught. Instead, it may be helpful to either decide on a regular schedule for reflection on lessons (e.g., once per fortnight) or select a sequence of lessons with a particular class that might present specific challenges.

Student learning	Lesson features		
	Teacher expectations	Teacher actions	Student actions
Engagement and involvement			
• attitude to learning			
Learning processes			
• how students learn			
Progress			
• how well students learn			
Student context			
• social environment for learning			

Figure 17.1 **Reflection card**

REVIEW AND REFLECT : *Before the lesson*, record your *expectations* in the first column of the reflection card by responding to the following questions (adapted from Artzt and Armour-Thomas, 2002):

Engagement and involvement

What do you know about your students in terms of their prior knowledge, achievements and experiences, attitudes and interests? How will you use your knowledge of your students to engage them in the lesson?

Learning process

What pedagogical approaches might be suitable for this topic and this class? How have you decided which approach to use? What types of tasks, and what resources and modes of representation (symbols, diagrams, manipulatives or digital), have you considered? What forms of teacher–student and student–student interaction have you planned?

Progress

What are the learning goals? What are the key mathematical ideas in the content and tasks for the lesson? What is the level of difficulty? How should tasks be sequenced? How does the content connect the mathematics students have already learned with content to be taught in future? What solutions might students produce? What methods of solution, problem solving and/or reasoning could students use? What difficulties do you anticipate the students might have and how have you prepared for these?

Social context

How will you establish a positive social and intellectual climate? What administrative and organisational routines will be important in this lesson?

During the lesson, pay attention to students' questions, observe them working on the task(s), interact with them about their work on the task(s) and collect samples of student work. You might also collect student self-assessments or exit tickets.

After the lesson, record the *actions* you took to ensure students were engaged in the lesson (*engagement and involvement*), to help them learn in the ways you envisaged (*learning process*), to help them achieve the progress you wanted (*progress*), and to establish a productive learning environment (*social context*).

Record the *student actions* you observed and analyse student *work samples* in each of these areas. That is, how did you know that the students were (or were not) engaged, that they were (or were not) learning in the ways you had expected, that they were (or were not) making the progress you anticipated, and that the learning environment was (or was not) as you desired?

Your self-analysis concludes by answering the following questions:

- Do you think your expectations for this lesson were appropriate? Why/ why not?
- Do you think that your actions were consistent with your expectations? Why/ why not? If not, what could you have done differently?
- What actions do you plan on keeping and taking in future lessons as a result of your reflection? What actions do you plan to modify or change?

Feedback from students

In addition to collecting and analysing student self-assessments or exit tickets (see Chapter 6), students can also provide valuable feedback on the effectiveness of your teaching. Good teachers monitor their students for informal feedback during lessons, such as by noticing students' level of interest and understanding, and adjust their teaching actions accordingly. More formal feedback can also be obtained from students through the use of surveys or questionnaires.

> **REVIEW AND REFLECT** : Close to the end of his first semester of secondary mathematics teaching, a beginning teacher emailed fellow graduates with the following request:
>
> I want to get some feedback from my students about how they perceive my teaching strengths and weaknesses. Does anyone have a questionnaire that they have used or can you suggest some questions that I could ask?
>
> What advice would you give him? What questions would you suggest he ask his students? Why do you suggest these questions?

One of the authors of this book recalls seeking feedback from a class by designing a survey based on the needs students expressed in the first lesson of the year. In this lesson, the teacher wanted to establish her expectations of students, but decided first to ask them what they needed from her in order to develop the idea that teacher–student relationships should be based on mutual respect and involve mutual obligations. She asked students to complete the sentence 'I need a teacher who …' Their responses included:

- explains things simply
- I can get along well with
- treats me like a human being
- helps when needed
- is understanding
- I can have a joke with

- makes class interesting and fun
- teaches!
- is encouraging
- tolerates 'dumb' questions
- allows students to help each other.

The teacher recorded these responses because they corresponded closely to her own personal goals, and at the end of Term 1 she administered a survey that asked students to rate the extent to which she had demonstrated these qualities, using a five-point scale ranging from 'Never' to 'Always'. Not surprisingly, students were impressed that the teacher had taken their opinions seriously and had tried to meet their expectations.

Teachers can also seek regular written feedback from students to help identify difficulties they may be experiencing, and thus plan more effective methods of instruction (see Chapter 6 for discussion of assessment as learning, and self-assessment practices). The IMPACT procedure (Clarke, 1988) provides a way of discovering students' concerns and opinions by administering a simple questionnaire during class (see Figure 17.2).

Name: ....

Class: ....

Teacher: ....

Date: ....

Write down the two most important things you have learnt in maths during the past month.

Write down at least one sort of problem which you have continued to find difficult.

What would you most like more help with?

How do you feel in maths classes at the moment? (Circle the words that apply to you.)

a) Interested	b) Relaxed	c) Worried
d) Successful	e) Confused	f) Clever
g) Happy	h) Bored	i) Rushed

j) Write down one word of your own...

What is the biggest worry affecting your work in maths at the moment?

How could we improve maths classes?

Figure 17.2 **The IMPACT procedure**

Administration should be regular (e.g., once per fortnight), and students' responses can be stored in a class folder in order to identify trends. Teachers have found this procedure very useful, but its success depends on respecting the confidentiality of student responses and acting on these responses where appropriate to improve students' experiences of learning mathematics.

REVIEW AND REFLECT : Below are some sample student responses using the IMPACT procedure.

- *What is the biggest worry affecting your work in maths at the moment?*
- Homework, because at home hardly anyone knows what to do because it is just as new to them as it is to me.
- *Write down at least one sort of problem that you have continued to find difficult.*
- Algebra a bit, because I don't understand why we don't just use numbers. It would be simpler.
- *How do you feel in maths classes at the moment?*
- Bored. Angry. (If you're wondering why I'm angry, it's because I don't like being bored.)
- *How could we improve maths classes?*
- By using some other method of learning instead of these boring textbooks. Have less work and more learning.

What information do these comments provide about the students' concerns and classroom learning experiences? If you were the teacher, what actions would you take to follow up on these comments?

Feedback from colleagues

Beginning teachers of mathematics and other teachers who lack formal qualifications in mathematics education often find it helpful to work with a mentor—a trusted colleague who is willing to observe lessons and to be observed in his or her classroom, to share resources, and to help less experienced teachers to develop their own teaching style (Zagorski, 2004). Observing lessons taught by others remains a powerful way to learn about teaching throughout your career. It is at the heart of professional learning programs such as Communities of Inquiry (Jaworski, 2008) and Japanese Lesson Study (Takahashi, n.d.; Takahashi & Yoshida, 2004). The relationship between observer and observed need not be hierarchical (e.g., expert and novice), and teachers who are equally experienced but still consider themselves learners will benefit from mutual observations and sharing of practice (Tanner & Jones, 2000).

Zagorski (2004) lists the following qualities to look for in a mentor:

- a knowledgeable teacher who is committed to the profession
- a teacher who has a positive attitude toward the school, colleagues and students, and is willing to share his or her own struggles and frustrations, avoiding the naysayer who constantly complains in staff meetings
- a teacher who is accepting of beginning teachers, showing empathy and acceptance without judgment
- a teacher who continuously searches for better answers and more effective solutions to problems rather than believes that he or she already has the only right answer to every question and the best solution to every problem
- a teacher who leads and attends workshops and who reads or writes for professional journals
- an open, caring, and friendly individual who has good communication skills
- someone who shares your teaching style, philosophy, grade level or subject area
- a teacher who is following the path you want to follow, someone with whom you can relate and with whom you share mutual respect
- someone who is aware of his or her own biases and opinions, and encourages you to listen to advice but also to form your own opinions. (p. 6)

In preparing to be observed, it is important to be specific about the kind of feedback desired for the particular lesson and how this feedback can contribute to achieving longer-term goals for improving practice. This should include noticing the actions of students in order to analyse the connections between teacher planning and actions and student engagement and learning. The reflection card shown in Figure 17.1 and the reflective questions that accompany it can help the beginning teacher identify lesson objectives, pedagogical strategies, anticipated student solutions, thinking and difficulties, and can also provide a focus for post-lesson discussion with the mentor. This discussion could explore options for developing specific teaching strategies or for finding solutions to problems experienced with a particular group of students.

> **REVIEW AND REFLECT :** Discuss Zagorski's (2004) list of qualities with a partner and rank them in order of importance to you. Add any other qualities you agree are important. Compare your list with those produced by others in your class.

Collaborative professional learning

Collaboration is regarded as central to all professional learning (Loucks-Horsley et al., 2003). Collaborative structures can bring together teachers within a school, sometimes called 'professional learning teams', or across schools to work towards a common learning goal (Jaworski, 2008). Professional networks, action research and lesson study are three common ways of engaging teachers in collaborative professional learning experiences.

Professional networks, action research and lesson study

Loucks-Horsley et al. (2003) define a professional network as 'an organised professional community that has a common theme or purpose' (p. 146). Networks typically bring together teachers and other educators across school boundaries—for example, through school–university partnerships, school clusters (including those that link primary and secondary schools), mathematics teacher associations or partnerships with community organisations. Often, these networks are formed for specific purposes, such as to improve teaching of particular subject matter or to support particular school or curricular reforms. For example, mathematics teachers from secondary schools in a geographically defined district might form a network in order to investigate effective ways to implement a new curriculum. A challenge for networks is to keep members engaged, connected and informed, and electronic communication methods such as email lists and websites are increasingly used for this purpose.

Action research is another form of practice-based collaborative inquiry that provides teachers with opportunities for deep analysis and reflection on critical questions they face in their work. Action research is 'an ongoing process of systematic study in which teachers examine their own teaching and students' learning through descriptive reporting, purposeful conversation, collegial sharing, and critical reflection for the purpose of improving classroom practice' (Miller & Pine, 1990, cited in Loucks-Horsley et al., 2003, p. 162). Although there are many different forms of action research, this approach is often considered to have the following key elements:

- Teachers contribute to or formulate their own questions, and collect the data to answer these questions (e.g., by observing or videotaping lessons, interviewing students or teachers, conducting surveys, collecting student work).
- Teachers use an action research cycle, comprising planning, observing, acting and reflecting.
- Teachers are linked with sources of knowledge and stimulation from outside their schools, such as professional associations or university researchers.
- Teachers work collaboratively.
- Learning from research is documented and shared.

Japanese Lesson Study is a collaborative professional learning process, brought to the attention of the world at the time of the TIMSS 1999 Video Study (Hollingworth et al., 2003) when Japanese students recorded the highest scores in mathematics achievement. Japanese Lesson Study, which began more than 100 years ago when the public education system was established in Japan, is regularly used at all school levels to enable teachers to learn from each other and to research their practice to improve student learning. It involves a four-step process (Takahashi, n.d.; Takahashi & Yoshida, 2004). Teachers work collaboratively to:

- Identify and document a long-term goal for student learning (the school improvement goal for mathematics).
- Research and plan a lesson for a particular topic and learning goal, related to the longer-term goal.

- Teach the planned research lesson that is observed by other members of the team and other teachers at the school, and in some instances from other schools, who carefully observe student actions, thinking, engagement and learning.
- Conduct a post-lesson discussion involving the teacher, lesson study planning team, and observers. The planning team will then meet to revise the lesson based on the feedback and discussion with observers.

The effectiveness of lesson study for teacher professional learning occurs because of the attention that the planning team and the observers pay to student thinking and learning, and the connections that they make between the task, teacher actions and student actions. Inquiry learning or problem-based approaches to learning is the pedagogical approach used for lesson study, as Japanese teachers aim for their students to understand, make sense of and create mathematics. Teachers from other parts of the world, when first introduced to lesson study, are often surprised at the depth of research that teachers undertake when planning the research lesson.

Working in professional communities

In recent years, there has been much interest in the idea of teacher professional communities as a means of supporting teacher networks that promote continuing inquiry into practice. Secada and Adajian (1997) define four characteristics of such communities: a shared sense of purpose in committing to common educational values and goals for student learning; coordinated effort to improve students' mathematics learning by examining curriculum across grade levels and by accessing parental support and community resources; collaborative professional learning to improve mathematics teaching practices; and collective control over important decisions affecting the mathematics program. Mathematics teacher professional communities with these characteristics can support experienced teachers learning to teach in new ways so they continually improve their practice (Stein et al., 1998).

REVIEW AND REFLECT : Browse the AITSL Illustrations of Practice for Standard 6: Engage in Professional Learning (www.aitsl.edu.au/australian-professional-standards-for-teachers/search?&fc=Standard!standard-6&fc=Scope!Iop&t=2)

and view videos of teachers discussing their professional learning practice. Find the professional learning policy and search for action research projects from your state's education department website.

Compare your experience of mentoring and collaboration for professional learning with these examples and the four characteristics of teacher professional communities outlined above. Which of these principles and practices would you like to build into your plan and practice for ongoing professional learning?

Understanding professional socialisation

Not only is learning to teach a lifelong process, but teachers also learn from experience in many different contexts. It can be challenging for beginning teachers to reconcile what they learn about teaching from their own schooling, the university pre-service program and practicum sessions, and initial professional experiences—especially if these produce conflicting images of mathematics teaching. This challenge is often associated with the perceived gap between the decontextualised knowledge provided by university-based teacher education and the practical realities of classroom teaching. As a result, many beginning teachers find it difficult to implement innovative approaches they may have learned about during their pre-service program while trying to cope with the demands of the early years of full-time teaching. It is common for beginning teachers to give up their innovative ideas in the struggle to survive, and instead conform to institutional norms of traditional practices. Teachers at later stages of their careers may experience similar challenges if they are required to implement new teaching approaches, curricula or assessment techniques.

Instead of viewing teachers as being passively moulded by the forces of professional socialisation and reliant on short-term coping strategies for survival, we prefer to take the view that they can take action to shape their own development. To show what this might look like, we present a theoretical model for teachers' learning (more fully discussed by Goos, 2005), illustrated by a case study of a beginning teacher.

A theory of teacher learning and development

Researchers have identified a range of factors that influence teacher learning and development. Rather than considering each separately, it is helpful to organise these factors into three 'zones of influence'. The first zone represents *teacher knowledge and beliefs*, and suggests the *potential for development*. This zone includes teachers' disciplinary knowledge, PCK (knowledge of how to represent concepts and to use examples and analogies, as described by Shulman, 1986), and beliefs about their discipline and how it is best taught and learned. The second zone represents the *professional context*, which defines the teaching actions *allowed*. Elements of the context may include curriculum and assessment requirements, access to resources, organisational structures and cultures, and teacher perceptions of student background, ability and motivation. The third zone represents the *sources of assistance* available to teachers in *promoting* specific teaching actions, such as that offered by a pre-service teacher education course, supervised practicum experience, professional colleagues and mentors in the school, or formal professional development activities.

To understand teacher learning, we need to investigate relationships between these three zones (represented by the overlapping circles in Figure 17.3). Professional learning is most effective when teachers experience enough challenge to disturb the balance between their existing beliefs and practices, but also enough support to think through the

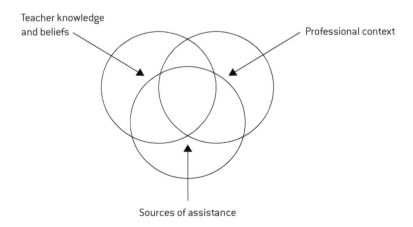

Figure 17.3 **Relationships between the three 'zones of influence' in teacher learning**

dissonance experienced and develop either a new repertoire of practice or a new way of interpreting their context that fits with their new understanding.

A case study of learning to teach

The case study in the box below illustrates how the 'three zone' model can be used to understand the professional socialisation of beginning teachers.

Adam's pre-service practicum

Adam was enrolled in a pre-service course that emphasised technology as a pedagogical resource. His practicum placement was in a large suburban school that had recently received funding to refurbish the mathematics classrooms and buy technology resources. All senior students had their own graphics calculators, and there were sufficient class sets of these calculators for use by junior classes. Some of these changes had been made in response to new senior mathematics syllabuses that mandated the use of computers or graphics calculators in teaching and assessment programs.

Adam had previously worked as a software designer. Although he had not used a graphics calculator before starting the pre-service course, he quickly became familiar with its capabilities and incorporated this and other technologies into his mathematics lessons, with the encouragement of his supervising teacher. However, at this stage, Adam had only ever used technology in his teaching, or observed its use by other teachers, as a tool for saving time in plotting graphs and performing complicated calculations, or for checking work done first by hand.

Table 17.1 identifies relevant aspects of Adam's *knowledge and beliefs, professional context* and *sources of assistance,* and indicates that all of these were likely to positively influence his professional socialisation and learning. Adam's practicum experience could be represented by the relationships (overlap) between the zones shown in Figure 17.3 (above).

Table 17.1 Adam's practicum experience

Knowledge and beliefs
- ✓ Skilled/experienced in using technology
- ✓ Student-centred pedagogical beliefs
- ✓ Developing PCK regarding technology integration

Professional context
- ✓ Syllabuses mandated use of technology
- ✓ Access to well-equipped classrooms, computers and graphics calculators

Sources of assistance
- ✓ Pre-service course emphasised technology
- ✓ Supervising teacher encouraged use of technology

Adam's first year of teaching

After graduation, Adam was employed by the same school where he had completed his practicum. Although the school environment and mathematics teaching staff were the same, Adam's experience of teaching changed dramatically. By now, he had developed a much more flexible teaching approach, and he encouraged students to use their graphics calculators as a learning tool to explore new mathematical concepts and model real-world situations. However, he discovered that many of the other mathematics teachers were unenthusiastic about using technology and favoured very structured lessons that left students with few opportunities to investigate mathematical ideas for themselves. Since he disagreed with this approach, conflicting pedagogical beliefs became a source of friction in the staffroom, and this was often played out in arguments where Adam was accused of not teaching in the 'right' way. He realised that as a pre-service teacher, he had not noticed the 'politics of teaching' because he had the luxury of focusing on a small number of classes and his relationship with a single supervising teacher. He now found himself in a more complex situation that required him to defend his instructional decisions while negotiating harmonious relationships with several colleagues who did not share his beliefs about learning.

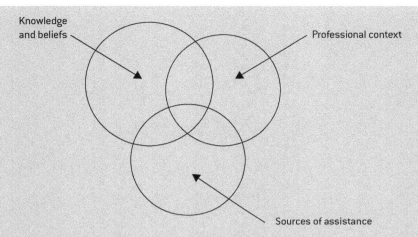

Figure 17.4 **Adam's first year of teaching**

As Adam's knowledge of technology integration had developed since the practicum, his knowledge and beliefs are now represented by a larger circle in Figure 17.4. However, the extent of overlap between his sources of assistance and the other two zones has decreased, because his teacher colleagues did not hold views compatible with either the teaching context (high access to technology) or Adam's potential for development.

How should Adam respond to this situation so that the overlap between the zones is restored?

Adam realised that this was the way he was comfortable teaching, and he refused to be drawn into arguments with other teachers. He interpreted his technology-rich context as *affording* his preferred teaching approach, and decided to pay attention only to those sources of assistance within the mathematics department that were consistent with his own beliefs and goals and the approaches that he experienced in his pre-service course.

Adam's second year of teaching

The following year, Adam was transferred to a different school where there was limited access to computer laboratories and only one class set of graphics

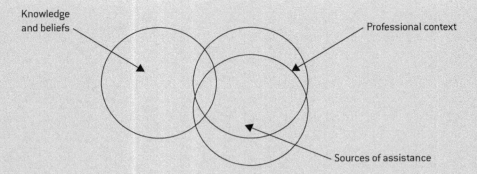

Figure 17.5 **Adam's second year of teaching**

calculators. The students were poorly motivated and unruly, and the school administration provided little support in managing the learning environment. None of the mathematics teachers was interested in using technology, and they preferred the same kind of teacher-centred, textbook-oriented teaching approaches as Adam's colleagues in his previous school. His experience as a second-year teacher is represented by Figure 17.5. The other teachers, through their lack of interest in technology, promoted approaches that were consistent with this technology-poor environment, but not with Adam's own beliefs and aspirations.

What actions might Adam take to increase the overlap between the three zones?

Since no other teacher wanted to use the class set of graphics calculators, Adam found he had unlimited access to the calculators for all of his classes. In other words, reinterpreting the context increased its overlap with his own beliefs and goals. He also joined the mathematics teachers' professional association and started to attend their professional development workshops and conferences. This brought him into contact with many other like-minded teachers and ideas for developing his teaching, and made available new sources of assistance that met his need for professional growth.

> **REVIEW AND REFLECT :** Analyse your own professional experience using the approach illustrated above. Begin by creating a table like Table 17.1 to list aspects of your knowledge and beliefs, professional context and sources of assistance. Use this information to draw a diagram like those in Figures 17.4 and 17.5. Describe the relationships between the three zones. What action could you take to reinterpret your teaching context or seek assistance from other sources in order to increase the degree of alignment between them?
>
> Compare your analysis and zone diagram with that of a peer who has professional experience in a different school.

Planning for continuing professional learning

There is no doubt that mathematics teachers need to know mathematics and constantly update their mathematical knowledge, especially when curriculum change brings new topics into school mathematics. However, mathematical content knowledge alone is not enough. Many professional development activities in mathematics education combine two core kinds of teacher knowledge—*mathematical knowledge* and *pedagogical content knowledge* (see Chapter 1 for an elaboration of PCK)—by immersing teachers in situations that require both mathematical and pedagogical problem solving, thus encouraging teachers to reflect on their practice and on their own understanding of the mathematics they teach (Zaslavsky, Chapman & Leiken, 2003).

Research on mathematics teachers' professional development (Darling-Hammond & Rothman, 2011; Darling-Hammond et al., 2009; Jaworski, 2008) concludes that effective professional development:

- focuses on deepening teachers' mathematics content knowledge and PCK
- addresses teachers' classroom work, their students' learning and the challenges that they encounter in their school settings
- is inquiry-focused and includes opportunities for practice, research and reflection
- is embedded in teachers' work and takes place during the school day
- is sustained over time

- is founded on, and develops a sense of, collegiality, collaboration and reciprocal forms of expertise, knowing and experience among teachers, between teachers and principals, and between teachers and education partners.

As mentioned earlier in this chapter, 'engaging in professional learning' is one of the Australian Professional Standards for Teachers (AITSL, 2014) and teachers are required to provide evidence of this practice when seeking promotion and, in some states, to maintain their registration. Instead of participating randomly in professional development activities to comply with renewal of registration policies, it is better to create a longer-term plan for career development and seek out professional learning opportunities with the characteristics described above to help achieve desired goals. Since the Australian Professional Standards for Teachers apply to all stages of a teaching career, it provides a framework about what teachers should know, understand and be able to do, and how this knowledge, understanding and practices should be developed over a career. There are seven main standards clustered into three categories of standings (see Table 17.2), with multiple elements in each of these standards (see AITSL, 2014).

Unlike the generic standards developed by teacher registration authorities, the professional standards published by the AAMT (2006) relate to the specialised work of teaching mathematics, describe characteristics of best teaching practice rather than competence, and provide a framework for teachers' career-long professional growth. The *Standards for Excellence in Teaching Mathematics in Australian Schools* (AAMT, 2006) were collaboratively

Table 17.2 Australian Professional Standards for Teachers

Professional Knowledge	Professional Practice	Professional Engagement
Know students and how they learn. Know the content and how to teach it.	Plan for and implement effective teaching and learning. Create and maintain supportive and safe learning environments. Assess, provide feedback and report on student learning.	Engage in professional learning. Engage professionally with colleagues, parents/carers and the community.

Source: AITSL (2014).

developed by mathematics teachers and researchers from Monash University to describe what mathematics teachers who are doing their job well should know and do. The AAMT Standards are arranged into the three domains, as shown in Table 17.3.

Teachers can use the standards statements for each of these domains, and the elaborations in each of these frameworks, to audit their professional knowledge, attributes and practice, and identify areas for further development. The domains and their related standards also provide a way of organising a professional portfolio that can be maintained and updated throughout your teaching career. A portfolio is more than a scrapbook of lesson plans and classroom artefacts; instead, it assembles evidence of professional learning in the form of annotated materials that relate explicitly to a professional standards framework (Campbell, 2004; Reese, 2004). Annotations should explain *why* each item was selected for inclusion and *how* each item is linked to one or more elements of the portfolio's organising framework. For example, one reason for including a particular lesson plan might be that use of collaborative learning strategies and a task that applied mathematics to a real-world problem succeeded in engaging students who had previously showed little interest in learning. The annotations could explain how this lesson provides evidence of knowledge of students (AAMT Standard 1.1; AITSL Standards 1.3, 1.5), knowledge of students' learning of mathematics (AAMT Standard 1.3; AITSL Standards 1.2), and planning for learning (AAMT Standard 3.2; AITSL Standards 3.1, 3.2, 3.4). Some thought also needs to be given to the format of the portfolio to ensure that the contents are appropriately indexed and cross-referenced, even when the materials included in the portfolio change over time to

Table 17.3 Domains of the AAMT Standards

Domain 1 Professional knowledge	Domain 2 Professional attributes	Domain 3 Professional practice
1.1 Knowledge of students	2.1 Personal attributes	3.1 The learning environment
1.2 Knowledge of mathematics	2.2 Personal professional development	3.2 Planning for learning
1.3 Knowledge of students' learning of mathematics	2.3 Community responsibilities	3.3 Teaching in action 3.4 Assessment

reflect continuing development as a teacher. Electronic portfolios have several advantages in this regard: they are portable rather than bulky, easy to modify, and the contents can be hyperlinked to the organising framework in multiple ways to illustrate the richness of professional experience.

REVIEW AND REFLECT : Download the AAMT Standards from www.aamt.edu.au/ standards and read the elaborations for each domain and standard. Download the AITSL Standards from www.aitsl.edu.au/australian-professional-standards-for-teachers/standards/list. Discuss with peers how you could select material for inclusion in a professional portfolio organised around these two sets of standards. Materials could include annotated lesson plans and student work samples; teaching resources you have created or selected together with an indication of their purpose and effectiveness; sample pages from websites used in your teaching; photographs of your classroom or student work; feedback from students and colleagues on your teaching; messages from parents; journal articles and other samples of professional reading; and evidence of participation in professional development activities and how this influenced your teaching.

Make a list of materials you have already collected in your pre-service course and classify these as providing evidence against one or more of the ten standards. Identify any standards for which you presently have little evidence of accomplishment: these represent opportunities for planning further professional learning.

Conclusion

A range of professional development opportunities is available specifically for mathematics teachers. Joining the local mathematics teachers' professional association provides access to journals, newsletters, workshops, and information about new curricula and assessment policies. (Details of state- and territory-based associations affiliated with the AAMT can be found on the AAMT website.) Annual conferences organised by these professional associations offer a wide choice of sessions for teachers of mathematics across all levels of schooling, as well as opportunities to interact with colleagues from other schools. Professional journals and resources (books and web-based materials) published by these associations are also an excellent source of practical teaching ideas and summaries of research findings that can be used to improve students' learning. As knowledge is continually changing, it is also worth considering postgraduate study to keep abreast of developments in mathematics education and to explore areas of interest in greater depth than is possible in a pre-service program. Finally, participation in university-based research and teacher development projects can be an energising experience that encourages teachers to analyse and reflect on their own practice and students' learning. Many of the examples of practice we have presented in this book have come from projects in which we have worked in partnership with teachers interested in trying out new approaches. Involvement in research grounded in classrooms allowed them to develop their personal professional knowledge while contributing to knowledge about mathematics teaching and learning more generally through publication of journal articles and presentation of conference papers. Taking this path has introduced many teachers to the exciting world of education research, and reinforced their commitment to lifelong learning.

Recommended reading

Artzt, A. & Armour-Thomas, E. (2002). *Becoming a reflective mathematics teacher: A guide for observations and self-assessment.* Mahwah, NJ: Lawrence Erlbaum.

Australian Association of Mathematics Teachers. (2006). *Standards for excellence in teaching mathematics in Australian schools.* Retrieved from www.aamt.edu.au/standards

Australian Institute for Teaching and School Leadership. (2014). *Australian professional standards for teachers.* Retrieved from www.aitsl.edu.au/australian-professional-standards-for-teachers

Campbell, D. M. (2004). *How to develop a professional portfolio: A manual for teachers.* Boston, MA: Pearson, Allyn and Bacon.

Geiger, V., Muir, T. & Lamb, J. (2015). Video-stimulated recall as a catalyst for teacher professional learning. *Journal of Mathematics Teacher Education.* doi: 10.1007/s10857-015-9306-y

Killen, R. (2013). Becoming a reflective teacher. In *Effective teaching strategies: Lessons from research and practice* (6th ed., Chapter 5). South Melbourne: Cengage Learning.

REFERENCES

AAMT *see* Australian Association of Mathematics Teachers.

ACARA *see* Australian Curriculam Assessment and Reporting Authority.

ACER *see* Australian Council for Eduactional Research.

Adler, J. & Davis, Z. (2006). Opening another black box: Researching mathematics for teaching in mathematics teacher education. *Journal for Research in Mathematics Education, 37*(4), 270–296.

Ainley, J., Kos, J. & Nicolas, M. (2008). *Participation in science, mathematics and technology in Australia.* Camberwell, Vic.: ACER.

AITSL *see* Australian Institute for Teaching and School Leadership.

Alejandre, S. (2005). The reality of using technology in the classroom. In W. Masalski & P. Elliott (Eds), *Technology-supported mathematics learning environments* (pp. 137–150). Reston, VA: NCTM.

Anthony, B. & Walshaw, M. (2002). Swaps and switches: Students' understanding of commutativity. In B. Barton, K. Irwin, M. Pfannkuch & M. Thomas (Eds), *Mathematics in the South Pacific* (Proceedings of the 25th annual conference of the Mathematics Education Research Group of Australasia, p. 91–99). Sydney: MERGA.

Anthony, G., Rawlins, P. & Winsley, J. (2002). Accelerated learning in New Zealand secondary school mathematics. *Australasian Journal for Gifted Education, 11*(2), 11–17.

Anthony, G. & Walshaw, M. (2007). *Effective pedagogy in mathematics/pāngarau: Best evidence synthesis iteration [BES].* Wellington: New Zealand Ministry of Education.

Anton, H., Bivens, I, & Davis, S. (2005). *Calculus.* Hoboken, NJ: Wiley.

Arcarvi, A. (2003). The role of visual representations in the learning of mathematics. *Educational Studies in Mathematics, 52*, 215–241.

Arnold, P. M. (2013). *Statistical investigative questions: An enquiry into posing and answering investigative questions from existing data* (PhD thesis, The University of Auckland). Retrieved from https://researchspace.auckland.ac.nz/bitstream/handle/2292/21305/whole.pdf?sequence=2

Arnold, S. (2001). Thinking irrationally. *The Australian Mathematics Teacher,* 57(1), 38–41.

Artzt, A. & Armour-Thomas, E. (2002). *Becoming a reflective mathematics teacher: A guide for observations and self-assessment.* Mahwah, NJ: Lawrence Erlbaum Associates Inc.

Asp, G., Dowsey, J., Stacey, K. & Tynan, D. (1995). *Graphic algebra: Explorations with a function grapher.* Melbourne, Vic.: Curriculum Corporation.

Attard, C. (2010). Students' experiences of mathematics during the transition from primary to secondary school. In L. Sparrow, B. Kissane & C. Hurst. (Eds), *Shaping the future of mathematics education* (Proceedings of the 33rd annual conference of the Mathematics Education Research Group of Australasia, pp. 53–60). Adelaide: MERGA.

Attard, C. & Orlando, J. (2014). Early career teachers, mathematics and technology: Device conflict and emerging mathematical knowledge. In J. Anderson, M. Cavanagh & A. Prescott (Eds), *Curriculum in focus: Research guided practice* (Proceedings of the 37th annual conference of the Mathematics Education Research Group of Australasia, pp. 71–78). Sydney: MERGA.

Atweh, B. (2011). Equity and quality in mathematics as ethical issues. In B. Atweh, M. Graven, W. Secada & P. Valero (Eds), *Mapping equity and quality in mathematics education* (pp. 563–576). Dordrecht, The Netherlands: Springer.

Atweh, B. & Ala'i, K. (2012). Socially response-able mathematics education: Lessons from three teachers. In J. Dindyal, L. P. Cheng & S. F. Ng (Eds), *Mathematics education: Expanding horizons* (Proceedings of the 35th annual conference of the Mathematics Education Research Group of Australasia, pp. 99–105). Singapore: MERGA.

Atweh, B. & Brady, K. (2009). Socially response-able mathematics education: Implications of an ethical approach. *Eurasia Journal of Mathematics, Science & Technology,* 5(3), 267–276.

Atweh, B. & Goos, M. (2011). The Australian mathematics curriculum: A move forward or back to the future? *Australian Journal of Education,* 55(3), 214–228.

Atweh, B., Goos, M., Jorgensen, R. & Siemon, D. (Eds). (2012). *Engaging the Australian curriculum: Mathematics—Perspectives from the field.* Mathematics Education Research Group of Australasia. Retrieved from www.merga.net.au/node/223

Atweh, B., Graven, M., Secada, W. & Valero, P. (2011). *Quality and equity agendas in mathematics education.* Dordrecht, The Netherlands: Springer.

Atweh, B., Vale, C. & Walshaw, M. (2012). Equity, diversity, social justice and ethics: Common concerns or divergent agendas? In B. Perry, T. Lowrie, T. Logan, A. MacDonald & J. Greenlees (Eds), *Research in mathematics education in Australasia 2008–2011* (pp. 39–65). Rotterdam: Sense Publishers B.V.

Australian Association of Mathematics Teachers (AAMT). (2006). *Standards for excellence in teaching mathematics in Australian schools.* Retrieved from www.aamt.edu.au/standards

Australian Association of Mathematics Teachers (AAMT). (2014). *AAMT position paper on digital learning in school mathematics.* AAMT.

Australian Bureau of Statistics. (2006). *CensusAtSchool archive.* Retrieved from www.abs.gov.au/websitedbs/cashome.nsf/Home/Home

Australian Bureau of Statistics. (2012). *1301.0—Year Book Australia, 2012.* Retrieved from www.abs.gov.au/ausstats/abs@.nsf/mf/1301.0

Australian Council for Educational Research. (1997). *Progressive achievement test in mathematics.* Camberwell, Vic.: ACER.

Australian Curriculum, Assessment and Certification Authorities (ACACA). (1996). *Guidelines for assessment quality and equity.* Retrieved from http://acaca.bostes.nsw.edu.au/files/pdf/guidelines.pdf

Australian Curriculum, Assessment and Reporting Authority (ACARA). (2012a). *Australian curriculum: History.* Retrieved from www.australiancurriculum.edu.au/History/Rationale

Australian Curriculum Assessment and Reporting Authority (ACARA). (2013b). *National assessment program.* Retrieved from www.nap.edu.au/

Australian Curriculum, Assessment and Reporting Authority (ACARA). (2014). *Australian curriculum.* Retrieved from www.australiancurriculum.edu.au/#

Australian Curriculum, Assessment and Reporting Authority (ACARA). (2015). *Australian curriculum: Mathematics* (Version 8.1). Retrieved from www.australiancurriculum.edu.au/mathematics/curriculum/f-10?layout=1

Australian Education Council. (1991). *A national statement on mathematics for Australian schools.* Melbourne: Curriculum Corporation.

Australian Education Council. (1994). *Mathematics—A curriculum profile for Australian schools.* Carlton, Vic.: Curriculum Corporation.

Australian Institute for Teaching and School Leadership (AITSL). (2014). *Australian professional standards for teachers.* Retrieved from www.aitsl.edu.au/australian-professional-standards-for-teachers

Australian Mathematics Trust. (n.d.). Australian mathematics competition. Retrieved from www.amt.canberra.edu.au/

Averill, R. (2012). Caring teaching practices in multi-ethnic mathematics classrooms: Attending to health and well-being. *Mathematics Education Research Journal, 24*, 105–128.

Averill, R. & Clark, M. (2012). Respect in teaching and learning mathematics: Professionals who know, listen to, and work with students. *Set: Research Information for Teachers, 3*, 50–57.

Averill, R. & Clark, M. (2013). Respectful and responsive pedagogies for mathematics and statistics. In V. Steinle, L. Ball & C. Bardini (Eds), *Mathematics education: Yesterday, today and tomorrow* (Proceedings of the 36th annual conference of the Mathematics Education Research Group of Australasia, pp. 66–73). Melbourne: MERGA.

Ball, D. L., Thames, M. H. & Phelps, G. (2008). Content knowledge for teaching: What makes it special? *Journal of Teacher Education, 59*, 389–407.

Ball, L. & Stacey, K. (2001). New literacies for mathematics: A new view of solving equations. *The Mathematics Educator, 6*(1), 55–62.

Ball, L. & Stacey, K. (2003). What should students record when solving problems with CAS? In J. Fey, A. Cuoco, C. Kieran, L. McMullin & R. Zblek (Eds), *Computer algebra systems in secondary school mathematics education* (pp. 289–303). Reston, VA: NCTM.

Ball, L. & Stacey, K. (2005). Middle school CAS: Opportunities and challenges for teachers and students. In J. Mousley, L. Bragg & C. Campbell (Eds), *Mathematics – Celebrating achievement* (MAV annual conference, pp. 121–128). Melbourne: MAV.

Barber, J., Barakos, L. & Bergman, L. (2000). *Parent partners: Workshops to foster school/home/family partnerships.* Berkeley, CA: Lawrence Hall of Science, University of California.

Barkatsas, A. & Malone, J. (2005). A typology of mathematics teachers' beliefs about teaching and learning mathematics and instructional practices. *Mathematics Education Research Journal, 17*, 69–90.

Barnes, H. (2007). Effectively using new paradigms in the teaching and learning of mathematics: An action research in a multicultural South African classroom. http://dipmathath.unipa.it/grim/SiBarnes.pdf>

Barnes, M. (1991). *Investigating change: An introduction to calculus for Australian schools.* Melbourne: Curriculum Corporation.

Barnes, M. (2000a). Effects of dominant and subordinate masculinities on interactions in a collaborative learning classroom. In J. Boaler (Ed.), *Multiple perspectives on mathematics teaching and learning* (pp. 145–169). Westport, CT: Ablex.

Barnes, M. (2000b). 'Magical moments' in mathematics: Insights into the process of coming to know. *For the Learning of Mathematics,* 20(1), 33–43.

Barrow, J. (2013). *Mathletics.* London: Random House.

Bartolini Bussi, M. & Borba, M. (2010). The role of resources and technology in mathematics education. *ZDM,* 42(1), 1–4.

Batanero, C., Burrill, G. & Reading, C. (2011). *Teaching statistics in school mathematics—Challenges for teaching and teacher education.* Dordrecht, The Netherlands: Springer.

Bate, F. G., Day, L. & Macnish, J. (2013). Conceptualising changes to pre-service teachers' knowledge of how to best facilitate learning in mathematics: A TPACK inspired initiative. *Australian Journal of Teacher Education,* 38(5), 14–30.

Battista, M. T. (2007). The development of geometric and spatial thinking. In F. Lester (Ed.), *Second handbook of research on mathematics teaching and learning* (pp. 843–908). Reston, VA: NCTM.

Beatty, R. & Geiger, V. (2010). Technology, communication and collaboration: Re-thinking communities of inquiry, learning and practice. In C. Hoyles & J.-B. Lagrange (Eds), *Mathematics education and technology: Rethinking the terrain* (pp. 251–284). New York: Springer.

Beesey, C., Clarke, B., Clarke, D., Stephens, M. & Sullivan, P. (2001). *Effective assessment for mathematics: CSF Levels 4–6.* Melbourne: Longman.

Belencky, M. F., Clinchy, B. M., Goldberger, N. R. & Tarule, J. M. (1986). *Women's ways of knowing: The development of self, voice and mind.* New York: Basic Books Inc.

Ben-Zvi, D. & Garfield, J. (2004). Statistical literacy, reasoning, and thinking: Goals, definitions, and challenges. In D. Ben-Zvi & J. Garfield (Eds), *The challenge of developing statistical literacy, reasoning and thinking.* Dordrecht, The Netherlands: Kluwer Academic Publishers.

Berthelot, R. & Salin, M. (1998). The role of pupils' spatial knowledge in the elementary teaching of geometry. In C. Mammana & V. Villani (Eds), *Perspectives on the teaching of geometry for the 21st century* (pp. 71–78). Dordrecht: Kluwer.

Beswick, K. (2005). The beliefs/practice connection in broadly defined contexts. *Mathematics Education Research Journal,* 17, 39–68.

Beswick, K. (2007). Teachers' beliefs that matter in secondary classrooms. *Educational Studies in Mathematics,* 65, 95–120.

Beswick, K. (2007–8). Influencing teachers' beliefs about teaching mathematics for numeracy to students with mathematics learning difficulties. *Mathematics Teacher Education and Development,* 9, 3–20.

Beswick, K. (2011a). Make your own paint chart: A realistic context for developing proportional reasoning with ratios. *The Australian Mathematics Teacher,* 67(1), 6–12.

Beswick, K. (2011b). Positive experiences with negative numbers: Building on students' in and out of school experiences. *The Australian Mathematics Teacher,* 67(2), 31–40.

Beswick, K. (2012). Teachers' beliefs about school mathematics and mathematician's mathematics and their relationship to practice. *Educational Studies in Mathematics,* 79, 127–147.

Bicknell, B. (2009). Multiple perspectives on the education of mathematically gifted and talented students. (Unpublished doctoral dissertation). Massey University, Auckland, New Zealand.

Biehler, R., Ben-Zvi, D., Bakker, A. & Makar, K. (2013). Technology for enhancing statistical reasoning at the school level. In K. Clements, A. Bishop, C. Keitel, J. Kilpatrick & F. Leung (Eds), *Third international handbook of mathematics education* (pp. 643–689). New York: Springer.

Biggs, J. & Collis, K. (1991). Developmental learning and the quality of intelligent behaviour. In H. Rowe (Ed.), *Intelligence: Reconceptualisation and measurement* (pp. 57–76). Hillsdale, NJ: Erlbaum.

Billings, E.M., & Klanderman, D. (2000). Graphical representations of speed: Obstacles pre-service K–8 teachers experience. *School Science and Mathematics,* 100, 440.

Bishop, A. (1994). Cultural conflicts in mathematics education: Developing a research agenda. *For the Learning of Mathematics,* 14(2), 15–18.

Bishop, A. (2001). What values do you teach when you teach mathematics? *Teaching Children Mathematics,* 7(6), 346–352.

Black, R. (2007). Crossing the divide. *The Education Foundation.* (ERIC Document No. ED501899).

Blaisdell, R. (2012). *Student understanding in the concept of limit in calculus: How student responses vary.* Paper presented at the 15th annual conference on Research in Undergraduate Mathematics Education, Portland, OR, 23–25 February 2012.

Blum, W. & Kaiser, G. (1984). Analysis of applications and of conceptions for an application-oriented mathematics instruction. In J. Berry, D. N. Burghes, I. D. Huntley, D. J. G. James & A. O. Moscardini (Eds), *Teaching and applying mathematical modelling* (pp. 201–214). Chichester, UK: Ellis Horwood.

Boaler, J. (1997a). *Experiencing school mathematics: Teaching styles, sex and setting.* Buckingham, England: Open University Press.

Boaler, J. (1997b). Reclaiming school mathematics: The girls fight back. *Gender and Education,* 9(3), 285–306.

Boaler, J. (1998). Open and closed mathematics: Student experiences and understanding. *Journal for Research in Mathematics Education*, 29, 41–62.

Boaler, J. (2000). Mathematics from another world: Traditional communities and the alienation of learners. *Journal of Mathematical Behavior*, 18(4), 1–19.

Boaler, J. (2002a). *Experiencing school mathematics: Traditional and reform approaches to teaching and their impact on student learning*. Mahwah, NJ: Lawrence Erlbaum Associates.

Boaler, J. (2002b). Learning from teaching: Exploring the relationship between reform curriculum and equity. *Journal for Research in Mathematics Education*, 33, 239–258.

Boaler, J. (2008a). *The elephant in the room: Helping children learn and love their most hated subject—and why it's important for America*. New York: Viking.

Boaler, J. (2008b). Promoting 'relational equity' and high mathematics achievement through an innovative mixed ability approach. *British Educational Research Journal*, 34(2), 167–194.

Boaler, J. & Staples, M. (2008). Creating mathematical futures through an equitable teaching approach: The case of Railside School. *Teachers College Record*, 110(3), 608–645.

Board of Studies New South Wales (2002). *Mathematics Years 7–10 Syllabus*. Sydney: Author

Board of Studies, Teaching and Educational Standards New South Wales. (n.d.). *Assessment*. Retrieved from http://syllabus.bostes.nsw.edu.au/mathematics/mathematics-k10/syllabus-assessment/

Booker, G., Bond, D., Sparrow, L. & Swan, P. (2004). *Teaching primary mathematics* (3rd ed.). Sydney: Pearson Education Australia.

Booker, G. & Windsor, W. (2010). Developing algebraic thinking: Using problem-solving to build from number and geometry in the primary school to the ideas that underpin algebra in high school and beyond. *Procedia Social and Behavioural Sciences*, 8, 411–419.

Borba, M. C. & Villarreal, M. E. (2005). *Humans-with-media and the reorganisation of mathematical thinking: Information and communication technologies, modelling, experimentation and visualization*. New York: Springer.

Bourke, P. (2002). Pinwhell aperiodic tiling. Retrieved from http://locals.wasp.uwa.edu.au/~pbourke/texture_coulour/nonperiodic/ (12 April, 2007).

Brader-Araje, L. (2004). Calling home: Keeping in contact with students' families. In M. Chappell, J. Choppin & J. Salls (Eds), *Empowering the beginning teacher of mathematics in high school* (pp. 52–53). Reston VA: NCTM.

Brendefur, J., Hughes, G. & Ely, R. (2015). A glimpse into secondary students' understanding of functions. *International Journal for Mathematics Teaching and Learning* (January), 1–22.

Brinkworth, P. & Scott, P. (2001). The place of mathematics: The London Underground. *The Australian Mathematics Teacher*, 57(1), 2–5.

Brinkworth, P. & Scott, P. (2002). The place of mathematics: San Francisco—the Golden Gate to geometry. *The Australian Mathematics Teacher*, 58(3), 2–5.

Brodie, K. (2010). *Teaching mathematical reasoning in secondary school classrooms*. New York: Springer.

Brown, D. (2015). *Tracker* (computer software). Available to download from www.cabrillo.edu/~dbrown/tracker/index.html

Brown, G. & Quinn, R. (2006). Algebra students' difficulty with fractions: An error analysis. *The Australian Mathematics Teacher*, 62(4), 28–40.

Brown, J. & Stillman, G. (2006). Defining moments in the graphing calculator solution of a cubic function task. *Nordic Studies in Mathematics Education*, 11(3), 57–84.

Brown, R. (2011). Actualising potential in the classroom: Moving from practicing to be numerate towards engaging in the literate practice of mathematics. In C. Wyatt-Smith, J. Elkins & S. Gunn (Eds.), *Multiple perspectives on difficulties in learning literacy and numeracy* (pp. 275–295). Dordrecht: Springer.

Buckskin, P. (2000). *Achieving educational equality for Indigenous Australians: A pathway of hope.* Paper presented at the 2000 Curriculum Corporation Conference, 18–19 May 2000.

Burke, M., Eickson, D., Lott, J. W. & Obert, M. (2001). *Navigating through algebra in grades 9–12.* Reston, VA: NCTM.

Burnett, J., Irons, C. & Turton, A. (2005). *GEO paper polygons* (rev. ed.). Brisbane: ORIGO Publications.

Burns, B. & Brade, G. (2003). Using the geoboard to enhance measurement instruction in the secondary mathematics classroom. In D. H. Clements (Ed.), *Learning and teaching measurement: 2003 yearbook* (pp. 256–270). Reston, VA: NCTM.

Burton, L. (1995). Moving towards a feminist epistemology of mathematics. In P. Rogers & G. Kaiser (Eds), Equity in mathematics education: Influences of feminism and culture (pp. 209–226). London: Falmer Press.

Bush, W. & Greer, A. (Eds). (1999). *Mathematics assessment: A practical handbook for grades 9–12.* Reston, VA: NCTM.

Byers, T. (2009). The BASICS intervention program for at-risk students. *The Australian Mathematics Teacher*, 65(1), 6–11.

Cai, J. (2003). Investigating parental roles in students' learning of mathematics from a cross-national perspective. *Mathematics Education Research Journal*, 15, 87–106.

Cai, J., Cirillo, M., Pelesko, J. A, Borromeo Ferri, R., Borba, M., Geiger, V., … Kwon, O.-N. (2014). Mathematical modeling in school education: Mathematical, cognitive, curricular, instructional and teacher education perspectives. *Proceedings of the Joint Meeting of PME 38 and PME-NA 36*, 1, 145–172.

Cai, J. & Howson, G. (2013). Towards an international mathematics curriculum. In M. A. Clements, A. Bishop, C. Keitel, J. Kilpatrick & F. Leung (Eds), *Third international handbook of mathematics education* (pp. 949–974). New York: Springer.

Cai, J. & Sun, W. (2002). Developing students' proportional reasoning: A Chinese perspective. In B. Litwiller & G. Bright (Eds), *Making sense of fractions, ratios and proportions: 2002 yearbook* (pp. 195–205). Reston, VA: NCTM.

Cairney, T. (2000). Beyond the classroom walls: The rediscovery of the family and community as partners in education. *Educational Review*, 52(2), 163–74.

Callingham, R. (2008). Dialogue and feedback: Assessment in the primary classroom. *Australian Primary Mathematics Classroom*, 13(3), 18–21.

Callingham, R. (2011). Assessing statistical understanding in middle schools: Emerging issues in a technology-rich environment. *Technological Innovations in Statistics Education*, 5(1), 1–12.

Callingham, R. & McIntosh, A. (2002). Mental computation competence across Years 3 to 10. In B. Barton, K. Irwin, M. Pfannkuch & M. Thomas (Eds), *Mathematics education in the South Pacific* (Proceedings of the 25th annual conference of the Mathematics Education Research Group of Australasia, pp. 155–162). Sydney: MERGA.

Callingham, R. & Watson, J. (2004). A developmental scale of mental computation with part whole numbers. *Mathematics Education Research Journal*, 16(2), 69–86.

Cameron, S. & Ball, L. (2015). CAS or pen-and-paper: Factors that influence students' choices. In M. Marshman, V. Geiger & A. Bennison (Eds), *Mathematics education in the margins* (Proceedings of the 38th annual conference of the Mathematics Education Research Group of Australasia, pp. 141–148). Sunshine Coast: MERGA.

Campbell, D. M. (2004). *How to develop a professional portfolio: A manual for teachers*. Boston, MA: Pearson, Allyn and Bacon.

Carlson, M., Jacobs, S., Coe, E., Larsen, S. & Hsu, E. (2002). Applying covariational reasoning while modeling dynamic events: A framework and a study. *Journal for Research in Mathematics Education*, 33(5), 352–378.

Carlson, M. P., Madison, B. & West, R. D. (2015). A study of students' readiness to learn calculus. *International Journal of Research in Undergraduate Mathematics Education*, 1(2), 209–233.

Carlson, M., Oehrtman, M., & Thompson, P.W. (2007). Foundational reasoning abilities that promote coherence in student's understanding of function. In M. Carlson & C. Ramussen (Eds), *Making the connection: Research and teaching in undergraduate mathematics* (pp. 150–171). Washington, DC: Mathematical Association of America.

Carmichael, C. (2014). Gender, parental beliefs and children's mathematics performance: Insights into a longitudinal study of Australian children. In J. Anderson, M. Cavanagh & A. Prescott (Eds), *Curriculum in focus: Research guided practice* (Proceedings of the 37th annual conference of the Mathematics Education Research Group of Australasia, pp. 119–126). Sydney: MERGA.

Carpenter, T. & Levi, L. (2000). *Developing conceptions of algebraic reasoning in the primary grades* (Research Report 00-2). Madison, WI: National Centre for Improving Student Learning and Achievement in Mathematics and Science. Retrieved from www.wcer.wisc.edu/ncisl

Carroll, J. (2006). *Working with primary school students who struggle with mathematics.* Paper presented at the annual conference of the Australian Association for Research in Education, Adelaide, 27–30 November.

Castro Sotos, A., Vanhoof, S., Van den Noortgate, W. & Onghena, P. (2007). Students' misconceptions of statistical inference: A review of the empirical evidence from research on statistics education. *Educational Research Review*, 2(2), 98–113.

Cavanagh, M. (2008). Trigonometry from a different angle. *The Australian Mathematics Teacher*, 64(1), 25–30.

Chan, L. K. S. & Dally, K. (2000). Review of literature. In W. Louden, L. K. S. Chan, J. Elkins, D. Greaves, H. House, M. Milton, … C. Van Kraayenoord (Eds), *Mapping the territory—Primary students with learning difficulties: Literacy and numeracy* (Vol. 2, pp. 171–358). Canberra, Australia: Department of Education, Training, and Youth Affairs.

Chapman, A. (2001). Maths talk is boys' talk: Constructing masculinity in school mathematics. In W. Martino & B. Meyenn (Eds), *What about the boys? Issues of masculinity in schools* (pp. 199–210). Buckingham, England: Open University.

Chernoff, E. J. & Sriraman, B. (Eds). (2014). *Probabilistic thinking: Presenting plural perspectives.* Berlin: Springer Science.

Chick, H. (2013). Busting myths. *The Australian Mathematics Teacher*, 69(4), 16–22.

Chick, H., Baker, M., Pham, T. & Cheng, H. (2006). Aspects of teachers' pedagogical content knowledge for decimals. In J. Novotna, H. Moraova, M. Kratka & N. Stehlikova (Eds), *Proceedings of the 30th conference of the International Group for the Psychology of Mathematics Education* (Vol. 2, pp. 297–304). Prague: PME.

Clarke, D. (1988). *The mathematics curriculum and teaching program professional development package: Assessment alternatives in mathematics.* Canberra: Curriculum Development Centre.

Clarke, D. & Clarke, B. (2008). Is time up for ability grouping? *EQ Australia* (Autumn), 31–33.

Clarke, D., Roche, A., Cheeseman, J. & Sullivan, P. (2014). Encouraging students to persist when working on challenging tasks: Some insights from teachers, *The Australian Mathematics Teacher, 70(1)*, 3–11.

Clarke, D. & Stephens, M. (1996). The ripple effect: The instructional impact of the systematic introduction of performance assessment in mathematics. In M. Birenbaum & F. Douchy (Eds), *Alternatives in assessment of achievements, learning processes and prior knowledge* (Chapter 3). New York: Springer.

Clarke, D. J. (1996a). Assessment. In A. Bishop, K. Clements, C. Keitel, J. Kilpatrick & C. Laborde (Eds), *International handbook of mathematics education* (Vol. 1, pp. 327–370). Dordrecht, The Netherlands: Kluwer.

Clarke, D. J. (1996b). The triple jump. *The Australian Mathematics Teacher*, 52(2), 4–7.

Clarke, D. J. (1997). *Constructive assessment in mathematics: Practical steps for classroom teachers.* Berkeley, CA: Key Curriculum Press.

Clarke, D. J., Goos, M. & Morony, W. (2007). Problem solving and working mathematically: An Australian perspective. *Zentralblatt für Didaktik der Mathematik* (International Reviews on Mathematical Education), 39(5–6), 475–490.

Clarke, D.M. (nd). Assessment for teaching and learning. Retrieved at http//cme.open.ac.uk/ Clarke_on_tasks1.htm (12 April 2007)

Clarke, L. & Bana, J. (2001). Phantom classmates: A case study of talented mathematics students learning via telematics. In J. Bobis, B. Perry & M. Mitchelmore (Eds), *Numeracy and beyond* (Proceedings of the 24th annual conference of the Mathematics Education Research Group of Australasia, pp. 163–169). Sydney: MERGA.

Clatworthy, N. & Galbraith, P. (1987). Mathematical modelling: Innovation in senior mathematics. *Australian Senior Mathematics Journal*, 1(2), 38–49.

Clatworthy, N. & Galbraith, P. (1989). Mathematical modelling: an attack on conformity. *Australian Senior Mathematics Journal*, 3(2), 38–49.

Clausen-May, T. (2005). Teaching maths to pupils with different learning styles. London: Chapman Publishing.

Clements, M. A., Grimison, L. & Ellerton, N. (1989). Colonialism and school mathematics in Australia 1788–1988. In N. Ellerton & M. A. Clements (Eds), *School mathematics: The challenge to change* (pp. 50–78). Geelong, Vic.: Deakin University Press.

Clements, R. R. (1989). *Mathematical modelling: A case study approach.* Cambridge: Cambridge University Press.

Coad, L. (2006). Paper folding in the middle school classroom and beyond. *The Australian Mathematics Teacher, 62*(1), 6–13.

Cobb, P. (1994). Where is the mind? Constructivist and sociocultural perspectives on mathematical development. *Educational Researcher, 23*(7), 13–20.

Coe, E. (2007). *Modeling teachers' ways of thinking about rate of change* (PhD thesis, Arizona State University). Retrieved from http://pat-thompson.net/PDFversions/Theses/2007Ted.pdf

Collins, J. (2001). *The education of gifted and talented children.* Canberra: Commonwealth of Australia.

Confrey, J. & Kazak, S. (2006). A thirty-year reflection on constructivism in mathematics education in PME. In A. Gutierrez & P. Boero (Eds), *Handbook of research on the psychology of mathematics education: Past, present and future* (pp. 305–345). Rotterdam: Sense Publishers.

Cooper, B. & Dunne, M. (2000). *Assessing children's mathematical knowledge: Social class, sex and problem-solving.* Buckingham: Open University Press.

Cooper, C., Dole, S., Geiger, V. & Goos, M. (2012). Numeracy in society and environment. *The Australian Mathematics Teacher, 68*(1), 16–20.

Corbett, M. (2009). Rural schooling in mobile modernity: Returning to the places I've been. *Journal of Research in Rural Education, 24*(7). Retrieved from http://jrre.psu.edu/articles/24-7.pdf

Crozier, G. (2000). *Parents and schools: Partners or protagonists?* Stoke on Trent: Trentham Books.

Csikszentmihalyi, M., Rathune, K. & Whalen, S. (1997). *Talented teenagers: The roots of success and failure.* Cambridge: Cambridge University Press.

Cunningham, K. C. (2006). Mathematics as a subject in the school curriculum. In M. Stephens (Ed.), *Master classes in mathematics* (Centenary publication of the Mathematical Association of Victoria, pp. 26–39). Brunswick, Vic.: MAV (original work published in 1956).

Curriculum Branch, Education Department of Western Australia. (1984). *Applying mathematics.* Perth: Author.

Curriculum Corporation. (1991). *A national statement on mathematics for Australian schools.* Carlton, Vic.: Curriculum Corporation and Australian Education Council.

D'Ambrosio, U. (1998). *Ethnomathematics: The art and technique of explaining and knowing.* (P. B. Scott, Trans.). Las Cruces, NM: International Study Group of Ethnomathematics (original work published in 1990).

Darling-Hammond, L. & Rothman, R. (2011). *Teacher and leader effectiveness in high-performing education systems.* Washington, DC: Alliance for Excellent Education.

Darling-Hammond, L., Wei, R. C., Andree, A., Richardson, N. & Orphanos, S. (2009). *Professional learning in the learning profession: A status report on teacher development in the United States and abroad.* Stanford, CA: National Staff Development Council and The School Redesign Network at Stanford University.

Daroczy, G., Wolska, M., Meurers, W. D. & Nuerk, H.-C. (2015). Word problems: A review of linguistic and numerical factors contributing to their difficulty. *Frontiers in Psychology,* 6, 348.

Davis, R. B. (1990). Discovery learning & constructivism. In R. B. Davis, C. A. Maher & N. Noddings (Eds), *Constructivist views on the teaching and learning of mathematics* (pp. 19–29). Reston, VA: NCTM.

Day, L. (2013). Using statistics to explore cross-curricular and social issues opportunities. *The Australian Mathematics Teacher,* 69(4), 3–7.

Day, M. (1997). From the experiences of women mathematicians: A feminist epistemology for mathematics (PhD thesis, Massey University, New Zealand).

De Bock, D., Van Dooren, W. & Verschaffel, L. (2005). Not everything is proportional: Task design and small-scale experiment. In H. Chick & J. Vincent (Eds), *Proceedings of the 29th conference of the International Group for the Psychology of Mathematics Education* (Vol. 1, pp. 97–102). Melbourne: PME.

delos Santos, A. & Thomas, M. (2003). Representational ability and understanding of derivative. In N. A. Pateman, B. J. Dougherty & J. Zilliox (Eds), *Proceedings of the 27th Conference of PME* (Vol. 2, pp. 325–332). Honolulu, HI: University of Hawaii.

delos Santos, A. & Thomas, M. (2005). The growth of schematic thinking about derivative. In P. Clarkson, A. Downton, D. Gronn, M. Horne, A. McDonough, R. Pierce & A. Roche (Eds), *Building connections: Research, theory and practice* (Proceedings of the 28th Conference of The Mathematics Education Research Group of Australasia, Melbourne, Vol. 1, pp. 377–384). Sydney: MERGA.

Department for Education, United Kingdom. (2013). Mathematics programme of study for key stage 4 (Draft). https://media.education.gov.uk/assets/files/pdf/m/mathematics%20-%20key%20stage%204%2004-02-13.pdf

Department of Education and Early Childhood Development. (2013). *Fractions and decimals online interview.* Retrieved from www.education.vic.gov.au/school/teachers/teachingresources/discipline/maths/continuum/Pages/fracdecint.aspx

Department of Education and Training, New South Wales. (1998). *Count me in too: Professional development package.* Sydney: Author.

Department of Education and Training, New South Wales. (2002). *Counting on: Re-connecting conceptual development.* Sydney: Author.

Department of Education and Training, Victoria. (2013a). *Assessment advice.* Retrieved from www.education.vic.gov.au/studentlearning/assessment/preptoyear10/assessadvice/default.htm

Department of Education and Training, Victoria. (2013b). *Scaffolding numeracy in the middle years: Assessment materials for multiplicative thinking.* Retrieved from www.education.vic.gov.au/school/teachers/teachingresources/discipline/maths/assessment/Pages/scaffoldnum.aspx

Department of Education and Training, Victoria. (2014a). *Fractions and decimals online interview.* Melbourne: State Government, Victoria. Retrieved from www.education.vic.gov.au/school/teachers/teachingresources/discipline/maths/continuum/Pages/fracdecint.aspx

Department of Education and Training, Victoria. (2014b). *Assessment for common misunderstandings.* Melbourne: State Government, Victoria. Retrieved from www.education.vic.gov.au/school/teachers/teachingresources/discipline/maths/assessment/Pages/misunderstandings.aspx

Department of Education and Training, Victoria. (2016). *Framework for improving student outcomes.* Melbourne: State Government, Victoria.

Department of Education, Queensland. (2004). *The shape we're in.* Retrieved from http://education.qld.gov.au/corporate/newbasics/pdfs/yr9rt7a.pdf (12 April 2007).

Department of Education, Queensland. (2016). *Assessment.* Retrieved from http://education.qld.gov.au/curriculum/framework/p-12/index.html

Department of Employment, Education, Training and Youth Affairs. (1997). *Numeracy = everyone's business* (Report of the Numeracy Education Strategy Development Conference). Adelaide: AAMT.

Dharmadasa, K., Nakos, A., Bament, J., Edwards, A. & Reeves, H. (2014). Beyond *NAPLAN* testing ... Nurturing mathematical talent. *The Australian Mathematics Teacher*, 70(2), 22–27.

Dickerson, D. & Doerr, H. (2014). High school mathematics teachers' perspectives on the purposes of mathematical proof in school mathematics. *Mathematics Education Research Journal*, 26(4), 711–733.

Diezmann, C., Faragher, R., Lowrie, T., Bicknell, B. & Putt, I. (2004). Exceptional students in mathematics. In B. Perry, G. Anthony & C. Diezmann (Eds), *Research in mathematics education in Australasia 2000–2003* (pp. 175–196). Flaxton, Qld: Post Pressed.

Diezmann, C. & Lowrie, T. (2009). An instrument for assessing primary students' knowledge of information graphics in mathematics. *Assessment in Education: Principles, Policy and Practice*, 16(2), 131–147.

Diezmann, C., Stevenson, M. & Fox, J. (2012). Supporting exceptional students to thrive mathematically. In B. Perry, T. Lowrie, T. Logan, A. McDonald & J. Greenlees (Eds), *Research in mathematics education in Australasia 2008–2011*. Rotterdam: Sense Publishers.

Diezmann, C., Thornton, C. & Watters, J. (2003). Addressing the needs of exceptional students through problem solving. In F. Lester & R. Charles (Eds), *Teaching mathematics through problem solving* (pp. 169–182). Reston, VA: NCTM.

Diezmann, C., Watters, J. & English, L. (2001). Implementing mathematical investigations with young children. In J. Bobis, B. Perry & M. Mitchelmore (Eds), *Numeracy and beyond* (Proceedings of the 24th annual conference of the Mathematics Education Research Group of Australasia, pp. 170–177). Sydney: MERGA.

Dimarco, S. (2009). Crossing the divide between teacher professionalism and national testing in middle school mathematics? In R. Hunter, B. Bicknell & T. Burgess (Eds), *Crossing divides* (Proceedings of the 32nd annual conference of the Mathematics Education Research Group of Australasia, pp. 673–676). Palmerston North, NZ: MERGA.

Dole, S. (2003). Questioning numeracy programs for at-risk students in the middle years of schooling. In L. Bragg, C. Campbell, G. Herbert & J. Mousley (Eds), *Mathematics education research: Innovation, networking, opportunity* (Proceedings of the 26th annual conference of the Mathematics Education Research Group of Australasia, pp. 278–285). Sydney: MERGA.

Dole, S., Cooper, T.J., Baturo, A.R. & Conpolia, Z. (1997). Year 8, 9 and 10 students' understanding and access of percent knowledge. In F. Biddulph & K. Carr (Eds), *People in mathematics education* (Proceedings of the 20th annual conference of the Mathematics Education Research Group of Australasia, p. 147–54). Rotorua: MERGA.

Doorman, M., Drijvers, P., Gravemeijer, K., Boon, P. & Reed, H. (2012). Tool use and the development of the function concept: From repeated calculations to functional thinking. *International Journal of Science and Mathematics Education*, 10(6), 1243–1267.

Downton, A. (2009). It seems to matters not whether it is partitive or quotitive division when solving one step division problems. In R. Hunter, B. Bicknell & T. Burgess (Eds), *Crossing divides* (Proceedings of the 32nd annual conference of the Mathematics Education Research Group of Australasia, Vol. 1). Palmerston North, NZ: MERGA.

Drake, M. (2013). How heavy is my rock? An exploration of students' understanding of the measurement of weight. In V. Steinle, L. Ball & C. Bardini (Eds), *Mathematics education: Yesterday, today and tomorrow* (Proceedings of the 36th annual conference of the Mathematics Education Research Group of Australasia, pp. 250–257). Melbourne: MERGA.

Dreyfus, T., Hershkowitz, R. & Schwarz, B. (2001). The construction of abstract knowledge in interaction. *Cognitive Science Quarterly*, 1, 307–368.

Drouhard, J.-P. & Teppo, A. R. (2004). Symbols and language. In K. Stacey, H. Chick & M. Kendal (Eds), *The teaching and learning of algebra: The 12th ICMI study* (pp. 227–264). Norwell, MA: Kluwer.

Duncan, B. (2011). Numeracy in household safety: How do you measure a slope of 10°? *The Australian Mathematics Teacher*, 67(4), 27–31.

Duncan, B. & Fitzallen, N. (2013). Developing box plots while navigating the maze of data representations. *The Australian Mathematics Teacher*, 69(4), 8–14.

Duval, R. (1998). Geometry from a cognitive point of view. In C. Mammana & V. Villani (Eds), *Perspectives on the teaching of geometry for the 21st century* (pp. 37–52). Dordrecht: Kluwer.

Earl, L. M. (2012). *Assessment as learning: Using classroom assessment to maximize student learning*. Thousand Oaks, CA: Corwin Press.

Edmonds-Wathen, C. (2014). Influences of indigenous language on spatial frames of reference in Aboriginal English. *Mathematics Education Research Journal*, 26(2), 169–192.

Education Services Australia. (n.d.). *Assessment for learning*. Retrieved from www.assessmentfor-learning.edu.au/professional_learning/

Todd Edwards, M. (2004). Fostering mathematical inquiry with explorations of facial symmetry. *Mathematics Teacher*, 94(4), 234–41

Ehnebuske, J. (1998), In the comfort of their own homes: Engaging families in mathematics. *Teaching Children Mathematics*, 4(6), 338–344.

Ell, F. & Meissel, K. (2011). Working collaboratively to improve the learning and teaching of mathematics in a rural New Zealand community. *Mathematics Education Research Journal*, 23(2), 169–187.

Ellis, A. B. (2011). Algebra in the middle school: Developing functional relationships through quantitative reasoning. In J. Kai & E. Knuth(Eds), *Early algebraization* (pp. 215–238). Berlin: Springer.

Eppstein, (2007). Tiling: The geometry junkyard. Retrieved from www.ics.uci.edu/~eppstein/junkyard/tiling.html

Epstein, J. (1995). School/family/community partnerships: Caring for the children we share. *Phi Delta Kappa*, 76(9), 701–712.

Epstein, J. (2001). *School, family, and community partnerships: Preparing educators and improving schools.* Boulder, CO: Westview.

Ernest, P. (1989). The impact of beliefs on the teaching of mathematics. In P. Ernest (Ed.), *Mathematics teaching: The state of the art* (pp. 249–253). New York: Falmer.

Evans, M., Jones, P., Leigh-Lancaster, D., Les, M., Norton, P. & Wu, M. (2008). The 2007 common technology free examination for Victorian Certificate of Education (VCE) Mathematical Methods and Mathematical Methods Computer Algebra System (CAS). In M. Goos, R. Brown & K. Makar (Eds), *Navigating currents and charting directions* (Proceedings of the 31st annual conference of the Mathematics Education Research Group of Australasia, pp. 331–336). Brisbane, Qld: MERGA.

Even, R. (2005). Using assessment to inform instructional decisions: How hard can it be? *Mathematics Education Research Journal*, 17(3), 45–61.

Ewing, B. (2004). 'Open your textbooks to page blah, blah, blah': So I just blocked off! In I. Putt, R. Faragher & M. McLean (Eds), *Mathematics education for the third millennium: Towards 2010* (Proceedings of the 27th annual conference of the Mathematics Education Research Group of Australasia, pp. 231–238). Sydney: MERGA.

Ewing, B. (2009). Torres Strait Island parents' involvement in their children's mathematics learning: A discussion paper. *First Peoples Child & Family Review*, 4(2), 129–134.

Falkner, K. P., Levi, L. & Carpenter, T. P. (1999). Children's understanding of equality: A foundation for algebra. *Teaching Children Mathematics*, 6(4), 56–60.

Farmer, J. (2005). The volume of a torus using cylindrical and spherical coordinates. *Australian Senior Mathematics Journal*, 19(2), 49–58.

Faulkner, P. (2004). Notes for applied mathematics in trigonometry and earth geometry/navigation. *Australian Senior Mathematics Journal*, 18(1), 55–9.

Fennema, E. (1995). Mathematics, gender and research. In B. Grevholm & G. Hanna (Eds), *Gender and mathematics education* (An ICMI Study, pp. 21–38). Lund: Lund University Press.

Fennema, E. & Sherman, J. (1977). Sex-related differences in mathematics achievement, spatial visualization, and affective factors. *American Educational Research Journal*, 14(1), 51–71.

Fennema, E. & Tartre, L. (1985). The use of spatial visualization in mathematics by boys and girls. *Journal of Research in Mathematics Education*, 16(3), 184–206.

Fernandez, S. H. A. A. & Healy, L. (2012). Representations of three-dimensional forms constructed by blind students: relations between 'seeing' and the 'knowing'. In S. J. Cho (Ed.), *Intellectual and attitudinal challenges* (Proceedings of the 12th International Congress on Mathematical Education). Seoul: ICMI.

Ferrini-Mundy, J., & Gaudard, M. (1992), Secondary school calculus: Preparation or pitfall in the study of college calculus? *Journal for Research in Mathematics Education*, 23(1), 56–71.

Finzer, W. (2005), Peanutbutter.ftm (data file). *Fathom dynamic data software.* Emeryville, CA: KCP Technologies.

Fitzgerald, D. (2012). How can we use Google Earth in mathematics class. In J. Cheeseman (Ed.), *It's my maths: Personalised mathematics learning* (Proceedings of the Mathematical Association of Victoria Annual Conference, pp. 172–174). Brunswick: MAV.

Fitzsimons, G., Seah, W.T., Bishop, A. & Clarkson, P. (2001). Beyond numeracy: Values in the mathematics classroom. In J. Bobis, B. Perry & M. Mitchelmore (Eds), *Numeracy and beyond* (Proceedings of the 24th annual conference of the Mathematics Education Research Group of Australasia, pp. 202–209). Sydney: MERGA.

Flores, A. (2002). Profound understanding of the division of fractions. In B. Litwiller & G. Bright (Eds), *Making sense of fractions, ratios and proportions: 2002 yearbook* (pp. 237–246). Reston, VA: NCTM.

Flynn, P. & Asp, G. (2002). Assessing the potential suitability of 'show that' questions in CAS-permitted examinations. In B. Barton, K. C. Irwin, M. Pfannkuch & M. O. J. Thomas (Eds), *Mathematics education in the South Pacific* (Proceedings of the 25th annual conference of the Mathematics Education Research Group of Australasia, pp. 252–259). Sydney: MERGA.

Flynn, P. & McCrae, B. (2001). Issues in assessing the impact of CAS on mathematics examinations. In J. Bobis, B. Perry & M. Mitchelmore (Eds), *Numeracy and beyond* (Proceedings of the 24th annual conference of the Mathematics Education Research Group of Australasia, pp. 210–217). Sydney: MERGA.

Fogliati, V. J. & Bussey, K. (2013). Stereo-type threat reduces motivation to improve: Effects of stereo-type threat and feedback on women's intentions to improve mathematical ability. *Psychology of Woman Quarterly*, 37(3), 310–324.

Forgasz, H. J. (1995). Gender and the relationship between affective beliefs and perceptions of grade 7 mathematics classroom learning environments. *Educational Studies in Mathematics*, 28, 219–239.

Forgasz, H. J. (2002). Computers for the learning of mathematics: Equity considerations. In B. Barton, K. Irwin, M. Pfannkuch & M. Thomas (Eds), *Mathematics education in the South Pacific* (Proceedings of the 25th annual conference of the Mathematics Education Research Group of Australasia, Auckland, pp. 260–267). Auckland: MERGA.

Forgasz, H. J. (2006). Australian year 12 'intermediate' level mathematics enrolments 2000–2004: Trends and patterns. In P. Grootenboer, R. Zevenbergen & M. Chinnappan (Eds), *Identities, cultures and learning spaces* (Proceedings of the 29th annual conference of the Mathematics Education Research Group of Australasia, pp. 211–220). Adelaide: MERGA.

Forgasz, H. J. (2010). Streaming for mathematics in Victorian secondary schools. *The Australian Mathematics Teacher*, 66(1), 31–40.

Forgasz, H. J. (2012). Streaming for mathematics in Years 7–10 in Victoria: An issue for equity? *Mathematics Education Research Journal*, 21(1), 57–90.

Forgasz, H. J. & Hill, J. C. (2013). Factors implicated in high mathematics achievement. *International Journal of Science and Mathematics Education*, 11(2), 481–499.

Forgasz, H. J. & Leder, G. C. (2011). Equity and quality of mathematics education: Research and media portrayals. In B. Atweh, M. Graven, W., Secada & P. Valero (Eds), *Mapping equity and quality in mathematics education* (pp. 205–222). Netherlands: Springer.

Forgasz, H. J., Leder, G. & Halliday, J. (2013). The *Make It Count* Project: *NAPLAN* achievement evaluation. In V. Steinle, L. Ball & C. Bardini (Eds), *Mathematics education: Yesterday, today and tomorrow* (Proceedings of the 36th annual conference of the Mathematics Education Research Group of Australasia MERGA, pp. 298–305). Melbourne, Vic.: MERGA.

Forgasz, H. J., Leder, G.C. & Tan, H. (2014). Public views on the gendering of mathematics and related careers: International comparisons. *Educational Studies in Mathematics*, 87(3), 369–388.

Forgasz, H.J., Leder, G.C., & Vale, C. (2000). Gender and mathematics: changing perspectives. In K. Owens & J. Mousley (Eds), *mathematics education research in Australasia: 1996–1999* (pp.305-40). Sydney: MERGA.

Forman, E. A. (2003). A sociocultural approach to mathematics reform: Speaking, inscribing, and doing mathematics within communities of practice. In J. Kilpatrick, W. G. Martin & D. Schifter (Eds.), *A research companion to principles and standards for school mathematics* (pp. 333–352). Reston, VA: NCTM.

Forster, P. (2000a). Patterns on a graphics calculator: Using vector components or trigonometry. *The Australian Mathematics Teacher, 56*(1), 8–9.

Forster, P. (2000b). Process and object interpretations of vector magnitude mediated by use of the graphics calculator. *Mathematics Education Research Journal, 12*(3), 254–268.

Forster, P. A. & Mueller, U. (2001). Outcomes and implications of students' use of graphics calculators in the public examination of calculus. *International Journal of Mathematical Education in Science and Technology, 32*(1), 37–52.

Forster P. A. & Mueller, U. (2002). What effect does the introduction of graphics calculators have on the performance of boys and girls in assessment of tertiary entrance calculus? *International Journal of Mathematical Education in Science and Technology, 33*(6), 801–818.

Frankenstein, M. (1997). In addition to the mathematics: Including equity issues in the curriculum. In J. Trentacosta & M. J. Kenny (Eds), *Multicultural and gender equity in the mathematics classroom: The gift of diversity* (pp. 10–22). Reston, VA: NCTM.

Frankenstein, M. (2001). Reading the world with math: Goals for a critical mathematical literacy curriculum. In B. Lee (Ed.), *Mathematics shaping Australia* (Proceedings of the 18th biennial conference of the Australian Association of Mathematics Teachers, pp. 53–64). Adelaide: AAMT.

Fraser, N. (2013). *Fortunes of feminism: From state-managed capitalism to neoliberal crisis.* London, UK: Verso Books.

Frid, S. (2000a). Constructivism and reflective practice in practice: Challenges and dilemmas of a mathematics educator. *Mathematics Teacher Education and Development, 2*, 17–33.

Frid, S. (2000b). Using learning cycles in mathematics. *The Australian Mathematics Teacher, 56*(4), 32–37.

Frid, S. (2001). Food for thought. *The Australian Mathematics Teacher, 57*(1), 12–16.

Friedlander, A. & Arcavi, A. (2005). Folding perimeters: Designer concerns and student solutions. In H. Chick & J. Vincent (Eds), *Proceedings of the 29th conference of the International Group for the Psychology of Mathematics Education* (Vol. 1, pp. 108–114). Melbourne: PME.

Fujii, T. & Stephens, M. (2001). Fostering an understanding of algebraic generalisation through numerical expressions: The role of quasi-variables. In H. Chick, K. Stacey, J. Vincent & J. Vincent (Eds), *The future of the teaching and learning of algebra: Proceedings of the 12th ICMI Study Conference* (Vol. 1, pp. 258–264). Melbourne: University of Melbourne.

Funkhouser, J. & Gonzales, M. (1997). *Family involvement in children's education: Successful local approaches.* Washington, DC: US Department of Education.

Gadanidis, G. & Geiger, V. (2010). A social perspective on technology enhanced mathematical learning—from collaboration to performance. *ZDM—The International Journal in Mathematics Education, 42*(1), 91–104.

Gaffney, J. & Treilibs, V. (1982). *Mathematical modelling.* Adelaide: Education Department of South Australia.

Galbraith, P. (1987). Modelling—teaching modelling. *Australian Mathematics Teacher, 43*(4), 6–9.

Galbraith, P. (1995). Assessment in mathematics: Developments, innovations and challenges. In L. Grimson & J. Pegg (Eds), *Teaching secondary school mathematics: Theory into practice* (pp. 289–314). Sydney: Harcourt Brace.

Galbraith, P. (1996). Modelling comparative performance: Some Olympic examples. *Teaching Mathematics and its Applications, 15*(2), 67–77.

Galbraith, P. L. & Stillman, G. A. (2001). Assumptions and context: Pursuing their role in modelling activity. In F. J. Matos, S. K. Houston, W. Blum & S. P. Carreira (Eds), *Modelling and mathematics education: Applications in science and technology* (pp. 317–327). Chichester, UK: Horwood.

Galbraith, P. L. & Stillman, G. A. (2006). A framework for identifying blockages during transitions in the modelling process. *ZDM, 38*(2), 143–162.

Galbraith, P. L., Stillman, G. & Brown, J. (2006). Identifying key transition activities for enhanced engagement in mathematical modelling. In P. Grootenboer, R. Zevenbergen & M. Chinnappan (Eds), *Identities, cultures and learning spaces* (Proceedings of the 29th annual conference of the Mathematics Education Research Group of Australasia, pp. 237–245). Adelaide: MERGA.

Galbraith, P. L., Stillman, G. & Brown, J. (2010). Turning ideas into modeling problems. In R. Lesh, P. Galbraith, C. Haines & A. Hurford (Eds), *Modeling students' mathematical modeling competencies* (ICTMA 13, pp. 133–144). Springer.

Galbraith, P., Stillman, G., Brown, J. & Edwards, I. (2007). Facilitating middle secondary modelling competencies. In C. Haines, P., Galbraith, W., Blum & S. Khan (Eds), *Mathematical modelling (ICTMA12): Education, engineering and economics* (pp. 130–140). Chichester, UK: Horwood.

Gardner, H. (1993). *Multiple intelligences: The theory in practice.* New York: Basic Books.

Garfield, J. & Ben-Zvi, D. (2007). How students learn statistics revisited: A current review of research on teaching and learning statistics. *International Statistical Review,* 75(3), 372–396.

Garfield, J. & Ben-Zvi, D. (2008). *Developing students' statistical reasoning: Connecting research and teaching practice.* New York: Springer.

Garfield, J. & Ben-Zvi, D. (2009). Helping students develop statistical reasoning: Implementing a statistical reasoning learning environment. *Teaching Statistics,* 31(3), 72–77.

Geiger, V. (2006). Standing on the outside: A tale of how technology can engage those working on the margins of a community of inquiry. In P. Grootenboer, R. Zevenbergen & M. Chinnappan (Eds), *Identities, cultures and learning spaces* (Proceedings of the 29th Annual Conference of the Mathematics Education Research Group of Australasia, Canberra, pp. 246–253). Adelaide: MERGA.

Geiger, V. (2009). The master, servant, partner, extension-of-self framework in individual, small group and whole class contexts. In R. Hunter, B. Bicknell & T. Burgess (Eds), *Crossing divides* (Proceedings of the 32nd annual conference of the Mathematics Education Research Group of Australasia, Wellington, NZ, pp. 201–208). Wellington: MERGA.

Geiger, V. (2011). Factors affecting teachers' adoption of innovative practices with technology and mathematical modelling. In G. Kaiser, W. Blum, R. Borromeo Ferri & G. Stillman (Eds), *Trends in the teaching and learning of mathematical modelling* (pp. 305–314). New York: Springer.

Geiger, V., Forgasz, H., Calder, N., Tan, H. & Hill, J. (2012). Technology in mathematics education. In R. Perry & T. Lowry (Eds), *Research in mathematics education in Australasia 2008–2011* (pp. 111–142). Rotterdam: Sense.

Geiger, V., Makar K., Faragher, R. & Goos, M. (2012). Mathematical applications and computer algebra systems project (MAACAS). Research based teaching resources available at www.qamt.org/maacas-project

Geiger, V., McKinlay, J. & O'Brien, G. (Eds). (1997). *The AToMIC Project: Applications to mathematics incorporating calculators.* Brisbane: Queensland Association of Mathematics Teachers.

Geiger, V., McKinlay, J. & O'Brien, G. (Eds). (1999). *The Sub-ATOMIC Project: Subsequent applications to mathematics incorporating calculators.* Brisbane: Queensland Association of Mathematics Teachers.

Geiger, V., Muir, T. & Lamb, J. (2015). Video-stimulated recall as a catalyst for teacher professional learning, *Journal of Mathematics Teacher Education*. doi: 10.1007/s10857-015-9306-y

Gibbs, M., Goos, M., Geiger, V. & Dole, S. (2012). Numeracy in secondary school mathematics. *The Australian Mathematics Teacher*, 68(1), 29–35.

Giraldo, V., Carvalho, L. M. & Tall, D. (2002). Descriptions and definitions in the teaching of elementary calculus. *Proceedings of the British Society for Research into Learning Mathematics Conference*. 22(3), 37–42.

Godfrey, D. & Thomas, M. O. (2008). Student perspectives on equation: The transition from school to university. *Mathematics Education Research Journal*, 20(2), 71–92.

González-Calero, J. A., Arnau, D. & Laserna-Belenguer, B. (2015). Influence of additive and multiplicative structure and direction of comparison on the reversal error. *Educational Studies in Mathematics*, 89(1), 133–147.

Goodell, J. & Parker, L. (2001). Creating a connected, equitable mathematics classroom:Facilitating gender equity. In B. Atweh, H. Forgasz, & B. Nebres (Eds), *Sociocultural research on mathematics education: An international perspective* (pp. 411–431). Mahwah, NJ: Lawrence Erlbaum Associates.

Goos, M. (1999). Scaffolds for learning: A sociocultural approach to reforming mathematics teaching and teacher education. *Mathematics Teacher Education and Development*, 1, 4–21.

Goos, M. (Ed.). (2002). *Techno Maths: Technology enriched activities for the mathematics classroom*. Flaxton, Qld: Post Pressed.

Goos, M. (2004a). Home, school and community partnerships to support children's numeracy. *Australian Primary Mathematics Classroom*, 9(4), 18–20.

Goos, M. (2004b). Learning mathematics in a classroom community of inquiry. *Journal for Research in Mathematics Education*, 35, 258–291.

Goos, M. (2005). A sociocultural analysis of learning to teach. In H. Chick & J. Vincent (Eds), *Proceedings of the 29th conference of the International Group for the Psychology of Mathematics Education* (Vol. 3, pp. 49–56). Melbourne: University of Melbourne.

Goos, M. (2012). Digital technologies in the Australian curriculum: Mathematics—A lost opportunity? In B. Atweh, M. Goos, R. Jorgensen & D. Siemon (Eds), *Engaging the Australian National Curriculum: Mathematics—Perspectives from the field* (pp. 135–152). Online Publication: Mathematics Education Research Group of Australasia.

Goos, M. & Askin, C. (2005). Towards numeracy across the curriculum: Integrating mathematics and science in the middle years. In R. Zevenbergen (Ed.), *Innovations in numeracy teaching in the middle years* (pp. 125–141). Canberra: Australian Curriculum Studies Association.

Goos, M. & Bennison, A. (2008). Teacher professional identities and the integration of technology into secondary school mathematics. In *Proceedings of the annual conference of the Australian Association for Research in Education*. Australian Association for Research in Education.

Goos, M., Dole, S. & Geiger, V. (2011 fol). Improving numeracy education in rural schools: A professional development approach. *Mathematics Education Research Journal*, 23(2), 129–148.

Goos, M., Dole, S. & Geiger, V. (2012a). Auditing the numeracy demands of the Australian Curriculum. In J. Dindyal, L. Chen & S. F. Ng (Eds), *Mathematics education: Expanding horizons* (Proceedings of the 35th annual conference of the Mathematics Education Research Group of Australasia, pp. 314–321). Singapore: MERGA.

Goos, M., Dole, S. & Geiger, V. (2012b). Numeracy across the curriculum. *The Australian Mathematics Teacher*, 68(1), 3–7.

Goos, M., Galbraith, P. & Renshaw, P. (2000). A money problem: A source of insight into problem solving action. *International Journal for Mathematics Teaching and Learning*. Retrieved from www.cimt.plymouth.ac.uk/journal.pgmoney.pdf

Goos, M., Galbraith, P. & Renshaw, P. (2002). Socially mediated metacognition: Creating collaborative zones of proximal development in small group problem solving. *Educational Studies in Mathematics*, 49, 193–223.

Goos, M., Galbraith, P., Renshaw, P. & Geiger, V. (2000). Re-shaping teacher and student roles in technology enriched classrooms. *Mathematics Education Research Journal*, 12, 303–320.

Goos, M., Galbraith, P., Renshaw, P. & Geiger, V. (2003). Perspectives on technology mediated learning in secondary school mathematics classrooms. *Journal of Mathematical Behavior*, 22, 73–89.

Goos, M., Geiger, V. & Dole, S. (2011). Teachers' personal conceptions of numeracy. In B. Ubuz (Ed.), *Proceedings of the 35th conference of the International Group for the Psychology of Mathematics Education* (Vol. 2, pp. 457–464). Ankara, Turkey: PME.

Goos, M., Geiger, V. & Dole, S. (2014). Transforming professional practice in numeracy teaching. In Y. Li, E. Silver & S. Li (Eds), *Transforming mathematics instruction: Multiple approaches and practices* (pp. 81–102). New York: Springer.

Goos, M., Lincoln, D., Coco, A., Frid, S., Galbraith, P., Horne, M., … Gholam, M. (2004). *Home, school and community partnerships to support children's numeracy.* Canberra: Australian Government Department of Education, Science and Training.

Gough, J. (2001). Algebra demons: Common mistakes in algebra, *Vinculum*, 38(1), 14–15.

Gough, J. (2006). Assessment in the middle years. *Professional Voice*, 4(2), 27–32.

Gould, P. (2005). Really broken numbers. *Australian Primary Mathematics Classroom*, 10(3), 4–10.

Gould, P., Outhred, L. & Mitchelmore, M. (2006). One-third is three-quarters of one-half. In P. Grootenboer, R. Zevenbergen & M. Chinnappan (Eds), *Identities, cultures and learning spaces* (Proceedings of the 29th annual conference of the Mathematics Education Research Group of Australasia, pp. 262–269). Adelaide: MERGA.

Graham, L. & Pegg, J. (2010). Hard data to support the effectiveness of *Quicksmart* numeracy. *Learning Difficulties in Australian Bulletin*, 42(1), 11–13.

Gravemeijer, K., van Galen, F. & Keijzer, R. (2005). Designing instruction on proportional reasoning with average speed. In H. Chick & J. Vincent (Eds), *Proceedings of the 29th conference of the International Group for the Psychology of Mathematics Education* (Vol. 1, pp. 103–7). Melbourne: PME.

Gray, E. & Tall, D. (1994). Duality, ambiguity and flexibility: A proceptual view of simple arithmetic. *Journal for Research in Mathematics Education*, 26, 115–141.

Greer, B. (2014). Commentary on perspective II: Psychology. In E. Chernoff & B. Sriraman (Eds), *Probabilistic thinking: Presenting plural perspectives* (pp. 299–309). Berlin: Springer Science.

Grunewald, S. (2013). The development of modelling competencies by Year 9 students: Effects of a modelling project. In G. A. Stillman, G. Kaiser, W. Blum & J. P. Brown (Eds), *Teaching mathematical modelling: Connecting to research and practice* (Chapter 16). Dordrecht: Springer.

Gullberg, J. (1997). *Mathematics: From the birth of numbers*. New York, NY: W.W. Norton & Company.

Gunn, S. & Wyatt-Smith, C. (2011). Learning difficulties, literacy and numeracy: Conversations across the fields. In C. Wyatt-Smith, J. Elkins & S. Gunn (Eds), *Multiple perspectives on difficulties in learning literacy and numeracy* (pp. 17–48). Dordrecht: Springer.

Gutstein, E. (2003). Teaching and learning mathematics for social justice in an urban, Latino school. *Journal of Research in Mathematics Education*, 34(1), 37–73.

Gutstein, E. & Peterson, B. (Eds). (2006). *Rethinking mathematics: Teaching social justice by the numbers*. Milwaukee, WI: Re-thinking Schools Ltd.

Hancock, C., Kaput, J. & Goldsmith, L. (1992). Authentic inquiry into data: Critical barriers to classroom implementation. *Educational Psychologist*, 27(3), 337–364.

Handal, B., Handal, P. & Herrington, T. (2005). Evaluating online mathematics resources: A practical approach for teachers. In M. Coupland, J. Anderson & T. Spencer (Eds), *Making mathematics vital* (Proceedings of the 20th biennial conference of the Australian Association of Mathematics Teachers, pp. 153–165). Adelaide: AAMT.

Handal, B., Watson, K., Petocz, P. & Maher, M. (2013). Retaining mathematics and science teachers in rural and remote schools. *Australian & International Journal of Rural Education*, 23(3), 13–27.

Hannigan, A., Gill, O. & Leavy, A. M. (2013). An investigation of prospective secondary mathematics teachers' conceptual knowledge of and attitudes towards statistics. *Journal of Mathematics Teacher Education*, 16(6), 427–449.

Harding, A., Wood, L., Muchatuta, M., Edwards, B., Falzon, L., Gudlaugsdottir, S., ... Steyn, T. (2010). Try and catch the wind: Women who do doctorates at a mature age in their lives. In H. J. Forgasz, J. Rossi-Becker, K. Lee & O. B. Steinthorsdottir (Eds), *International perspective on gender in mathematics education* (pp. 1–11). Charlotte, NC: Information Age Publishing Inc.

Harradine, A., Batanero, C. & Rossman, A. (2011). Students and teachers' knowledge of sampling and inference. In C. Batanero, G. Burrill & C. Reading (Eds), *Teaching statistics in school mathematics—Challenges for teaching and teacher education* (pp. 235–246). Dordrecht, The Netherlands: Springer.

Harris, C. & Marsh, C. (2005). Analysing curriculum change: Some reconceptualised approaches. In C. Harris & C. Marsh (Eds), *Curriculum developments in Australia: Promising initiatives, impasses and dead-ends* (pp. 15–37). Canberra: Australian Curriculum Studies Association.

Harris, L. R. & Brown, G. T. (2008). New Zealand teachers' conceptions of the purpose of assessment: Phenomenographic analysis for teachers' thinking. In P. L. Jeffrey (Ed.), *Changing climates, education for sustainable futures* (Proceedings of the AARE 2008 International Education Research Conference). Brisbane: AARE.

Hashemi, N., Abu, M. S., Kashefi, H. & Rahimi, K. (2014). Undergraduate students' difficulties in conceptual understanding of derivation. *Procedia-Social and Behavioral Sciences*, 143, 358–366.

Hassan, I. & Mitchelmore, M. (2006), The role of abstraction in learning about rates of change. In P. Grootenboer, R. Zevnbergen & M. Chinnappan (Eds), *Identities, Cultures and Learning Spaces: Proceedings of the 29th Annual Conference of the Mathematics Education Group of Australasi*a, Vol. 1, pp. 278–285. Canberra, Australia: MERGA.

Hattie, J. & Timperley, H. (2009). The power of feedback. *Review of Educational Research*, 77(1), 81–112.

Hayes, D., Lingard, B. & Mills, M. (2000). Productive pedagogies. *Education Links*, 60, 10–13.

Healy, L. & Powell, A. (2013). Understanding and overcoming 'disadvantage' in learning mathematics. In K. Clements, A. Bishop, J. Kilpatrick, F. Leung & C. Keitel (Eds), *Third international handbook of mathematics education* (Vol. 1, pp. 69–100). New York: Springer.

Heid, K. (1995). *Algebra in a technological world.* Reston, VA: NCTM.

Heid, M. K. (1988). Resequencing skills and concepts in applied calculus using the computer as a tool. *Journal for Research in Mathematics Education,* 19, 3–25.

Heid, M. K., Thomas, M. O. & Zbiek, R. M. (2013). How might computer algebra systems change the role of algebra in the school curriculum?. In M. A. (Ken) Clements, C. Keitel, K. S. Leung, A. Bishop & J. Kilpatrick (Eds), *Third international handbook of mathematics education* (pp. 597–641). New York: Springer.

Hekimoglu, S. (2005). Romance of a mathematician: Celebrating St Valentine's Day in a mathematics class. *Australian Senior Mathematics Journal,* 19(1), 39–43.

Henningsen, M. & Stein, M. K. (1997). Mathematical tasks and student cognition: Classroom-based factors that support and inhibit high-level mathematical thinking and reasoning. *Journal for Research in Mathematics Education,* 28, 524–549.

Henry, M. & McAuliffe, R. (1994). Drying out. *Mathstralia* (pp. 41–48). Melbourne: AAMT.

Herbel-Eisenmann, B. A. & Breyfogle, M. L. (2005). Questioning our patterns of questioning. *Mathematics Teaching in the Middle School,* 10(9), 484–489.

Herbert, S. (2010). *An investigation of middle secondary students' mathematical conceptions of rate* (PhD Thesis, University of Ballarat, Victoria).

Herbert, S. (2013). Challenging traditional sequence of teaching introductory calculus. *Computers in the Schools,* 30 (1–2), 172–190.

Herbert, S. & Pierce, R. (2011). What is rate? Does context or representation matter? *Mathematics Education Research Journal,* 23 (4), 455–477.

Herbert, S. & Pierce, R. (2012). Revealing educationally critical aspects of rate. *Educational Studies in Mathematics,* 81(1), 85–101.

Hershkowitz, R (1998). About reasoning in geometry. In C. Mammana & V. Villani (Eds), *Perspectives on the teaching of geometry for the 21st century* (pp. 29–37). Dordrecht, The Netherlands: Kluwer.

Hershkowitz, R. & Kieran, C. (2001). Algorithmic and meaningful ways of joining together representatives within the same mathematical activity: An experience with graphing calculators. In M. van den Heuvel-Panhuizen (Ed.), *Proceedings of the 25th international conference for the Psychology of Mathematics Education* (Vol. 1, pp. 96–107). Utrecht, The Netherlands: PME.

Hiebert, J. & Carpenter, T. P. (1992). Learning and teaching with understanding. In D. A. Grouws (Ed.), *Handbook of research on mathematics teaching and learning* (pp. 65–97). New York: Macmillan.

Hilton, A., Hilton, G., Dole, S. & Goos, M. (2013). Development and application of a two-tier diagnostic instrument to assess middle-years students' proportional reasoning. *Mathematics Education Research Journal*, 25, 523–545.

Hino, K. (2015). Comparing multiple solutions in the structured problem solving: Deconstructing Japanese lessons from learner's perspective. *Education Studies in Mathematics*, 90, 121–141.

Hobbs, L. (2013). Teaching 'out-of-field' as a boundary-crossing event: Factors shaping teacher identity. *International Journal of Science and Mathematics Education*, 11, 271–297.

Hodgson, B. & Leigh-Lancaster, D. (1990). Coastal navigation. In B. Hodgson & D. Leigh-Lancaster (Eds), *Space and number* (Jacaranda Senior Mathematics Series). Brisbane: Jacaranda.

Hogan, J. (2002). Mathematics and numeracy: Is there a difference? *Australian Mathematics Teacher*, 58(4), 14–15.

Hohenwarter, M., Hohenwarter, J., Kreis, Y. & Lavicza, Z. (2008). Teaching and learning calculus with free dynamic mathematics software GeoGebra. In *11th International Congress on Mathematical Education*. Monterrey, Nuevo Leon, Mexico.

Hohenwarter, M. & Preiner, J. (2007). Dynamic mathematics with GeoGebra. *AMC*, 10, 12.

Hollingsworth, H., Lokan, J. & McCrae, B. (2003). *Teaching mathematics in Australia: Results from the TIMSS 1999 Video Study*. Melbourne: ACER.

Holton, D. & Clarke, D. (2006). Scaffolding and metacognition. *International Journal of Mathematical Education in Science and Technology*, 37(2), 127–143.

Horne, M. (1998). Linking parents and school mathematics. In N. Ellerton (Ed.), *Issues in mathematics education: A contemporary perspective* (pp. 115–135). Perth: Mathematics, Science and Technology Education Centre, Edith Cowan University.

Hoyles, C. & Jones, K. (1998). Proof in dynamic geometry contexts. In C. Mammana & V. Villani (Eds), *Perspectives on the teaching of geometry for the 21st century* (pp. 121–8). Dordrecht, The Netherlands: Kluwer.

Hoyles, C. & Lagrange, J.-B. (2010). Introduction. In C. Hoyles & J.-B. Lagrange (Eds), *Mathematics education and technology—Rethinking the terrain* (pp. 1–11). New York: Springer.

Huff, D. (1954). *How to lie with statistics*. New York: W. W. Norton & Company.

Hufferd-Ackles, K., Fuson, K. & Sherin, M. (2004). Describing levels and components of a math-talk learning community. *Journal for Research in Mathematics Education*, 35, 81–116.

Hunting, R., Doig, B., Pearn, C. & Nugent, E. (1995). *Initial clinical assessment procedure mathematics tasks—Level B Short Version (upper primary)*. Unpublished paper, Institute of Mathematics Education, La Trobe University.

Huwaldt, J. (2015). Plot digitizer (computer software). Available to download from http://plot-digitizer.sourceforge.net.

Hyde, H. (2004). The measurement of digitised motion. *The Australian Mathematics Teacher*, 60(3), 14–15.

Inoue, N. (2011). Zen and the art of neriage: Facilitating consensus building in mathematics inquiry lessons through lesson study. *Journal of Mathematics Teacher Education*, 14(1), 5–23.

Irwin, K. C. (2001). Using everyday knowledge of decimals to enhance understanding. *Journal for Research in Mathematics Education*, 32(4), 399–420.

Ives, B. & Hoy, C. (2003). Graphic organisers applied to higher-level secondary mathematics. *Learning Disabilities Research and Practice*, 18(1), 36–51.

Jaafar, R. (2016). Writing-to-learn activities to provoke deeper learning in calculus. *PRIMUS*, 26(1), 67–82.

Jablonka, E. (2003). Mathematical literacy. In A. Bishop, M. A. Clements, C. Keitel, J. Kilpatrick & F. Leung (Eds), *Second international handbook of mathematics education* (pp. 75–102). Dordrecht, The Netherlands: Kluwer.

Jackson, K. & Cobb, P. (2010). *Refining a vision of ambitious mathematics instruction to address issues of equity.* Paper presented at the National Council of Teachers of Mathematics Research Pre-Session in San Diego, CA, April 2010.

James, D., Jurich, S. & Estes, S. (2001). *Raising minority academic achievement: A compendium of education programs and practices* (Report). Washington, DC: American Youth Policy Forum. Retrieved from www.aypf.org/publicatons/index.html

Jaworski, B. (2008). Building and sustaining inquiry communities in mathematics teaching development: Teachers and didacticians in collaboration. In K. Krainer & T. Wood (Eds), *Participants in mathematics teacher education: Individuals, teams, communities and networks* (pp. 309–330). Rotterdam: Sense.

Johnston-Wilder, S. & Mason, J. (2005). *Developing thinking in geometry.* London: Open University.

Jones, G., Langrall, C. & Mooney, E. (2007). Research in probability. In F. Lester (Ed.), *Second handbook of research on mathematics teaching and learning* (pp. 909–955). Reston, VA: NCTM.

Jordan, C., Ozorco, E. & Arerett, A. (2001). *Emerging issues in school, family and community connections. Report.* Austin; TX: Southwest Educational Development Laboratory.

Jorgensen, R. (2012a). Exploring scholastic mortality among working class and indigenous students: A perspective from Australia. In B. Herzelman, J. Choppin, D. Wagner &

D. Pimm (Eds), *Equity in discourse for mathematics education: Theories, practices and policies* (pp. 35–49). Dordrecht: Springer.

Jorgensen, R. (2012b). Curriculum leadership: Reforming and reshaping successful practice in remote and regional indigenous education. In J. Dindyal, L. P. Cheng & S. F. Ng (Eds), *Proceedings of the 35th annual conference of the Mathematics Education Research Group of Australasia* (pp. 370–377). Singapore: MERGA.

Jorgensen, R. (2014). Social theories of learning: A need for a new paradigm in mathematics education. In J. Anderson, M. Cavanagh & A. Prescott (Eds), *Proceedings of the 37th annual conference of the Mathematics Education Research Group of Australasia* (pp. 311–318). Sydney: MERGA.

Jorgensen, R., Gates, P. & Roper, V. (2014). Structural exclusion through school mathematics: Using Bourdieu to understand mathematics as a social practice. *Educational Studies in Mathematics*, 87(2), 1–19.

Jorgensen, R. & Lowrie, T. (2013). Socio-economic status and rurality as scholastic mortality: Exploring failure at mathematics. In M. Berger, K. Brodie, V. Frith & K. le Roux (Eds), *Proceedings of the 7th international Mathematics Education and Society Conference* (Vol. 2, pp. 340–347). Cape Town: MES7.

Jungwirth, H. (1991). Interaction and gender—Findings of a microethnographical approach to classroom discourse. *Educational Studies in Mathematics*, 22, 263–284.

Jungwirth, H. (2003). What is a gender-sensitive mathematics classroom? In L. Burton (Ed.), *Which way social justice in mathematics education?* (pp. 3–26). Westport, CT: Praeger.

Juter, K. (2005). Limits of functions: Traces of students' concept images. *Nordic Studies in Mathematics Education*, 10 (3/4), 65–82.

Kaiser, G. & Rogers, P. (1995). Introduction: Equity in mathematics education. In P. Rogers & G. Kaiser (Eds), *Equity in mathematics education. Influences of feminism and culture* (pp. 1–10). London: Falmer Press.

Kaiser, G. & Sriraman, B. (2006). A global survey of international perspectives on modelling in mathematics education. *ZDM*, 38(3), 302–310.

Kaput, J. (1999). Teaching and learning a new algebra. In T. Romberg & E. Fennema (Eds), *Mathematics classroom that promotes understanding* (pp. 133–155). Hillsdale: Lawrence Erlbaum.

Kaput, J., Noss, R. & Hoyles, C. (2008). Developing new notations for a learnable mathematics in the computational era. In L. D. English (Ed.), *Handbook of international research in mathematics education* (2nd ed., pp. 693–715). New York: Routledge.

Kaput, J. & Schorr, R. Y. (2002). Changing representational infrastructures changes most everything: The case of SimCalc, algebra & calculus. In M. K. Heid & G. Blume (Eds), *Research on the impact of technology on the teaching and learning of mathematics* (pp. 47–75). Mahwah, NJ: Erlbaum.

Kastberg, S. (2003). Using Bloom's Taxonomy as a framework for classroom assessment. *Mathematics Teacher*, 96(6), 402–405.

Kastberg, S., D'Ambrosio, B. & Lynch-Davis, K. (2012). Understanding proportional reasoning. *The Australian Mathematics Teacher*, 68(3), 32–40.

Katz, V. (2006). Stages in the history of algebra with implications for teaching. *Educational Studies in Mathematics*, 66(2), 185–201.

Katz, Y. (2000). The parent–school partnership: Shared responsibility for the education of children. *Curriculum and Teaching*, 15(2), 95–102.

Keith, N. (1999). Whose community schools? New discourses, old patterns. *Theory into practice*, 38(4), 225–234.

Kendal, M. (2002). *Teaching and learning introductory differential calculus* (PhD thesis, The University of Melbourne). Retrieved from http://thesis.lib.unimelb.edu.au/

Kendal, M. & Stacey, K. (1999). Teaching trigonometry. *The Australian Mathematics Teacher*, 54(1), 34–39.

Kendal, M. & Stacey, K. (2003). Tracing learning of three representations with the differentiation competency framework. *Mathematics Education Research Journal*, 15(1), 22–41.

Kendal, M. & Stacey, K. (2004). Algebra: A world of difference. In K. Stacey, H. Chick & M. Kendal (Eds), *The teaching and learning of algebra: The 12th ICMI study* (pp. 329–346). Norwell, MA: Kluwer.

Kennedy, D. (1983). *Little sparrow: A portrait of Sophia Kovalevsky*. Athens, OH: Ohio University Press.

Kenney, R. H. (2014). Investigating a link between pre-calculus students' uses of graphing calculators and their understanding of mathematical symbols. *International Journal for Technology in Mathematics Education*, 21(4), 157–166.

Kieran, C. (2004). The core of algebra: Reflections on its main activities. In K. Stacey, H. Chick & M. Kendal (Eds), *The future of the teaching and learning of algebra: The 12th ICMI study* (pp. 21–33). Norwell, MA: Kluwer.

Kieran, C. & Yerushalmy, M. (2004). Research on the role of technological environments in algebra learning and teaching. In K. Stacey, H. Chick & M. Kendal (Eds), *The future of the teaching and learning of algebra: The 12th ICMI study* (pp. 99–152). Norwell, MA: Kluwer.

Killen, R. (2013). Becoming a reflective teacher. In *Effective teaching strategies: Lessons from research and practice* (6th ed., Chapter 5). South Melbourne: Cengage Learning.

Kilpatrick, J., Swafford, J. & Findell, B. (2001). *Adding it up: Helping children learn mathematics.* Washington, DC: National Academy Press.

Kissane, B. (2001). Algebra and technology: Emerging issues. In B. Lee (Ed.), *Mathematics shaping Australia* (Proceedings of the 18th biennial conference of the Australian Association of Mathematics Teachers, pp. 120–130). Adelaide: AAMT.

Kissane, B. (2009). What does the Internet offer for students? In C. Hurst, M. Kemp, B. Kissane, L. Sparrow & T. Spencer (Eds), *Mathematics: It's mine* (Proceedings of the 22nd biennial conference of the Australian Association of Mathematics Teachers, pp. 135–144). Adelaide: AAMT.

Kissane, B., Bradley, J. & Kemp, M. (1994). Graphics calculators, equity, and assessment. *Australian Senior Mathematics Journal, 8*(2), 31–43.

Klaoudatous, N. (1994). Modelling-orientated teaching (a theoretical development for teaching mathematics through the modelling process). *International Journal of Mathematics Education, Science and Technology, 25*(1), 69–79.

Klenowski, V. (2009). Assessment for learning revisited: An Asia-Pacific perspective (Editorial). *Assessment in Education: Principles, Policy & Practice, 16*(3), 263–268.

Klenowski, V. (2011). Assessment for learning in the accountability era: Queensland Australia. *Studies in Educational Evaluation, 37*(1), 78–83.

Kline, M. (1979). *Mathematics: An introduction to its spirit and use. Readings from Scientific American* (pp. 46–48). San Francisco: W.H. Freeman and Co.

Knuth, E. J., Alibali, M. W., McNeil, N. M., Weinberg, A. & Stephens, A. C. (2005). Middle school students' understanding of core algebraic concepts: Equivalence & variable. *ZDM, 37*(1), 68–76.

Knuth, E. J., Stephens, A. C., McNeil, N. M. & Alibali, M. W. (2006). Does understanding the equal sign matter? Evidence from solving equations. *Journal for Research in Mathematics Education, 37*(4), 297–312.

Konold, C. & Harradine, A. (2014). Contexts for highlighting signal and noise. In T. Wassong, D. Frischemeier, P. Fischer, R. Hochmuth & P. Bender (Eds), *Using tools for learning mathematics and statistics* (pp. 237–250). Wiesbaden, Germany: Springer Spektrum.

Konold, C. & Pollatsek, A. (2002). Data analysis as the search for signals in noisy processes. *Journal for Research in Mathematics Education, 33*(4), 259–289.

Kouropatov, A. & Dreyfus, T. (2014). Learning the integral concept by constructing knowledge about accumulation. *ZDM, 46*(4), 533–548.

Krainer, K. (2001). Teachers' growth is more than the growth of individual teachers: The case of Gisela. In F. Lin & T. Cooney (Eds), *Making sense of mathematics teacher education* (pp. 271–293). Dordrecht, The Netherlands: Kluwer.

Kronberg, L. & Plunkett, M. (2008). Curriculum differentiation: An innovative Australian secondary school program to extend academic talent. *The Australian Journal of Gifted Education, 17*(2), 23–32.

Kronberg, L. & Plunkett, M. (2015). Providing an optimal school context for talent development: An extended curriculum program in practice. *The Australian Journal of Gifted Education, 24*(2), 61–69.

Küchemann, D. (1978). Children's understanding of numerical variables. *Mathematics in School, 7* (4), 23–26.

Lambert, R. (2015). Constructing and resisting disability in mathematics classrooms: A case study exploring the impact of different pedagogies. *Educational Studies in Mathematics, 89*, 1–15.

Lamon, S. (1999) *Teaching fractions and ratios for understanding: Essential content knowledge and instructional strategies for teachers*. London: Lawrence Erlbaum Associates.

Lampert, M. (2001). Teaching to deliberately connect content across lessons. *Teaching problems and the problems of teaching* (pp. 179–211). New Haven: Yale University Press.

Lannin, J., Ellis, A. & Elliott, R. (2011). *Developing essential understanding of mathematical reasoning for teaching mathematics in pre-kindergarten–Grade 8*. Reston, VA: NCTM.

Larkin, K. & Calder, N. (2015). Mathematics education and mobile technologies. *Mathematics Education Research Journal*. doi: 10.1007/s13394-015-0167-6

Lawson, D. A. & Tabor, J. H. (2001). APs, GPs and vCJD. *Teaching Mathematics and its Applications, 20*(1), 3–9.

Leder, G. C. (1993). Teacher/student interactions in the mathematics classroom: A different perspective. In E. Fennema & G. Leder (Eds), *Mathematics and gender* (pp. 149–168). Brisbane: Queensland University Press.

Leder, G. C. (2001). The Victorian certificate of education: A gendered affair? *Australian Educational Researcher, 28*(2), 53–66.

Leder, G. C. (2006). Catering for individual differences: Lessons learnt from the Australian Mathematics Competition. In P. Grootenboer, R. Zevenbergen & M. Chinnappan

(Eds.), *Identities, cultures and learning spaces* (Proceedings of the 29th annual conference of the Mathematics Education Research Group of Australasia, pp. 336–343). Adelaide: MERGA.

Leder, G. C., Brew, C. & Rowley, G. (1999). Gender differences in mathematics achievement here today and gone tomorrow? In G. Kaiser, E. Luna & I. Huntley (Eds), *International comparisons in mathematics education* (Studies in Mathematics Series 11, pp. 213–224). London: Falmer.

Leder, G. & Forgasz, H.(2000). Mathematics and gender: Beliefs they are a changin'. In J. Bana & A. Chapman (Eds), *Mathematics education beyond 2000* (Proceedings of the 23rd annual conference of the Mathematics Education Research Group of Australasia, pp. 370–376). Sydney: MERGA.

Leder, G. C. & Forgasz, H. (2000). Affect and mathematics education. In A. Gutiérrez & P. Boero (Eds), *Handbook of research on the psychology of mathematics education* (pp. 403–427). Rotterdam, The Netherlands: Sense Publishers.

Leder, G. C., Forgasz, H. J. & Jackson, G. (2014). Mathematics, English and gender issues: Do teachers count? *Australian Journal of Teacher Education, 39*(9), 18–34.

Leder, G. C. & Taylor, P. (2010). Are Raelene, Marjorie and Betty still in the race? *The Australian Mathematics Teacher, 66*(4), 17–24.

Lee, K. H. (2005). Mathematically gifted students' geometrical reasoning and informal proof. In H. L. Chick & J. L. Vincent (Eds), *Proceedings of the 29th conference of the International Group for the Psychology of Mathematics Education* (Vol. 3, pp. 241–248). Melbourne: PME.

Lehrer, R. & Schauble, L. (2000). Modeling in mathematics and science. In R. Glaser (Ed.), *Advances in instructional psychology: Educational design and cognitive science* (Vol. 5, pp. 101–159). Mahweh NJ: Lawrence Erlbaum Associates.

Leigh-Lancaster, A. (2004). What shape will be unfolded? *Vinculum, 41*(3), 19–20.

Leigh-Lancaster, A. & Leigh-Lancaster, D. (2002). AEMining for algebra Years 7–10. In C. Vale, J. Roumeliotis & J. Horwood (Eds), *Valuing mathematics in society* (pp. 251–262). Melbourne: MAV.

Leigh-Lancaster, D. (2004). Constructing quadrilaterals. *Vinculum, 41*(3), 15–17.

Leigh-Lancaster, D. (2010). The case of technology in senior secondary mathematics: Curriculum and assessment congruence? In C. Glascodine & K.-A. Hoad (Eds), *Teaching mathematics? Make it count: What research tells us about effective teaching and learning of mathematics* (Paper presented at the Australian Council for Educational Research Conference, pp. 43–60). Melbourne: ACER.

Leigh-Lancaster, D., Les, M. & Evans, M. (2010). Examinations in the final year of transition to mathematical methods computer algebra system (CAS). In L. Sparrow, B. Kissane & C. Hurst (Eds), *Shaping the future of mathematics education* (Proceedings of the 33rd annual conference of the Mathematics Education Research Group of Australasia, pp. 336–343). Freemantle: MERGA.

Leong, Y. H. & Yap, S. F., with Teo, M. L. Y., Thilogram, S., Irni, K. B., Quek, E. C. & Tang, K. L. K. (2010). Concretising factorisation of quadratic expressions. *The Australian Mathematics Teacher*, 66(3), 19–24.

Lerman, S. (1996). Intersubjectivity in mathematics learning: A challenge to the radical constructivist paradigm? *Journal for Research in Mathematics Education*, 27, 133–150.

Lerman, S. (2001). Cultural, discursive psychology: A sociocultural approach to studying the teaching and learning of mathematics. *Educational Studies in Mathematics*, 46, 87–113.

Lerman, S. (2006). Cultural psychology, anthropology and sociology: The developing 'strong' social turn. In J. Maass & W. Schloglmann (Eds), *New mathematics education research and practice* (pp. 171–188). Rotterdam, The Netherlands: Sense.

Lesh, R. (2000). Beyond constructivism: Identifying mathematical abilities that are most needed for success beyond school in an age of information. *Mathematics Education Research Journal*, 12(3), 177–195.

Lesh, R. (2003). How mathematizing reality is different from realizing mathematics. In S. Lamon, W. Parker & S. K. Houston (Eds), *Mathematical modelling: A way of life* (pp. 37–69). Chichester, UK: Horwood.

Lesh, R. & Zawojewski, J. (2007). Problem solving and modeling. In F. K. Lester, Jr. (Ed.), *Second handbook of research on mathematics teaching and learning* (pp. 763–804). Reston, VA: NCTM.

Leung, F. K., Park, K., Holton, D. & Clarke, D. (Eds). (2014). *Algebra teaching around the world*. New York: Springer.

Lins, R. & Kaput, J. (2004). The early development of algebraic reasoning: The current state of the field. In K. Stacey, H. Chick & M. Kendal (Eds), *The future of the teaching and learning of algebra: The 12th ICMI study* (pp. 47–70). Norwell, MA: Kluwer.

Livy, S. & Vale, C. (2011). First year pre-service teachers' mathematical content knowledge: Methods of solution to a ratio question. *Mathematics Teacher Education and Development*, 13(1), 22–43.

Lloyd, G.M. & Wilson, M. (1998), Supporting innovation: The impact of a teacher's conceptions of functions on his implementation of a reform curriculum. *Journal for Research in Mathematics Education*, 29(3), 248–74.

Lo, M. L. & Marton, F. (2012). Towards a science of the art of teaching: Using variation theory as a guiding principle of pedagogical design. *International Journal for Lesson and Learning Studies*, 1(1), 7–22.

Lo, M. P. (2012). *Variation theory and the improvement of teaching and learning.* Goteborg: Acta Universitatis Gothoburgensis.

Lokan, J., Ford, P. & Greenwood, L. (1996). *Mathematics and science on the line: Australian junior secondary students' performance in the Third International Mathematics and Science Study.* Melbourne: ACER.

Lokan, J., Ford, P. & Greenwood, L. (1997). *Maths & science on the line: Australian middle primary students' performance in the Third International Mathematics and Science Study.* Melbourne: ACER.

Lokan, J., Greenwood, L. & Cresswell, J. (2001). *15-up and counting, reading, writing, reasoning ... How literate are Australia's students?* Program for International Student Assessment (PISA). Melbourne: ACER.

Lopez, G. (2001). The value of hard work: Lessons on parental involvement from an (im) migrant household. *Harvard Educational Review*, 71(3), 416–437.

Loucks-Horsley, S., Love, N., Stiles, K., Mundry, S. & Hewson, P. (2003). *Designing professional development for teachers of science and mathematics* (2nd ed.). Thousand Oaks, CA: Corwin Press.

Lovitt, C. & Clarke, D. M. (1988). *Mathematics curriculum and teaching program: Professional development package. Activity bank* (Vols 1 & 2). Canberra: Curriculum Development Centre.

Lovitt, C. & Clarke, D. M. (1992). *Mathematics curriculum and teaching program activity bank* (Vols 1 & 2). Canberra: Curriculum Corporation.

Lovitt, C., Clarke, D. M. & Stephens, M. (1986). Mathematics curriculum and teaching program—An invitation for you to be involved. In K. V. Swinson (Ed.), *Proceedings of the 11th biennial conference of the Australian Association of Mathematics Teachers* (pp. 4–5). Brisbane: AAMT.

Lowe, I. (1984). The R.I.M.E. teacher development project. In P. Costello, S. Ferguson, K. Slinn., M. Stephens, D. Trembath & D. Williams (Eds), *Proceedings of the 10th biennial conference of the Australian Association of Mathematics Teachers* (pp. 210–217). Blackburn: AAMT.

Lowe, I. (1988). *Mathematics at work: Modelling your world* (Vol. 1). Canberra: Australian Academy of Science.

Lowe, I. (1991). *Mathematics at work: Modelling your world* (Vol. 2). Canberra: Australian Academy of Science.

Lowe, I. (2013). Mathematical understanding. *Vinculum*, 50(4), 8–11.

Lowrie, T., Greenlees, J. & Logan, T. (2012). Assessment beyond all: The changing nature of assessment. In B. Perry et al. (Eds) *Research in mathematics education in Australasia 2008–2011* (pp. 143–165). Rotterdam: Sense Publishers B.V.

Lowrie, T., Jorgensen, R. & Logan, T. (2013). Navigating and decoding dynamic maps: Gender preferences and engagement differences within-and outside-of game experiences. *Australasian Journal of Educational Technology, 29*(5), 626–639.

Maaβ, K. (2006). What are modelling competencies? *ZDM*, 38(2), 113–142.

MacGregor, M. (1990). Writing in natural language helps students construct algebraic equations. *Mathematics Education Research Journal*, 2(2), 1–10.

MacGregor, M. & Moore, R. (1991). *Teaching mathematics in the multicultural classroom: A resource for teachers and teacher educators*. Melbourne: University of Melbourne.

Mack, J. & Wilson, R. (2014). Declines in high school mathematics and science participation: Evidence of students' and future teachers' disengagement with maths. *International Journal of Innovation in Science and Mathematics Education*, 22(7), 35–48.

Mackrell, K. & Johnston-Wilder, P. (2005). Thinking geometrically: Dynamic geometry. In S. Johnston-Wilder & D. Pimm (Eds), *Teaching secondary mathematics with ICT* (pp. 81–100). Maidenhead: Open University Press.

Makar, K (2004). *Developing statistical inquiry: Prospective secondary mathematics and science teachers' investigations into equity and fairness through analysis of accountability data* (PhD thesis, University of Texas at Austin). Retrieved from http://iase-web.org/Publications.php?p=Dissertations

Makar, K. (2011). Learning over time: Pedagogical change in teaching mathematical inquiry. In J. Clark, B. Kissane, J. Mousley, T. Spencer & S. Thornton (Eds), *Mathematics: Traditions and [new] practices* (Proceedings of the 23rd biennial conference of the Australian Association of Mathematics Teachers and the 34th annual conference of the Mathematics Education Research Group of Australasia, pp. 27–37). Adelaide: AAMT & MERGA.

Makar, K (2013). Predict! Teaching statistics using informal statistical inference. *The Australian Mathematics Teacher*, 69(4), 34–40.

Makar, K. & Fielding-Wells, J. (2011). Teaching teachers to teach statistical investigations. In C. Batanero, G. Burrill & C. Reading (Eds), *Teaching statistics in school mathematics—Challenges for teaching and teacher education* (pp. 347–358). Dordrecht, The Netherlands: Springer.

Malloy, C. (2002). Democratic access to mathematics through democratic education: An introduction. In L. D. English (Ed.), *Handbook of international research in mathematics education* (pp. 17–25). Mahwah, NJ: Lawrence Erlbaum Associates.

Marsh, C. (1994). *Producing a national curriculum: Plans and paranoia.* Sydney: Allen & Unwin.

Martin, T. (2000). Calculus students' ability to solve geometric related-rates problems. *Mathematics Education Research Journal*, 12, 74.

Martinovic, D. & Karadag, Z. (2012). Dynamic and interactive mathematics learning environments: The case of teaching the limit concept. *Teaching mathematics and its applications*. Oxford, UK: Oxford University Press.

Mason, J. (1996). Expressing generality and roots of algebra. In N. Bednarz, C. Kieran & L. Lee (Eds), *Approaches to algebra: Perspectives for research and teaching* (pp. 65–86). Dordretch, The Netherlands: Kluwer.

Mason, J. (2004). *Doing ≠ construing and doing + discussing ≠ learning: The importance of the structure of attention.* Plenary and regular lectures presented at the 10th international congress on mathematical education, Copenhagen, Denmark, 4–11 July.

Mason, J. (2010). Attention and intention in learning about teaching through teaching. In R. Leikin & R. Zazkis (Eds), *Learning through teaching mathematics: Development of teachers' knowledge and expertise in practice* (pp. 23–47). Rotterdam, The Netherlands: Springer.

Mason, J., Graham, A. & Johnston-Wilder, S. (2005). *Developing thinking in algebra.* London, UK: Sage.

Mason, J., Graham, A., Pimm, D. & Gowar, N. (1987). *Routes to/roots of algebra* (Australian ed.). Milton Keynes: Open University.

Mason, J. & Spence, M. (1998). Towards a psychology of knowing-to. In C. Kanes, M. Goos & E. Warren (Eds), *Teaching mathematics in new times* (Proceedings of the 21st annual conference of the Mathematics Education Research Group of Australasia, pp. 342–349). Brisbane: MERGA.

Mathematical Association of Victoria. (2006). *MAV Maths talent quest 2006: Information booklet.* Retrieved from www.mav.vic.edu.au

Maths300. (n.d.). Retrieved from www.maths300.com/

Matthews, C., Watego, L., Cooper, T. & Baturo, A. (2005). Does mathematics education in Australia devalue Indigenous culture? Indigenous perspectives and non-Indigenous reflections. In P. Clarkson, A. Downton, D. Gronn, M. Horne, A. McDonough, R. Pierce & A. Roche (Eds), *Building connections: Theory, research and practice* (Proceedings of the 28th annual conference of the Mathematics Education Research Group of Australasia, pp. 513–520). Sydney: MERGA.

Mawson, R. (2013). *Theorising teaching essay* (Unpublished document, Deakin University).

Maxwell, G. (1997). Future directions for competency based assessment. *Queensland Journal of Educational Research*, 13(3), 71–84.

McCarthy, J. (2002). Connecting numeracy across the curriculum in the outback. *The Australian Mathematics Teacher*, 58(4), 22–27.

McClain, K. & Cobb, P. (2001). An analysis of development of sociomathematical norms in one first-grade classroom. *Journal for Research in Mathematics Education*, 32, 236–266.

McDonough, A. & Sullivan, P. (2008). Focussing Year 8 students on self-regulating their learning of mathematics. In M. Goos, R. Brown & K. Makar (Eds), *Navigating currents and charting directions* (Proceedings of the 31st annual conference of the Mathematics Education Research Group of Australasia, pp. 337–343). Brisbane, Qld: MERGA.

McGaw, B. (2004). Australian mathematics learning in an international context. Paper presented at the 27th annual conference of the Mathematics Education Research Group of Australasia, Townsville, 27–30 June.

McIntosh, A. (2002). Common errors in mental computation of students in Grades 3–10. In B. Barton, K. Irwin, M. Pfannkuch & M. Thomas (Eds), *Mathematics education in the South Pacific* (Proceedings of the 25th annual conference of the Mathematics Education Research Group of Australasia, pp. 457–472). Sydney: MERGA.

McIntosh, A. & Dole, S. (2004). *Mental computation: A strategies approach.* Hobart: Department of Education, Tasmania.

McIntosh, A., Reys, B. & Reys, R. (1997). *Number SENSE: Simple effective number sense experiences, Grades 6–8.* Parsippany, NJ: Dale Seymour.

McKimmie, T. (2002). Gender and textbooks. *Vinculum*, 39(4), 18–23.

McKinlay, J. (2000). Graphics calculators and assessment in mathematics education. In W. Morony & M. Stephens (Eds), *Students, mathematics and graphics calculators into the new millennium* (pp. 81–91). Adelaide: AAMT.

McLeod, D. B. (1992). Research on affect in mathematics education: A reconceptualisation. In D. A. Grouws (Ed.), *Handbook of research on mathematics teaching and learning* (pp. 575–596). New York: Macmillan.

McMillan, J. (2004). *Classroom assessment: Principles and practice for effective instruction* (3rd ed.). Boston: Pearson Education.

McMullin, L. (1999). The scrambler, or a family of vectors at the amusement park. *Mathematics Teacher*, 92(1), 64–69.

McMurchy-Pilkington, C., Trinick, T. & Meaney, T. (2013). Mathematics curriculum development and indigenous language revitalisation: Contested spaces. *Mathematics Education Research Journal*, 25, 341–360.

McNeil, N. M. & Alibali, M. W. (2005). Knowledge change as a function of mathematics experience: All contexts are not created equal. *Journal of Cognition and Development*, 6(2), 285–306.

McQuade, V. (2013). A discussion of the statistical investigation process in the Australian Curriculum. *The Australian Mathematics Teacher*, 69(4), 32–33.

Meaney, T. & Fairhall, U. (2003). Tensions and possibilities: Indigenous parents doing mathematics curriculum development. In L. Bragg, C. Campbell, G. Herbert & J. Mousley (Eds), *Mathematics education research: Innovation, networking, opportunity* (Proceedings of the 26th annual conference of the Mathematics Education Research Group of Australasia, pp. 507–514). Sydney: MERGA.

Meaney, T., McMurchy-Pilkington, C. & Trinick, T. (2008). Mathematics education and Indigenous students. In H. Forgasz, A. Barkatsas, A. Bishop, B. Clarke, S. Keast, W.-T. Seah, … S. Willis (Eds), *Research in mathematics education in Australasia 2004–2007* (pp. 119–139). Rotterdam, The Netherlands: Sense.

Meaney, T., Trinick, T. & Fairhall, U. (2013). One size does NOT fit all: Achieving equity in Māori mathematics classrooms. *Journal for Research in Mathematics Education*, 44(1), 255–263.

Mewborn, D. (2003). Teaching, teachers' knowledge, and their professional development. In J. Kilpatrick, W. G. Martin & D. Schifter (Eds), *A research companion to principles and standards for school mathematics* (pp. 45–52). Reston VA: NCTM.

Meyer, D. (2010). Math curriculum makeover. Retrieved from www.youtube.com/watch?v= BlvKWEvKSi8

Miedel, W. & Reynolds, A. (1999). Parent involvement in early intervention for disadvantaged children: Does it matter? *Journal of School Psychology*, 37(4), 379–402.

Miller, J. (2014). Young Australian Indigenous students' growing pattern generalisations: The role of gesture when generalizing. In J. Anderson, M. Cavanagh & A. Prescott (Eds), *Proceedings of the 37th annual conference of the Mathematics Education Research Group of Australasia* (pp. 461–468). Sydney: MERGA.

Mills, C. & Gale, T. (2004). Parent participation in disadvantaged schools: Moving beyond attributions of blame. *Australian Journal of Education*, 48, 268–281.

Ministerial Council on Education, Employment, Training and Youth Affairs (MCEETYA). (2008). *Melbourne Declaration on educational goals for young Australians.* Retrieved from www.curriculum.edu.au/verve/_resources/National_Declaration_on_the_Educational_Goals_for_Young_Australians.pdf

Ministerial Council on Education, Employment, Training and Youth Affairs (MCEETYA). (2009). *MCEETYA four-year plan, 2009–2012: A companion document for the Melbourne declaration on educational goals for young Australians.* Retrieved from www.scseec.edu.au/site/DefaultSite/filesystem/documents/Reports%20and%20publications/Publications/National%20goals%20for%20schooling/MCEETYA_Four_Year_Plan_%282009-2012%29.pdf

Mirra, A. (Ed.). (2004). *A family's guide: Fostering your child's success in school mathematics.* Reston, VA: NCTM.

Mitchelmore, M. (1995). Number and measurement. In L. Grimison & J. Pegg (Eds), *Teaching secondary school mathematics: Theory into practice* (pp. 50–86). Sydney: Harcourt Brace.

Mitchelmore, M. & McMaster, H. (2009). *Working mathematically in a rural context.* Australian School Innovation in Science, Technology and Mathematics Project. Retrieved from www.wmrural.net/activities.php

Mitchelmore, M. & White, P. (1998). Development of angle concepts: A framework for research. *Mathematics Education Research Journal*, 10(3), 4–27.

Mitchelmore, M. & White, P. (2000). Development of angle concepts by progressive abstraction and generalisation. *Educational Studies in Mathematics*, 41, 209–238.

Mojica-Casey, M., Dekkers, J. & Thrupp, R. (2014). Research guided practice: Student online experiences during mathematics class in the middle school. In J. Anderson, M. Cavanagh & A. Prescott (Eds.), *Curriculum in focus: Research guided practice* (Proceedings of the 37th annual conference of the Mathematics Education Research Group of Australasia, pp. 469–476). Sydney: MERGA.

Moll, L. C., Amanti, C., Neff, D. & Gonzales, N. (1992). Funds of knowledge for teaching: Using a qualitative approach to connect homes and classrooms. *Theory into Practice*, 31(2), 132–141.

Moloney, K. & Stacey, K. (1997). Changes with age in students' conceptions of decimal notation. *Mathematics Education Research Journal*, 9(5), 25–38.

Monaghan, J. (1991). Problems with the language of limits. *For the Learning of Mathematics*, 11(3), 20–24.

Moore, T. J., Miller, R. L., Lesh, R. A., Stohlmann, M. S. & Kim, Y. R. (2013). Modeling in engineering: The role of representational fluency in students' conceptual understanding. *Journal of Engineering Education*, 102(1), 141–178.

Morgan, C. & Watson, A. (2002). The interpretative nature of teachers' assessment of students' mathematics: Issues for equity. *Journal of Research in Mathematics Education*, 33(2), 78–110.

Mousley, J. (n.d.). Do quadrilaterals tessellate? *Top Drawer Teachers: Resources for teachers of mathematics*. Australian Association of Mathematics Teachers. Retrieved from http://topdrawer.aamt.edu.au/Reasoning/Big-ideas/Mathematical-truth/Truth-of-propositions/Do-quadrilaterals-tessellate

Mueller, U. & Forster, P. (2000). A comparative analysis of the 1996–1999 calculus TEE papers. In J. Bana & A. Chapman (Eds), *Mathematics education beyond 2000* (Proceedings of the 23rd annual conference of the Mathematics Education Research Group of Australasia, pp. 465–473). Sydney: MERGA.

Muir, T. (2005). When near enough is good enough, *The Australian Primary Classroom*, 10(2), 9–14.

Muir, T. & Chick, H. (2014). Flipping the classroom: A case study of mathematics methods class. In J. Anderson, M. Cavanagh & A. Prescott (Eds), *Curriculum in focus: Research guided practice* (Proceedings of the 37th annual conference of the Mathematics Education Research Group of Australasia) (pp. 485–492). Sydney, NSW: MERGA.

Muir, T. & Geiger, V. (2015). The affordances of using a flipped classroom approach in the teaching of mathematics: A case study of a Grade 10 mathematics class. *Mathematics Education Research Journal*. doi 10.1007/s13394-015-0165-8

Mulligan, J., Papic, M., Prescott, A. & Mitchelmore, M. (2006). Improving early numeracy through a Pattern and Structure Mathematics Awareness Program (PASMAP). In P. Grootenboer, R. Zevenbergen & M. Chinnappan (Eds), *Identities, cultures and learning spaces* (Proceedings of the 29th annual conference of the Mathematics Education Research Group of Australasia, pp. 376–383). Adelaide: MERGA.

Munoz, M. & Mullet, S. (1998). Evolution of the intuitive mastery of the relationship between base, exponent, and number magnitude in high-school students, *Mathematical Cognition*, 4(1), 67–77.

National Council of Teachers of Mathematics (NCTM). (1989). *Curriculum and evaluation standards for school mathematics.* Reston, VA: National Council of Teachers of Mathematics.

National Council of Teachers of Mathematics (NCTM). (2000). *Principles and standards for school mathematics.* Reston, VA: National Council of Teachers of Mathematics.

National Council of Teachers of Mathematics (NCTM). (2011). Technology in teaching and learning mathematics: A position of the National Council of Teachers of Mathematics. Retrieved from www.nctm.org/uploadedFiles/Standards_and_Positions/Position_Statements/Technology_(with%20references%202011).pdf

National Curriculum Board. (2009). *Shape of the Australian curriculum: Mathematics.* Retrieved from www.acara.edu.au/verve/_resources/Australian_Curriculum_-_Maths.pdf

NCTM *see* National Council of Teachers of Mathematics (1980). An agenda for action: Recommendations for school mathematics of the 1980s. Reston, VA: NCTM

Neiderer, K. & Irwin, K. C. (2001). Using problem solving to identify mathematically gifted children. In M. van den Heuvel-Pranhuizen (Eds), *Proceedings of the 25th conference of the International Group for the Psychology of Mathematics Education* (Vol. 3, pp. 431–438). Utrecht: PME.

Neiderer, K., Irwin, K. C. & Reilly, I. I. (2003). Identification of mathematically gifted students in New Zealand. *High Ability Studies*, 14(1), 71–84.

Neumann, D. L., Hood, M. H. & Neumann, M. M. (2012). An evaluation of computer-based interactive simulations in the assessment of statistical concepts. *International Journal for Technology in Mathematics Education,* 19(1), 17–23.

Nolan, C. & Herbert, S. (2015). Introducing linear functions: An alternative statistical approach. *Mathematics Education Research Journal,* 27(4), 401–421. doi: 10.1007/s13394-015-0147-x

Norris, N. & Dixon, R. (2011). Two exceptional-gifted students with Asperger syndrome. *The Australasian Journal of Gifted Education,* 20(2), 34–45.

Northcote, N. & McIntosh, A. (1999). What mathematics do adults really do in everyday life? *Australian Primary Mathematics Classroom,* 4(1), 19–21.

Noss, R. & Hoyles, C. (1996). *Windows on mathematical meanings.* Dordrecht, The Netherlands: Kluwer.

Noss, R., Hoyles, C. & Pozzi, S. (2000). Working knowledge: Mathematics in use. In A. Bessot & J. Ridgeway (Eds), *Education for mathematics in the workplace* (pp. 17–35). Dordrecht, The Netherlands: Kluwer.

Noura, K. (2005). How to teach mathematics using technology. In J. Mousley, L. Bragg & C. Campbell (Eds), Mathematics—Celebrating Achievement (pp. 256–264). Melbourne: Mathematical Association of Victoria.

Oates, G. (2009). Relative values of curriculum topics in undergraduate mathematics in an integrated technology environment. In R. Hunter, B. Bicknell & T. Burgess (Eds), *Crossing divides* (Proceedings of the 32nd annual conference of the Mathematics Education Research Group of Australasia, pp. 419–426). Palmerston North, NZ: MERGA.

Ocean, J. & Miller-Reilly, B. (1997). Black holes and beginning teachers: A connected approach to teaching mathematics. *The Australian Mathematics Teacher*, 53(4), 17–20.

Op't Eynde, P., de Corte, E. & Verschaffel, L. (2002). Framing students' mathematics-related beliefs. In G. Leder, E. Pehkonen & G. Törner (Eds), *Beliefs: A hidden variable in mathematics education?* (pp. 13–37). Dordrecht, The Netherlands: Kluwer Academic Publishers.

Organisation for Economic Cooperation and Development (OECD). (2000). *Programme for international student assessment* (PISA). Retrieved from www.pisa.oecd.org/

Orton, A. (1983). Students' understanding of differentiation. *Educational Studies in Mathematics*, 14, 235–250.

Orton, A. (1985). When should we teach calculus? *Mathematics in School* (March), 11–15.

Owens, K. (2010). Changing our perspective on space: Place mathematics as a human endeavour. In L. Sparrow, B. Kissane & C. Hurst (Eds), *Shaping the future of mathematics* (Proceedings of the 33rd annual conference of the Mathematics Education Research Group of Australasia, pp. 454–461). Freemantle, WA: MERGA.

Owens, K. (2015). Changing the teaching of mathematics for improved Indigenous education in a rural Australian city. *Journal of Mathematics Teacher Education*, 18(1), 53–78.

Padula, J., with Lam, S. & Schmidtke, M. (2001). Syntax and word order: Important aspects of mathematical English. *The Australian Mathematics Teacher*, 57(4), 31–35.

Padula, J., with Lam, S. & Schmidtke, M. (2002). Mathematical English: Some insights for teachers and students. *The Australian Mathematics Teacher*, 58(3), 40–44.

Pagliaro, C. (2006). Mathematics education and the deaf learner. In D. F. Moores & D. S. Martin (Eds), *Deaf learners: Developments in curriculum and instruction* (pp. 29–40). Washington, DC: Gallaudet University Press.

Pagni, D. (1998). Giving meaning to multiplication and division of fractions. *The Australian Mathematics Teacher*, 54(1), 11–15.

Parameswaran, R. (2007). On understanding the notion of limits and infinitesimal quantities. *International Journal of Science and Mathematics Education*, 5(2), 193–216. doi: 10.1007/s10763-006-9050-y

Parliament of the Commonwealth of Australia. (2000). Inquiry into the education of boys. (Media release). House of Representatives Standing Committee on Employment, Education and Workplace Relations: Commonwealth of Australia.

Parliament of the Commonwealth of Australia. (2002). *Boys: Getting it right* (Report on the Inquiry into the Education of Boys). House of Representatives Standing Committee on Education and Training. Canberra: Commonwealth of Australia.

Parsons, J. & Reilly, Y. (2012). *Mathematics in the inclusive classroom* (Book 1). Albert Park, Vic.: Teaching Solutions.

Pearn, C. (2007). Using paper folding, fraction walls, and number lines to develop understanding of fractions for students from Years 5–8. *The Australian Mathematics Teacher, 63*(4), 31–36.

Pearn, C. & Stephens, M. (2004). Why you have to probe to discover what Year 8 students really think about fractions. In I. Putt, R. Faragher & M. McLean (Eds), *Mathematics education for the third millennium: Towards 2010* (Proceedings of the 27th annual conference of the Mathematics Education Research Group of Australasia, pp. 430–437). Sydney: MERGA.

Pedemonte, P. (2007). How can the relationship between argumentation and proof be analysed? *Educational Studies in Mathematics, 66*, pp. 23–41.

Pegg, J. (2003). Assessment in mathematics: A developmental approach. In M. Royer (Ed.), *Mathematical cognition* (pp. 227–259). Greenwich, CN: Information Age.

Pegg, J. (2009). Lessons learnt: Implications of four large-scale SiMMER projects for rural education in Australia. In T. Lyons, J. Y. Choi & G. McPhan (Eds), *Innovation for equity in rural education* (Proceedings of the international symposium for Innovation in Rural Education, ISFIRE, pp. 33–44). Armidale, NSW: University of New England.

Pegg, J., Graham, L. & Bellert, A. (2005). The effect of improved automaticity of basic number skills on persistently low-achieving pupils. In H. L. Chick & J. L. Vincent (Eds), *Proceedings of the 29th conference of the International Group for the Psychology of Mathematics Education* (Vol. 4, pp. 49–56). Melbourne, Australia: PME.

Pegg, J., Lyons, T., Panizzon, D., Parnell, A., Brown, T., Reading, C. & Graham, L. (2005). *Issues in teaching and learning science, ICT, and mathematics in rural and regional Australia: A national survey.* SiMERR National Research Centre. Retrieved from http://simerr.une.edu.au/pages/projects/1nationalsurvey/Abridged%20report/Abridged_Full.pdf

Pegg, J. & Panizzon, D. (2004). Addressing changing assessment agendas: Experiences of secondary mathematics teachers in rural NSW. In I. Putt, R. Faragher & M. McLean (Eds),

Mathematics education for the third millennium: Towards 2010 (Proceedings of the 27th annual conference of the Mathematics Education Research Group of Australasia, Townsville, Vol. 2, pp. 438–445). Sydney: MERGA.

Pegg, J. & Panizzon, D. (2007–08). Addressing changing assessment agendas: Impact of professional development on secondary mathematics teachers in NSW, *Mathematics Teacher Education and Development*, 9, 66-80.

Pegg, J., Panizzon, D. & Inglis, M. (2003). Helping teachers incorporate changing agendas in assessment into their primary and secondary classrooms. Retrieved from http://scs.une.edu.au/CF/Papers/pdf/Pegg.pdf

Peled, I. & Suzan, A. (2011). Pedagogical, mathematical, and epistemological goals in designing cognitive conflict tasks for teacher education. In O. Zaslavsky & O. Sullivan (Eds), *Constructing knowledge for teaching secondary mathematics: Tasks to enhance prospective and practicing teacher learning* (pp. 73–88). New York: Springer.

Pena, D. (2000). Parent involvement: Influencing factors and implications. *The Journal of Educational Research*, 94(1), 42–54.

Peressini, D. (1997). Parental involvement in the reform of mathematics education. *The Mathematics Teacher*, 90(6), 421–427.

Peressini, D. (1998). The portrayal of parents in the school mathematics reform literature: Locating the context for parental involvement. *Journal for Research in Mathematics Education*, 29, 555–572.

Pérez Carreón, G., Drake, C. & Barton, A. (2005). The importance of presence: Immigrant parents' school engagement experiences. *American Educational Research Journal*, 42, 465–498.

Perry, B. & Howard, P. (2001). Arithmetic thinking strategies and low achieving junior high school students in Australia. In R. Speiser, C. A. Maher & C. N. Walter (Eds), *Proceedings of the 23rd annual meeting of the Psychology of Mathematics Education—North America Group* (pp. 273–280). Columbus, OH: ERIC Clearing House.

Perry, B. & Howard, P. (2002). A systematic program for students who are experiencing difficulty with mathematics as they transition from elementary to high school in Australia. In B. Barton, K. Irwin, M. Pfannkuch & M. Thomas (Eds), *Mathematics education in the South Pacific* (Proceedings of the 25th annual conference of the Mathematics Education Research Group of Australasia, pp. 543–550). Sydney: MERGA.

Perry, B., Howard, P. & Tracey, D. (1999). Head mathematics teachers' beliefs about the learning and teaching of mathematics. *Mathematics Education Research Journal*, 11, 39–53.

Perso, T. (1992). Making the most of errors. *The Australian Mathematics Teacher*, 48(2), 12–14.

Perso. T. (2003). *Improving Aboriginal numeracy: A book for education systems, school administrators, teachers and teacher educators.* Perth, WA: Mathematics, Science and Technology Education Centre, Edith Cowan University.

Peters, C., Geiger, V., Goos, M. & Dole, S. (2012). Numeracy in health and physical education. *The Australian Mathematics Teacher*, 68(1), 21–27.

Pfannkuch, M. & Ben-Zvi, D. (2011). Developing teachers' statistical thinking. In C. Batanero, G. Burrill & C. Reading (Eds), *Teaching statistics in school mathematics—Challenges for teaching and teacher education* (pp. 323–333). Dordrecht, The Netherlands: Springer.

Pfannkuch, M. & Brown, C. M. (1996). Building on and challenging students' intuitions about probability: Can we improve undergraduate learning. *Journal of Statistics Education*, 4(1), 39–53.

Pfannkuch, M., Budgett, S., Parsonage, R. & Horring, J. (2004). *Comparison of data plots: Building a pedagogical framework.* Paper presented at the International Conference on Mathematics Education, Bergen, Norway.

Pfannkuch, M. & Ziedins, I. (2014). A modelling perspective on probability. In E. Chernoff & B. Sriraman (Eds), *Probabilistic thinking: Presenting plural perspectives* (pp. 101–116). Berlin: Springer Science.

Piaget, J. (1954). *The construction of reality in the child.* New York: Basic Books.

Picker, S. & Berry, J. 2001. Investigating pupils' images of mathematicians. In M. van der Heuvel-Pantuizen (Ed.), *Proceedings of the 25th conference of the International Group for the Psychology of Mathematics Education* (Vol.4, pp.49–56). Utrecht, The Netherlands: PME.

Pierce, R. & Ball, L. (2009). Perceptions that may affect teachers' intention to use technology in secondary mathematics classes. *Educational Studies in Mathematics*, 71(3), 299–317

Pierce, R. & Stacey, K. (2001). A framework for algebraic insight. In J. Bobis, B. Perry & M. Mitchelmore (Eds), *Numeracy and beyond.* (Proceedings of the 24th annual conference of the Mathematics Education Research Group of Australasia, Vol. 2, pp. 418–425). Sydney: MERGA.

Pierce, R. & Stacey, K. (2008). Using pedagogical maps to show the opportunities afforded by CAS for improving the teaching of mathematics. *Australian Senior Mathematics Journal*, 22(1), 6–12.

Pierce, R. & Stacey, K. (2013). Teaching with new technology: Four 'early majority' teachers. *Journal of Mathematics Teacher Education*, 16(5), 323–347.

Pierce, R., Stacey, K. & Ball, L. (2005). Mathematics from still and moving images. *The Australian Mathematics Teacher*, 61(3), 26–31.

Pirie, S. E. B. & Kieren, T. (1994). Growth in mathematical understanding: How can we characterise it and how can we represent it? *Educational Studies in Mathematics*, 26, 165–190.

Pirie, S. E. B. & Martin, S. (2000). The role of collecting in the growth of mathematical understanding. *Mathematics Education Research Journal*, 12, 127–146.

Pitman, A. (1989). Mathematics education reform in its social, political and economic contexts. In N. Ellerton & M. A. Clements (Eds), *School mathematics: The challenge to change* (pp. 101–119). Geelong, Vic.: Deakin University Press.

Plummer, F. (1999). Rich assessment tasks: Exploring quality assessment for the School Certificate. *SCAN*, 18(1), 14–19.

Pollak, H. (1997). Solving problems in the real world. In L. A. Steen (Ed.), *Why numbers count: Quantitative literacy for tomorrow's America*. New York: College Entrance Examination Board.

Porzio, D. (1997), 'Effects of different instructional approaches on calculus student's understanding of the relationship between slope, rate of change, and the first derivative', in Dossey, J., A., Swafford, J. O., Parmantie, M., Dossey, A.E (Eds), *Proceedings of the Annual Conference of the North American Chapter of International Group for the Psychology of Mathematics Education* (pp. 37–44), Bloomington-Normal, IL: NA-PME.

Prinsley, R. & Baranyai, K. (2015). *STEM skills in the workplace: What do employers want?* (Occasional Paper Issue 9). Office of the Chief Scientist, Australian Government. Retrieved from www.chiefscientist.gov.au/wp-content/uploads/OPS09_02Mar2015_Web.pdf

Przenioslo, M. (2004). Images of the limit of function formed in the course of mathematical studies at the university. *Educational Studies in Mathematics*, 22(1), 1–36.

Pugalee, D., Frykholm, J., Johnson, A., Slovin, H., Malloy, C. & Reston, R. (2002). *Navigating through geometry in Grades 6–8*. Reston, VA: National Council of Mathematics Teachers.

Puig, L. (2011). Researching the history of algebraic ideas from an educational point of view. In V. Katz & C. Tzanakis (Eds), *Recent developments on introducing a historical dimension in mathematics education* (pp. 29–42). Washington, DC: The Mathematical Association of America.

Queensland Board of Senior Secondary School Studies (2001). *Mathematics B senior syllabus*. Brisbane: Author.

Queensland Board of Teacher Registration. (2005). *Numeracy in teacher education: The way forward in the 21st century* (Report of the Numeracy in Preservice Teacher Education Working Party). Brisbane: Author.

Queensland Curriculum and Assessment Authority. (2015). *A-Z of senior moderation of achievement in school-based assessment.* Retrieved from www.qcaa.qld.edu.au/downloads/senior/snr_qa_mod_a-z.pdf

Quinlan, C. (2004). Sparking interest in trigonometry. *The Australian Mathematics Teacher, 60*(3), 17–20.

Quinnell, L. (2014). Enhancing the teaching and learning of mathematical visual images. *The Australian Mathematics Teacher, 70*(1), 18–25.

Radford, L. (1995). Before the other unknowns were invented: Didactic inquiries on the methods and problems of medieval Italian algebra. *For the Learning of Mathematics, 15*(3), 28–38.

Radford, L. & Grenier, M. (1996). On the dialectical relationships between symbols and algebraic ideas. In L. Puig & A. Gutierrez (Eds), *Proceedings of the 20th international conference for the Psychology of Mathematics Education* (Vol. 4, pp. 179–186). Valencia, Spain: PME.

Rasmussen, C., Marrongelle, K., & Borba, M. (2014), Research on calculus: what do we know and where do we need to go?. *ZDM, 46*(4), 507–515.

Rasmussen, D., Rasmussen, S. & Bennett, D. (Eds). (1995). *Teaching geometry with The Geometer's Sketchpad: Teaching notes and sample activities.* Berkeley, CA: Key Curriculum Press.

Raymond, A. (1997). Inconsistency between a beginning elementary school teacher's mathematical beliefs and teaching practice. *Journal for Research in Mathematics Education, 28,* 550–576.

Reay, D. (1998). *Class work: Mothers' involvement in their children's schooling.* London: University College Press.

Reese, S. (2004). Teacher portfolios: Displaying the art of teaching. *Techniques: Connecting Education and Careers, 79*(5), 18.

Reid, A. (2005). The politics of national curriculum collaboration: How can Australia move beyond the railway gauge metaphor? In C. Harris & C. Marsh (Eds), *Curriculum developments in Australia: Promising initiatives, impasses and dead-ends* (pp. 39–51). Canberra: Australian Curriculum Studies Association.

Reilly, Y. & Parsons, J. (2011). Delivering differentiation in the fully inclusive middle years classroom. In L. Bragg (Ed.), *Mathematics is multi-dimensional* (Proceedings of the 48th annual conference of the Mathematical Association of Victoria). Brunswick: MAV.

Reilly, Y., Parsons, J. & Bortolot, E. (2009). Reciprocal teaching in mathematics. In *Mathematics of prime importance* (Proceedings of the 46th annual conference of the Mathematical Association of Victoria). Brunswick: MAV.

Reilly, Y., Parsons, J. & Bortolot, E. (2010). An effective numeracy program for the middle years. In M. Westbrook et al. (Eds), *New curriculum, new opportunities* (Proceedings of the 46th annual conference of the Mathematical Association of Victoria). Brunswick: MAV.

Reys, R., Lindquist, M. M., Lambdin, D. V., Smith, N. L., Rogers, A., Falle, J., ... Bennett, S. (2012). *Helping children learn mathematics* (1st Australian ed.). Milton, Qld: John Wiley and Sons.

Richards, J. (1991). Mathematical discussions. In E. von Glaserfeld (Ed.), *Radical constructivism in mathematics education* (pp. 13–51). Dordrecht, The Netherland: Kluwer Academic Publishers.

Ridgway, J., Nicholson, J. & McCusker, S. (2011). Developing statistical literacy in students and teachers. In C. Batanero, G. Burrill & C. Reading (Eds), *Teaching statistics in school mathematics— Challenges for teaching and teacher education* (pp. 311–322). Dordrecht, The Netherlands: Springer.

Roche, A. & Clarke, D. (2011). Some lessons learned from the experience of assessing teacher pedagogical content knowledge in mathematics. In J. Clark, B. Kissane, J. Mousley, T. Spencer & S. Thornton (Eds), *Mathematics: Traditions and [new] practices* (Proceedings of the 34th annual conference of the Mathematics Education Research Group of Australasia, pp. 658–688). Adelaide: MERGA.

Rogers, E. M. & Shoemaker, F. F. (1971). *Communication of innovations: A cross-cultural approach.* New York: Free Press.

Roorda, G. Vos, P. & Goedhart, M. (2007). Derivatives in applications: Exploring students' understanding. Paper presented at the CERMES Conference, Cyprus. Retrieved from http://ermeweb.free.fr/CERME%205/WG13/13_Roorda.pdf

Rossman, A., Chance, B. & Locke, R. (2001). *Workshop statistics: Discovery with data and Fathom.* Emeryville, CA: Key College Press.

Rothman, S. (2002). Achievement in literacy and numeracy by 14 year olds, 1975–1998 (LSAY Research Project No. 29). Melbourne: ACER.

Rowland, D. R., & Jovanoski, Z. (2004), Student interpretations of the terms in first-order ordinary differential equations in modelling contexts, *International Journal of Mathematical Education in Science & Technology*, 35, 503–516.

Rowland, T., Turner, F., Thwaites, A., & Huckstep, P. (2009). *Developing primary mathematics teaching: Reflecting on practice with the knowledge quartet.* London: Sage Publications.

Rozek, C. S., Hyde, J. & Svodoba, R. C. (2015). Gender differences in the effects of a utility-value intervention to help parents motivate adolescents in mathematics and science. *Journal of Educational Psychology*, 107(1), 195–206.

Runnesson, U., Kulberg, A. & Maunula, T. (2011). Sensitivity to student learning: A possible way to enhance teachers' and students' learning? In O. Zaslavsky & O. Sullivan (Eds), *Constructing knowledge for teaching secondary mathematics: Tasks to enhance prospective and practicing teacher learning* (pp. 263–278). New York: Springer.

Ryan, J. (1991). Calculus: With some respect to time. *Australian Senior Mathematics Journal*, 5(2), 88–97.

Sadler, D. R. (1987). Specifying and promulgating achievement standard. *Oxford Review of Education*, 13(2), 191–209.

Sadler, D. R. (1996). Criteria and standards in student assessment. Paper presented at the HERDSA conference, Perth, WA, 7–8 July.

Sarra, C. (2012). *Stronger smarter institute: Changing the tide of low expectations in Indigenous education.* Retrieved from http://strongersmarter.com.au/

Saxe, J. (2002). Children's developing mathematics in collective practices: A framework for analysis. *The Journal of Learning Sciences*, 11(2–3), 275–300.

Schliemann, A. D., Carraher, D. W. & Brizuela, B. M. (2007). *Bringing out the algebraic character of arithmetic: From children's ideas to classroom practice.* Mahwah, NJ: Lawrence Erlbaum.

Schneider, M. (1992), A way to analyse several difficulties the pupils meet with in calculus. In *Proceedings of the 7th International Congress of Mathematical Education* (Working Group on Students' Difficulties in Calculus, pp. 31–34), Québec: Les Presses de l'Université Laval.

Schoenfeld, A. H. (1988). When good teaching leads to bad results: The disasters of 'well-taught' mathematics courses. *Educational Psychologist*, 23, 145–166.

Schoenfeld, A. H. (1992). Learning to think mathematically: Problem solving, metacognition and sense making in mathematics. In D. A. Grouws (Ed.), *Handbook of research on mathematics teaching and learning* (pp. 334–370). New York: Macmillan.

Scott, P. (2011). The square root of 2, *The Australian Mathematics Teacher*, 67(1), 3–5.

Secada, W. & Adajian, L. (1997). Mathematics teachers' change in the context of their professional communities. In E. Fennema & B. Nelson (Eds), *Mathematics teachers in transition* (pp. 193–219). Mahwah, NJ: Erlbaum.

Serow, P. & Inglis, M. (2010). Templates in action. *The Australian Mathematics Teacher*, 66(4), 10–16.

Sfard, A., Forman, E. A. & Kieran, C. (2001). Learning discourse: Sociocultural approaches to research in mathematics education. *Educational Studies in Mathematics*, 46(1–3), 1–11.

Shan, S. & Bailey, P. (1994). *Multiple factors: Classroom mathematics for equality and justice.* Oakhill: Trentham Books.

Sharma, Y. (2013). Mathematical giftedness: A creative scenario. *The Australian Mathematics Teacher*, 69(1), 15–24.

Shaughnessy, M. (2007). Research on statistics. In F. Lester (Ed.), *Second handbook of research on mathematics teaching and learning* (pp. 909–955). Reston, VA: NCTM.

Shaughnessy, M. & Pfannkuch, M. (2002). How faithful is Old Faithful? *The Mathematics Teacher*, 95(4), 252–259.

Shell Centre for Mathematical Education. (1985). *The language of functions and graphs*. Nottingham, UK: Joint Matriculation Board and Shell Centre for Mathematical Education.

Shephard, L. (2001). The role of classroom assessment in teaching and learning. In V. Richardson (Ed.), *Handbook of research on teaching* (pp. 1066–1099). Washington, DC: American Educational Research Association.

Shield, M. & Dole, S. (2002). Investigating textbook presentations of ratio and proportion. In B. Barton, K. Irwin, M. Pfannkuch & M. Thomas (Eds) *Mathematics education in the South Pacific* (Proceedings of the 25th annual conference of the Mathematics Education Research Group of Australasia, pp. 608–616). Auckland: MERGA.

Shield, M. & Dole, S. (2008). Proportion in middle school mathematics: It's everywhere. *The Australian Mathematics Teacher*, 64(3), 7–9.

Shield, M. & Dole, S. (2013). Assessing the potential of textbooks to promote deep learning. *Educational Studies in Mathematics*, 82, 183–199.

Shimizu, Y. (2010). A task-specific analysis of explicit linking in lesson sequences in three Japanese mathematics classrooms. In Y. Shimizu, B. Kaur, R. Huang & D. J. Clarke (Eds), *Mathematical tasks in classrooms around the world* (pp. 87–102). The Netherlands: Sense Publications.

Shulman, L. (1986). Those who understand: Knowledge growth in teaching. *Educational Researcher*, 15(2), 4–14.

Siemon, D. & Virgona, J. (2002). Reflections on the middle years numeracy research project— Is it a case of too much too soon, for too many? In B. Barton, K. Irwin, M. Pfannkuch & M. Thomas (Eds), *Mathematics education in the South Pacific* (Proceedings of the 25th annual conference of the Mathematics Education Research Group of Australasia, pp. 617–624). Sydney: MERGA.

Siemon, D., Virgona, J. & Corneille, K. (2001). *The middle years numeracy research project: 5–9* (Final Report). Retrieved from www.sofweb.vic.edu.au/mys/research/MYNRP/index.htm

Sinicrope, R., Mick, H. & Kolb, J. (2002). Fraction division interpretations. In B. Litwiller & G. Bright (Eds), *Making sense of fractions, ratios and proportions: 2002 yearbook* (pp. 162–175). Reston, VA: NCTM.

Skemp, R. (1987). *The psychology of learning mathematics.* Hillsdale, NJ: Erlbaum.

Skovmose, O. & Valero, P. (2002). Democratic access to powerful mathematics in a democratic country. In L. English (Ed.), *Handbook of international research in mathematics education* (pp. 383–408). Mahwah, NJ: Lawrence Erlbaum Assoc.

Slee, R. (2011). *The irregular school: Exclusion, schooling and inclusive education.* Oxon, UK: Routledge.

Small, M. (2011). *One, two, infinity.* Retrieved from www.onetwoinfinity.ca/

Sriraman, B. (2003). Mathematical giftedness, problem solving, and the ability to formulate generalisations. *The Journal of Secondary Gifted Education,* 14(3), 151–165.

Stacey, K. (1991). Teaching mathematical modelling. In J. O'Reilly & S. Wettenhall (Eds), *Mathematics: Inclusive dynamic exciting active stimulating* (Proceedings of the 28th annual conference of the Mathematical Association of Victoria, pp. 221–227). Melbourne: MAV.

Stacey, K. (2003). The need to increase attention to mathematical reasoning. In H. Hollingsworth, J. Lokan & B. McCrae (Eds), *Teaching mathematics in Australia: Results from the TIMSS 1999 Video Study* (pp. 119–122). Melbourne: ACER.

Stacey, K. (2005). The place of problem solving in contemporary mathematics curriculum documents. *Journal of Mathematical Behavior,* 24, 341–350.

Stacey, K. & Chick, H. (2004). Solving the problem with algebra. In K. Stacey, H. Chick & M. Kendal (Eds), *The future of the teaching and learning of algebra: The 12th ICMI study* (pp. 1–20). Norwell, MA: Kluwer.

Stacey, K., Chick, H. & Kendal, M. (Eds). (2004). *The future of the teaching and learning of algebra: The 12th ICMI study.* Norwell, MA: Kluwer.

Stacey, K. & Groves, S. (1984). Problem solving: People and projects in Australia. In P. Costello, S. Ferguson, K. Slinn, M. Stephens, D. Trembath & D. Williams (Eds), Proceedings of the 10th biennial conference of the Australian Association of Mathematics Teachers (pp. 205–209). Blackburn: AAMT.

Stacey, K. & Price, E. (2005). Surds, spirals, dynamic geometry and CAS. In J. Mousley, L. Bragg & C. Campbell (Eds.), *Mathematics—Celebrating achievement* (MAV annual conference, pp. 298–306). Melbourne: MAV.

Stacey, K. & Vincent, J. (2009a). Finding the area of a circle: Didactic explanations in school mathematics. *The Australian Mathematics Teacher*, 65(3), 6–9.

Stacey, K. & Vincent, J. (2009b). Modes of reasoning in explanations in Australian eighth-grade mathematics textbooks. *Educational Studies in Mathematics*, 72, 271–288.

Stack, S. & Watson, J. (2013). Randomness, sample size, imagination and metacognition making judgements about differences in data sets. *The Australian Mathematics Teacher*, 69(4), 23–30.

Steen, L. (2001). The case for quantitative literacy. In L. Steen (Ed.), *Mathematics and democracy: The case for quantitative literacy* (pp. 1–22). Princeton, NJ: National Council on Education and the Disciplines.

Steen, L. (Ed.). (n.d.). *Mathematics and democracy: The case for quantitative literacy*. The National Council on Education and the Disciplines. Retrieved from www.maa.org/Ql/mathanddemocracy. html

Stein, M., Engle, R., Smith, M. & Hughes, E. (2008). Orchestrating productive mathematical discussions: Five practices for helping teachers move beyond show and tell. *Mathematical Thinking and Learning*, 10(4), 313–340.

Stein, M. K., Remillard, J. & Smith, M. (2007). How curriculum influences student learning. In F. K. Lester, Jr. (Ed.), *Second handbook of research on mathematics teaching and learning* (pp. 319–369). Charlotte, NC: Information Age Publishing.

Stein, M. K., Silver, E. & Smith, M. (1998). Mathematics reform and teacher development: A community of practice perspective. In J. Greeno & S. Goldman (Eds), *Thinking practices: A symposium on mathematics and science learning* (pp. 17–52). Mahwah, NJ: Erlbaum.

Steinle, V., Stacey, K. & Chambers, D. (2002). *Teaching and learning about decimals* (Version 2.1). Melbourne: University of Melbourne.

Stenhouse, L. (1975). *An introduction to curriculum research and development*. London: Heinemann.

Stephens, M. (1990). Discipline review of teacher education in mathematics and science: A response from the Australian Association of Mathematics Teachers. *Australian Senior Mathematics Journal*, 4(1), 40–51.

Stephens, M. (2000). *Identification and evaluation of teaching practices that enhance numeracy achievement*. Canberra: AusInfo.

Stephens, M. (2014). The Australian curriculum: Mathematics—How did it come about? What challenges does it present for teachers and for the teaching of mathematics? In Y. Li and G. Lappan (Eds), *Mathematics curriculum in school education* (pp. 157–176). New York: Springer.

Stephens, M., Clarke, D. & Wallbridge, M. (1994). Policy to practice: High stakes assessment as a catalyst for classroom change. In G. Bell, B. Wright, N. Lesson & J. Geake (Eds), *Challenges in mathematics education: Constraints on construction* (Proceedings of the 17th annual conference of the Mathematics Education Research Group of Australasia, pp. 571–580). Lismore: MERGA.

Stewart, S. (2006). A new elementary function for our curricula? *Australian Senior Mathematics Journal*, 19(2), 8–25.

Stewart, S. & Thomas, M. O. J. (2006). Student thinking about eigenvalues and eigenvectors: Formal, symbolic and embodied notions. In P. Grootenboer, R. Zevenbergen & M. Chinnappan (Eds), *Proceedings of the 29th annual conference of the Mathematics Education Research Group of Australasia, Canberra* (Vol. 2, pp. 487–495). Adelaide: MERGA.

Stiggins, R., Chappuis, J., Chappuis, S. & Arter, J. (2014). Assessment *for* learning and *of* learning. In *Classroom assessment for student learning* (Chapter 2, pp. 29–46). Upper Saddle River, NJ: Pearson.

Stillman, G. (1998). The emperor's new clothes: Teaching and assessment of mathematical applications at the senior secondary level. In P. Galbraith, W. Blum, G, Booker & I. D. Huntley (Eds), *Mathematical modelling: Teaching and assessment in a technology-rich world* (pp. 243–254). Chichester, UK: Horwood.

Stillman, G. (2001). The impact of school-based assessment on the implementation of a modelling/applications-based curriculum: An Australian example. *Teaching Mathematics and its Applications*, 20(3), 101–107.

Stillman, G. (2002). *Assessing higher order mathematical thinking through applications* (PhD thesis, University of Queensland).

Stillman, G. (2006). The role of challenge in engaging lower secondary students in investigating real world tasks. In E. Barbeau & P. Taylor (Eds), *Challenging mathematics in and beyond the classroom* (Proceedings of ICMI Study Conference 16). Retrieved from www.amt.edu.au/pdf/icmis16oaustillman.pdf

Stillman, G. (2007). Implementation case study: Sustaining curriculum change. In W. Blum, P. Galbraith, H.-W. Henn & M. Niss (Eds), *Modelling and applications in mathematics education* (New ICMI Studies Series No. 10, pp. 497–502). New York: Springer.

Stillman, G., Brown, J., Farragher, R., Geiger, V. & Galbraith, P. (2013). The role of textbook tasks in developing a socio-critical perspective on mathematical modelling. In G. Stillman, G. Kaiser, W. Blum & J. Brown (Eds), *Teaching mathematical modelling: Connecting to research and practice* (pp. 361–371). Dordrecht, The Netherlands: Springer.

Stillman, G., Kaiser, G., Blum, W. & Brown, J. P. (Eds). (2013). *Teaching mathematical modelling: Connecting to research and practice*. Dordrecht, The Netherlands: Springer.

Stillman, G. & Ng, K. E. D. (2013). Embedding authentic real world tasks into secondary mathematics curricula. In A. Damlamian, J. F. Rodrigues & R. Strasser (Eds), *Educational interfaces between mathematics and industry* (pp. 299–307). Cham, Switzerland: Springer.

Stutzman, R. & Race, K. (2004). EMRF: Everyday rubric grading. *Mathematics Teacher*, 97(1), 34–39.

Sullivan, P. & Clarke, D. (1991). *Communication in the classroom: The importance of good questioning*. Geelong: Deakin University Press.

Sullivan, P., Clarke, D. J., Clarke, D. M., Farrell, L. & Gerrard, J. (2013). Processes and priorities in planning mathematics teaching. *Mathematics Education Research Journal*, 25, 457–480.

Sullivan, P., Mousley, J. & Zevenbergen, R. (2006). Developing guidelines for teachers helping students experiencing difficulty in learning mathematics. In P. Grootenboer, R. Zevenbergen & M. Chinnappan (Eds), *Identities, cultures and learning spaces* (Proceedings of the 29th annual conference of the Mathematics Education Research Group of Australasia, pp. 496–503). Adelaide: MERGA.

Summers, A. (2003). *The end of equality: Work, babies, and women's choices in 21st century Australia*. Milsons Point, NSW: Random House.

Suurmatamm, C., Koch, M. & Arden, A. (2010). Teachers' assessment practices in mathematics: Classrooms in the context of reform. *Assessment in Education: Principles, Policy & Practice*, 17(4), 399–417.

Swan, K., Adamson, R. & Cocking, M. (1997). *Nelson maths 9: Maths for CSFII*. Melbourne: Nelson.

Swedosh, P., Dowsey, J., Caruso, N., Flynn, P. & Tynan, D. (2006). *MathsWorld: Mathematical methods units 1 & 2*. Melbourne: Macmillan.

Takači, D., Stankov, G. & Milanovic, I. (2015). Efficiency of learning environment using GeoGebra when calculus contents are learned in collaborative groups. *Computers & Education*, 82, 421–431.

Takahashi, A. (n.d.). *Implementing lesson study in North American schools and school districts*. Retrieved from http://hrd.apec.org/images/a/ae/51.2.pdf

Takahashi, A. (2006). Characteristics of Japanese mathematics lessons. Retrieved from www. criced.tsukuba.ac.jp/math/sympo_2006/takahashi.pdf

Takahashi, A. & Yoshida, M. (2004). Ideas for establishing lesson-study communities. *Teaching Children Mathematics* (May), 436–443.

Takahashi, T. (2012). Let's investigate solid figures (IMPLUS Lesson Plan). Tokyo: Gakugei University. Retrieved from www.impuls-tgu.org/en/library/geometry/page-87.html

Tall, D. (1985). Understanding the calculus. *Mathematics Teaching*, 110, 49–53.

Tall, D. (1987). Whither calculus? *Mathematics Teaching*, 117, 50–54.

Tall, D. (1991). *Advanced mathematical thinking*. Dordrecht, The Netherlands: Kluwer.

Tall, D. (1992). The transition to advanced mathematical thinking: Functions, limits, infinity and proof. In D. A. Grouws (Ed.), *Handbook of research on mathematics teaching and learning* (pp. 495–511). New York: Macmillan.

Tall, D. (1994). Computer environments for the learning of mathematics. In R. Biehler, R. Scholz, B. Straber & B. Winkelmann (Eds), *Didactics of mathematics as a scientific discipline* (pp. 189–199). Dordrecht, The Netherlands: Kluwer Academic Publishers.

Tall, D. (1996). Functions and calculus. In A. Bishop, K. Clements, C. Keitel, J. Kilpatrick & C. Laborde (Eds), *International handbook of mathematics education* (Vol. 1, pp. 289–325). Dordrecht, The Netherlands: Kluwer.

Tall, D. (1997). Functions and calculus. In A. J. Bishop, M. A. Clements, C. Keitel, J. Kilpatrick & C. Laborde (Eds), *International handbook of mathematics education* (pp. 289–325). Dordrtecht, The Netherlands: Kluwer.

Tall, D. (2011). Looking for the bigger picture. *For the Learning of Mathematics*, 31(2), 17–18.

Tall, D. & Vinner, S. (1981). Concept image and concept definition in mathematics with particular reference to limits and continuity. *Educational Studies in Mathematics*, 12(2), 151–169.

Tall, D. & Watson, A. (2001). Schemas and processes for sketching the gradient of a graph. Available from www.warwick.ac.uk/staff/David Tall/

Tan, J. (2012). Students' ways of knowing and learning mathematics and their ways of interacting with advanced calculators. In J. Dindyal, L. P. Cheng & S. F. Ng (Eds), *Proceedings of the 35th annual conference of the Mathematics Education Research Group of Australasia* (pp. 704–711). Singapore: MERGA.

Tanko, M. G. & Atweh, B. (2012). Developing mathematical knowledge through social justice pedagogy with young adult Arab women. In J. Dindyal, L. P. Cheng & S. F. Ng (Eds),

Mathematics education: Expanding horizons (Proceedings of the 35th annual conference of the Mathematics Education Research Group of Australasia, pp. 720–727). Singapore: MERGA.

Tanner, H. & Jones, S. (2000). *Becoming a successful teacher of mathematics.* London: Routledge Falmer.

Tate, W. F. & Rousseau, C. (2002). Access and opportunity: The political and social context of mathematics education. In L. English (Ed.), *International handbook of research in mathematics education* (pp. 271–300). Mahwah, NJ: Lawrence Erlbaum.

Teese, R. (2000). *Academic success and social power: Examinations and inequality.* Carlton South: Melbourne University Press.

Teese, R. (2013). *Academic success and social power: Examinations and inequality* (2nd ed.). Carlton South: Australian Scholarly Publishing.

Texas-Instruments (2006). TI-Nspire (Computer Software). Dallas, TX: Texas-Instruments.

Thomas, M. O. J., Monaghan, J. & Pierce, R. (2004). Computer algebra systems and algebra: Curriculum, assessment, teaching, and learning. In K. Stacey, H. Chick & M. Kendal (Eds), *The future of the teaching and learning of algebra: The 12th ICMI study* (pp. 155–186). Norwell, MA: Kluwer.

Thompson, P. (1994), 'Images of rate and operational understanding of the fundamental theorem of calculus', *Educational Studies in Mathematics*, 26(2), 229–274.

Thomson, S., De Bortoli, L. & Buckley, S. (2013). *PISA 2012: How Australia measures up. The PISA 2012 Assessment of Students' Mathematical, Scientific and Reading Literacy.* Melbourne: ACER.

Thomson, S., Hillman, K. & De Bortoli, L. (2013). *A teacher's guide to PISA mathematical literacy.* Melbourne: ACER.

Thomson, S., Hillman, K., & Wernet, N. (2012). *Monitoring Australian Year 8 student achievement internationally: TIMSS 2011.* Camberwell: ACER.

Treilibs, V. (Ed.). (1980–81). *Mathematics at work* (Vols. 1–6). Canberra, Australia: Australian Academy of Science.

Treilibs, V., Burkhardt, H. & Low, E. (1980). *Formulation processes in mathematical modelling.* Nottingham, UK: Shell Centre for Mathematical Education.

Trouche, L. & P. Drijvers (2010). Handheld technology for mathematics education: Flashback into the future. *ZDM*, 42(7), 667–681.

Tukey, J. (1977). *Exploratory data analysis.* Reading, MA: Addison-Wesley.

Turner, P. (2007). Reflections on numeracy and streaming in mathematics education. *The Australian Mathematics Teacher*, 63(2), 28–33.

Ubuz, B. (2007). Interpreting a graph and constructing its derivative graph: Stability and change in students' conceptions. *International Journal of Mathematics Education in Science and Technology*, 38(5), 609–637.

Ubuz, B. & Üstün, I. (2004). Figural and conceptual aspects in defining and identifying polygons. *Eurasian Journal of Educational Research*, 16, 15–26.

Ubuz, B., Üstün, I. & Erbas, A. K. (2009). Effect of dynamic geometry environment on immediate and retention level achievements of seventh grade students. *Eurasian Journal of Educational Research*, 35, 147–164.

United Nations Educational, Scientific and Cultural Organization (2003). *Gender and education for all: The leap to equality* (Summary Report). Paris: UNESCO.

United States College Board (2016). *AP calculator policy*. Retrieved from https://apstudent.collegeboard.org/apcourse/ap-calculus-ab/calculator-policy?calcab

Vacher, H. L. & Mylroie, J. E. (2001). Connecting with geology: Where's the end of the cave? *Mathematics Teacher*, 94(8), 640–649.

Vale, C. (n.d.). *Middle years numeracy intervention interview*. Unpublished paper, School of Education, Victoria University, Melbourne.

Vale, C. (1987a). *Negotiating and mathematics course*. Melbourne: Brooks Waterloo.

Vale, C. (1987b). *Up, up and away, Schools Resource Program: The participation of girls in maths and science, participation and equity program*. Melbourne: Ministry of Education.

Vale, C. (2002a). Girls back off mathematics again: The views and experiences of girls in computer based mathematics. *Mathematics Education Research Journal*, 14(3), 52–68.

Vale, C. (2002b). Numeracy intervention in the middle years. In C. Vale, J. Roumeliotis & J. Horwood (Eds), *Valuing mathematics in society* (pp. 284–293). Brunswick, Vic.: MAV.

Vale, C. (2003). Computers in mathematics: A super highway to social justice? In L. Burton (Ed.), *Which way social justice in mathematics education?* (pp. 277–301). Westport, CT: Praeger.

Vale, C. (2004). Numeracy intervention in junior secondary mathematics. In I. Putt, R. Faragher & M. McLean (Eds), *Mathematics education for the third millennium: Towards 2010* (Proceedings of the 27th annual conference of the Mathematics Education Research Group of Australasia, p. 631). Sydney: MERGA.

Vale, C. (2010). Gender mainstreaming: Maintaining attention on gender equality. In H. J. Forgasz, J. Rossi-Becker, K. Lee & O. B. Steinthorsdottir (Eds), *International perspective on gender in mathematics education* (pp. 111–143). Charlotte, NC: Information Age Publishing Inc.

Vale, C., Atweh, B., Averill, R. & Skourdoumbis, A. (2016). Equity, social justice and ethics in mathematics educations. In K. Makar, S. Dole, J. Visnovska, M. Goos, A. Bennison & K. Fry (Eds), *RiMEA 2011–2015*. Rotterdam, The Netherlands: Springer.

Vale, C., Davies, A., Hooley, N., Weaven, M., Davidson, K. & Loton, D. (2011). *Outcome evaluation of literacy and numeracy pilots in low SES school communities 2009–2010* (Final Report for the Department of Education and Early Childhood). Victoria University, Melbourne.

Vale, C., Forgasz, H. & Horne, M. (2004). Gender and mathematics: Back to the future? In B. Perry, C. Diezmann & G. Anthony (Eds), *Review of research in mathematics education in Australasia 2000–2003* (pp. 75–102). Sydney: MERGA.

Vale, C. & Leder, G. (2004). Student views of computer based mathematics in the middle years: Does gender make a difference? *Educational Studies in Mathematics*, 56(3), 287–312.

Vale, C., Weaven, M., Davies, A., Hooley, N., Davidson, K. & Loton, L. (2013). Growth in literacy and numeracy achievement: Evidence and explanations of a summer slowdown in low socio-economic schools. *Australian Educational Researcher*, 40(1), 1–25.

Vale, C., Widjaja, W., Herbert, S., Loong, E. & Bragg, L.A. (2016). Mapping variation in children's mathematical reasoning: The case of 'What else belongs?' *International Journal of Science and Mathematics Education*. DOI 10.1007/s10763-016-9725-y

van den Akker, J. (2003). Curriculum perspectives: An introduction. In J. van den Akker, W. Kuiper & U. Hameyer (Eds), *Curriculum landscapes and trends* (pp. 1–10). Dordrecht, The Netherlands: Kluwer Academic Publishers.

van Dooren, W., De Bock D., Janssens, D. & Verschaffel, L. (2005). Students' over reliance on linearity: An effect of school-like word problems? In H. Chick & J. Vincent, *Proceedings of the 29th conference of the International Group for the Psychology of Mathematics Education* (Vol. 4, pp. 265–272). Melbourne: PME.

van Hiele, P. (1986). *Structure and insight: A theory of mathematics education.* New York: Academic Press.

van Kraayernoord, C. E. & Elkins, J. (2004). Learning difficulties in numeracy in Australia, *Journal of Learning Difficulties*, 37(1), 32–41.

van Oers, B. (2001). Educational forms of initiation in mathematical culture. *Educational Studies in Mathematics*, 46, 59–85.

van Zoest, L., Jones, G. & Thornton, C. (1994). Beliefs about mathematics teaching held by preservice teachers involved in a first grade mentorship program. *Mathematics Education Research Journal*, 6, 37–55.

Victorian Curriculum and Assessment Authority. (2014). *Assessment principles*. Retrieved from www.insight.vic.edu.au/assessment-in-principle/assessment-principles

Victorian Curriculum and Assessment Authority. (2015). *On demand testing*. Retrieved from www.vcaa.vic.edu.au/Pages/prep10/ondemand/index.aspx

Vincent, J. (2000). *Computer enriched mathematics for Years 7 and 8 (CSF II Level 5)* (2nd ed.). Melbourne: MAV.

Vincent, J. (2005) *From shrine to university: A geometry journey along St Kilda Road and Swanston Street* (2nd ed.). Melbourne: MAV.

Vinner, S. (1989). The avoidance of visual considerations in calculus students. *Focus on Learning Problems in Mathematics*, 11(2), 149–156.

Vinner, S. & Dreyfus, T. (1989). Images and definitions for the concept of function. *Journal for Research in Mathematics Education*, 20(4), 356–366.

Visser, G. (2004). GridPic (Version 1.2) (Computer software). Melbourne: Gerard Inc. Retrieved from http://extranet.edfac.unimelb.edu.au/DSME/RITEMATHS

Vygotsky, L. S. (1978). *Mind in society*. Cambridge, MA: Harvard University Press.

Vygotsky, L. S. (1997). *Obras escogidas V—Fundamentos da defectología* (The Fundamentals of defectology). Madrid, Spain: Visor.

Walden, R. & Walkerdine, V. (1985). *Girls and mathematics: From primary to secondary schooling* (Bedford Way Papers 24). London: Heinemann.

Wallace, A. & Boylan, C. (2009). Reviewing the rural lens in education policy and practice. *Education in Rural Australia*, 19(2), 23–30.

Walshaw, M. & Siber, E. (2005). Inclusive mathematics: Catering for the 'learning-difficulties' student. In H. L. Chick & J. L. Vincent (Eds), *Proceedings of the 29th conference of the International Group for the Psychology of Mathematics Education* (Vol. 1, p. 290). Melbourne, Australia: PME.

Walter, J. & Gerson, H. (2007). Teachers' personal agency: Making sense of slope through additive structures. *Educational Studies in Mathematics*, 65(2), 203–233.

Warren, E. (2003). The role of arithmetic structure in the transition from arithmetic to algebra. *Mathematics Education Research Journal*, 15(2), 122–137.

Warren, E., Cooper, T. & Baturo, A. (2009). Bridging the educational gap: Indigenous and non-Indigenous beliefs, attitudes and practices in a remote Australian school. In J. Zajda & K. Freeman (Eds), *Race, ethnicity and gender in education* (pp. 161–184). New York, NY: Springer.

Watson, A. (2006a). *Raising achievement in secondary education*. Maidenhead, Berskshire: Open University Press.

Watson, A. (2006b). Some difficulties in informal assessment in mathematics. *Assessment in Education: Principles, Policy & Practice*, 13(3), 289–303.

Watson, A. & Mason, J. (2006). Seeing an exercise as a single mathematical object: Using variation to structure sense-making. *Mathematical Thinking and Learning*, 8(2), 91–111.

Watson, J. (2005). Is statistical literacy relevant for middle school students? *Vinculum*, 42(1), 3–10.

Watson, J. (2006). *Statistical literacy at school: Growth and goals*. Mahwah, NJ: Lawrence Erlbaum Associates.

Watson, J. (2014). What is 'typical' for different kinds of data? Examples from the Melbourne Cup. *The Australian Mathematics Teacher*, 70(2), 33–40.

Watt, H. (2005). Attitudes to use of alternative assessment methods in mathematics: A study with secondary mathematics teachers in Sydney, Australia. *Educational Studies in Mathematics*, 58(1), 21–44.

Watt, H. M. G. (2000). Exploring perceived personal and social gender stereotypes of maths with secondary students: An explanation for continued gender differences in participation? Paper presented at the annual meeting of the Australian Association for Research in Education, Sydney. Retrieved from www.aare.edu.au/00pap/wat00302.htm

Way, J. (n.d.). Fractions. *Top Drawer Teachers: Resources for teachers of mathematics*. Australian Association of Mathematics Teachers. Retrieved from http://topdrawer.aamt.edu.au/Fractions

Wenger, E. (1998). *Communities of practice: Learning, meaning, and identity*. Cambridge, UK: Cambridge University Press.

Wesslen, M. & Fernanez, S. (2005). Transformation geometry, *Mathematics Teaching*, *191*, 27–29.

Western Australian Board of Studies. (2014). *Assessment principles and reflective questions*. Retrieved from http://k10outline.scsa.wa.edu.au/home/assessment/principles-and-reflective-questions

Western Australian School Curriculum and Standards Authority (2013). *Year 12 grade descriptions*. Retrieved from www.scsa.wa.edu.au/internet/Senior_Secondary/School_Based_Assessment/guide_to_grades

Wettenhall, S. & O'Reilly, J. (1989). *Space to grow*. Melbourne: The Girls and Maths Special Interest Group, Mathematical Association of Victoria.

White, P. (n.d.). Mathematical modelling and the general mathematics syllabus. *NSW HSC Online*. Sydney: Curriculum Support Directorate. Retrieved from www.curriculumsupport.education.nsw.gov.au

White, P. (1995). An introduction to teaching calculus. In L. Grimson & J. Pegg (Eds), *Teaching secondary school mathematics: Theory into practice* (pp. 165–185). Sydney: Harcourt Brace.

White, P. & Mitchelmore, M. C. (1996). Conceptual knowledge in introductory calculus. *Journal for Research in Mathematics Education*, 27(1), 79–95.

White, P. & Mitchelmore, M. C. (2002). Teaching and learning mathematics by abstraction. In D. Tall & M. Thomas (Eds), *Intelligence, learning and understanding in mathematics: A tribute to Richard Skemp* (pp. 235–255). Flaxton, Queensland: Post Pressed.

Widjaja, W., Stacey, K. & Steinle, V. (2008). Misconceptions about density of decimals: Insights from preservice teacher's work. *Journal for Science and Mathematics Education in Southeast Asia*, 31(2), 117–131.

Widjaja, W., Stacey, K. & Steinle, V. (2011). Locating negative decimals on the number line: Insights into the thinking of pre-service primary teachers. *Journal of Mathematical Behaviour*, 30(1), 80–91.

Wienk, M. (2015). *Discipline profile of the mathematical sciences 2015*. Parkville, Vic.: Australian Mathematical Sciences Institute.

Wild, C. & Pfannkuch, M. (1999). Statistical thinking in empirical enquiry. *International Statistical Review*, 67(3), 223–265.

Wilkins, J. & Hicks, D. (2001). A s(t)imulating study of map projections: An exploration integrating mathematics and social studies. *Mathematics Teacher*, 94(8), 660–671.

Williams, A. (2013). A teacher's perspective of dyscalculia: Who counts? An interdisciplinary overview. *Australian Journal of Learning Disabilities*, 18(1), 1–16.

Williams, C. G. (1998). Using concept maps to assess conceptual knowledge of function. *Journal for Research in Mathematics Education*, 29(4), 414–421.

Williams, G. (2000). Collaborative problem solving and discovered complexity. In J. Bana & A. Chapman (Eds), *Mathematics education beyond 2000* (Proceedings of the 23rd annual conference of MERGA, Fremantle, Vol. 2, pp. 656–663). Sydney: MERGA.

Williams, G. (2002). Autonomous access, spontaneous pursuit, and creative execution: Insightful and creative mathematical problem-solving. In C. Vale, J. Roumeliotis & J. Horwood (Eds), *Valuing mathematics in society* (Proceedings of the 39th annual conference of the Mathematical Association of Victoria, pp. 331–347). Melbourne: MAV.

Williams, G. (2012). High mathematical performance on class tests is not a predictor of problem solving ability: Why? In J. Cheeseman (Ed.), *It's my maths: Personalised mathematics learning* (Proceedings of the 49th annual conference of the Mathematical Association of Victoria, pp. 288–296). Brunswick, Vic.: MAV.

Williams, S. (2001). Predications of the limit concept: An application of repertory grids. *Journal for Research in Mathematics Education*, 32(4), 341–367.

Willis, K., Geiger, V., Goos, M. & Dole, S. (2012). Numeracy for what's in the news and building an expressway. *The Australian Mathematics Teacher*, 68(1), 9–15.

Wilson, T. & Barkatsas, T. (2014). The effect of language, gender and age on *NAPLAN* numeracy data. In J. Anderson, M. Cavanagh & A. Prescott (Eds), *Curriculum in focus: Research guided practice* (Proceedings of the 37th annual conference of the Mathematics Education Research Group of Australasia, pp. 653–660). Sydney: MERGA.

Wong, K. Y. & Veloo, P. (2001). Situated sociocultural mathematics education: Vignettes from southeast Asian practices. In B. Atweh, H. Forgasz & B. Nebres (Eds), *Sociocultural research on mathematics education* (pp. 113–134). Mahwah, NJ: Lawrence Erlbaum.

Wood, D., Bruner, J. & Ross, G. (1976). The role of tutoring in problem solving. *Journal of Child Psychology and Psychiatry*, 17, 89–100.

Wood, T., Williams, G. & McNeal, B. (2006). Children's mathematical thinking in different classroom cultures. *Journal for Research in Mathematics Education*, 37, 222–255.

Wright, R. (1998). An overview of a research-based framework for assessing and teaching early number learning. In C. Kanes, M. Goos & E. Warren (Eds), *Teaching mathematics in new times* (Proceedings of the 21st annual conference of the Mathematics Education Research Group of Australasia, pp. 701–708). Brisbane: MERGA.

Wright, V. & Downton, A. (n.d.). Mental computation. *Top Drawer Teachers: Resources for teachers of mathematics*. Australian Association of Mathematics Teachers. Retrieved from http://top-drawer.aamt.edu.au/Mental-computation

Yackel, E. & Hanna, G. (2003). Reasoning and proof. In J. Kilpatrick, W. G. Martin & D. Schifter (Eds), *A research companion to principles and standards for school mathematics* (pp. 227–236). Reston, VA: NCTM.

Yerushalmy, M. (2000). Problem solving strategies and mathematical resources: A longitudinal view on problem solving in a function-based approach to algebra. *Educational Studies in Mathematics*, 43, 125–147.

Yerushalmy, M. & Gilead, S. (1999). Structures of constant rate word problems: A functional approach analysis. *Educational Studies in Mathematics*, 39(1–3), 185–203.

Yerushalmy, M. & Schwartz, J. L. (1993). Seizing the opportunity to make algebra mathematically and pedagogically interesting. In T. A. Romberg, E. Fennema & T. P. Carpenter

(Eds), *Integrating research on the graphical representation of functions* (pp. 41–68). Hillsdale, NJ: Earlbaum.

Young-Loveridge, J. (2005). Fostering multiplicative thinking using array-based materials. *The Australian Mathematics Teacher*, 61(3), 34–39.

Yunkaporta, T. & McGinty, S. (2009). Reclaiming Aboriginal knowledge at the cultural interface. *The Australian Educational Researcher*, 36(2), 55–72.

Zagorski, S. (2004). Choosing and working with a mentor. In M. Chappell, J. Choppin & J. Salls (Eds), *Empowering the beginning teacher of mathematics in high school* (pp. 6–7). Reston, VA: NCTM.

Zakaria, E. & Salleh, T. S. (2015). Using technology in learning integral calculus. *Mediterranean Journal of Social Sciences*, 6(5), 144.

Zandieh, M. (2000). A theoretical framework for analyzing student understanding of the concept of derivative. In E. Dubinsky, J. Kaput & A. H. Schoenfeld (Eds), *Research in collegiate mathematics education* (pp. 103–127). Providence, RI: American Mathematical Society.

Zandieh, M. J. & Knapp, J. (2006), Exploring the role of metonymy in mathematical understanding and reasoning: The concept of derivative as an example, *The Journal of Mathematical Behavior*, 25(1), 1–17.

Zaslavsky, O., Chapman, O. & Leiken, R. (2003). Professional development of mathematics educators: Trends and tasks. In A. Bishop, M. A. Clements, C. Keitel, J. Kilpatrick & F. Leung (Eds), *Second international handbook of mathematics education, Part 2* (pp. 877–917). Dordrecht, The Netherlands: Kluwer.

Zazkis, R. (2001). From arithmetic to algebra via big numbers. In H. Chick, K. Stacey, J. Vincent & J. Vincent (Eds), *The future of the teaching and learning of algebra* (Proceedings of the 12th ICMI Study Conference, Vol. 2, pp. 676–681). Melbourne: University of Melbourne.

Zbiek, R. & Conner, A. (2006). Beyond motivation: Exploring mathematical modeling as a context for deepening students' understanding of curricular mathematics. *Educational Studies in Mathematics*, 63, 89–112.

Zbiek, R. & Heid, M. K. (2008). Digging deeply into intermediate algebra: Using symbols to reason and technology to connect symbols and graphs. In C. Greenes & R. N. Rubenstein (Eds), *Algebra and algebraic thinking in school mathematics, 2008 yearbook of the National Council of Teachers of Mathematics* (pp. 247–261). Reston, VA: NCTM.

Zevenbergen, R. (2003a). Grouping by ability: A self-fulfilling prophecy? *The Australian Mathematics Teacher*, 59(4), 2–7.

Zevenbergen, R. (2003b). Teachers' beliefs about teaching mathematics to students from socially disadvantaged backgrounds: Implications for social justice. In L. Burton (Ed.), *Which way social justice in mathematics education?* (pp. 133–151). Westport, CT: Praeger.

Zevenbergen, R. (2004). Technologising numeracy: Intergenerational differences in working mathematically in new times. *Educational Studies in Mathematics*, 56, 97–117.

Zuber, E. N. & Anderson, J. (2013). The initial response of secondary mathematics teachers to a one-to-one laptop program. *Mathematics Education Research Journal*, 25(4), 279–298.

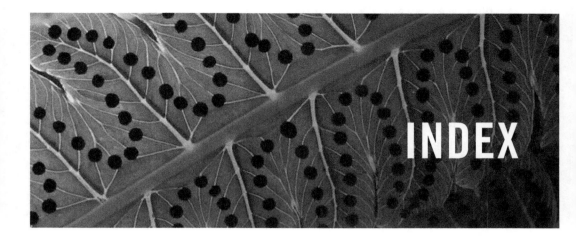

INDEX

addition 170
algebra 114, 119, 122, 199–225
  activities 200
  algebraic expressions 206–7
  algebraic insight 210
  approaches to teaching 199–200, 201–5
  ascendancy of the function concept 217–24
  Computer Algebra Systems (CAS) 209–11
  core algebraic ideas 211–7
  equivalence 211–3
  expressing generality 206–7
  families of functions 222–4
  function-graphing technologies 218–24
  functions as entities, manipulating 224
  generalisation problem 202–3
  generalising from number patterns 207
  generational activities 206–7
  primary years 200
  quasi-variable problem 203–5
  rate 215–7
  technology 209–11, 219–24
  transformational activities 207–9
  transition to algebraic thinking 201–5
  translation difficulties 205–6
  variables 213–5
area and volume 265–8
arithmetic 119–20
assessment *see* mathematics learning,
    assessment of

Australian Association of Mathematics Teachers
    (AAMT) 77, 116, 456–7
Australian Mathematics Competition 362–3

calculators 67, 76, 84, 95–7, 102–4, 224
calculus 302–37
  algebraic representations 303
  anti-differentiation 329–32
  approaches to teaching 321–3
  background required for successful learning
    303–13
  change, study of 307–9
  derivative 302–3
  developing rules of differentiation 329
  differential calculus 313–5, 323–32
  Differentiation Competency Framework 324
  difficulties with limit concept 316–7
  enactive experiences 302
  function 303–4
  graphical representation 303–4
  graphs 308–9, 314–5
  integral 303, 332–7
  integration 332–7
  investigative approach to teaching 323
  notation confusions 320–1
  numerical representation 303, 305–7
  obstacles to learning 315–21
  procepts 303
  rate, understanding of 318–9

rates of change 310–3
representational diversity, fluency and versatility
    303–7
Representational Framework of Knowing
    Derivative 327
secants, slope of 310
tangents, misconceptions with 319–20
technology 303–7, 321–3
visual-enactive approach 322
chance and data 274–300
    aggregate perspective of data 279–80
    challenges in understanding 272–81
    chance in primary years 276–7
    consuming statistical information 285–6
    contexts for teaching 275
    covariation 287–90
    creating statistical information 291–2
    data in primary years 276–7
    data-logging equipment 84
    graph interpretation 280
    integrating contextual knowledge 280–1
    intuition about randomness 277–8
    primary years 276–7
    probability 282–4, 290–1
    procedural knowledge 278
    secondary curriculum 282–300
    statistical inference, informal 297–300
    statistical information, utilising 285–6
    statistical investigations 279, 294–7
    statistical literacy 284–5
    statistical thinking, dispositions for 292–3
    statistics and mathematics, distinction 274
    technologies 274
    topic integration 275
    univariate descriptive statistics 286–7
classroom environment 39
classroom practice *see also* mathematics teaching
        practices
    impact of mathematical beliefs 4
    integration of technologies 76
    technology and 76, 95–101, 105–8
classroom scenarios 51–61
cognitive development 30–1, 120, 145
communities *see* parents and communities,
        working with

Computer Algebra Systems (CAS) calculators 67,
    76, 92, 94, 102–5, 195–7, 209–11, 303, 306, 314,
    316–7
congruence 230
constructivism 30–1
curriculum concepts 111–2 *see also* mathematics
    curriculum

data *see* chance and data
data-logging equipment 84
decimals 170–6, 185–6
derivatives 302, 303, 324–9
differential calculus *see* calculus
differentiation 302, 323–9, 329
division 176
dynamic geometry software 81–2, 239–42, 245–8

education in Australia 119, 120–1, 422–3
educational partnerships 419–33
equations 202, 208–9, 210, 219, 221–4
estimating and measuring 260–2
Euclidean geometry 119, 228, 229
exponential thinking 191–3

focusing 55–61
fractions 170–2, 176–9, 185–6
function
    calculus 303–8, 310–3
    change of 310
    concept as object of algebra 217–24
    exponential 84, 223
    families of 79, 222–4
    introductory calculus 306
    linear 78–9
    trigonometric 81, 84, 223
funnelling 52–5

gender equity and justice 395–414
    attitudes to mathematics 398–9
    classroom environment 403–6
    computer usage 403–5
    definitions 396–7
    differences in performance 230
    equal opportunity 396
    equal outcomes 396, 399

equal treatment 396–7
gender equity in practice 412–3
gender gaps 395–401
graphical interpretation 400
inclusive curriculum 408–12
intervention programs 402
male domain 401–3
mathematical achievement 397, 399–401
nature of classroom tasks 412
participation in mathematics 397–8
resources 403–6
response-able curriculum 412
spatial visualisation 400
stereotyping 401–12
teacher beliefs 406
teaching, in 406–7
technology use 399, 401
theories of gender equity and practice 406–7
geometry and spatial concepts 227–55
concepts, discussion of 237
construction 239–41
contexts for student activities 232
curriculum 229, 243
do-talk-record strategy 238
dynamic geometry software 80–2, 239–42,
245–8
Euclidean 228
geometric reasoning 227, 231
geometric tools 228, 239–40, 249
international studies of performance 230
invariance 230, 252
isometric transformations 248–51
language and communicating 236–8
lines, planes and angles 243–5
location and spatial reasoning 252–4
materials for activities 233–4
network analysis 253
non-isometric transformations 251
primary years 228–9
properties and relationships of figures 245–8
reasoning 241–3
rich tasks/real problems 246, 250, 252
secondary years 229–31
terms, understanding of 236, 237
transformations 228, 248–51

vectors 254–5
visual-spatial skills 228, 230, 361
gifted and talented students 360–7
graphics calculators 67, 76, 77, 79, 82, 84, 93–7, 102–4,
108
graphing software 208, 221, 223–4
graphs in calculus 308–9, 314–5

Indigenous students 375–7, 386–90
inquiry mathematics 27–30, 41–4
integers 193–5
integrals 302, 303
integration 302, 332–7
internet 76, 81, 82, 84, 86–7
irrational number 195–6
isometries 248

Japanese lesson sequence 50

Key Learning Areas (KLAs) 122

language in learning 377–8
learning 30–4
assessment *see* mathematics learning, assessment
of
constructivism 30–1
outcomes 122
practical activities, use of 33–4
sociocultural perspectives 31–2, 141
learning difficulties in mathematics
assessment tools 254–6
clinical interviews 354–5
defining and identifying students with 352–4
diagnostic tools 354–5
inquiry approaches 257
intervention programs 358–9
learning disabilities, distinction 352
learning obstacles 353–4
primary students 352
research findings 353
structural understanding, importance of 353
teaching strategies 358
teaching students with 356–9
lines, planes and angles 243–5
location, concepts and skills 227–8, 252–4

mathematical application problems 61–6
  border problems 62–3
  tapestry problems 64–6
  wrapper problems 63–4
mathematical beliefs 4
  nature of mathematics 4
  relationship between practices and 4–5
  student beliefs 8–9, 39–40
  teaching and learning 7
mathematical connections 46–74
  classroom scenario 51–61
  connecting content across lessons 49–51
  curriculum areas, across 48–9
  focusing 55–61
  funnelling 52–5
  making connections in lessons 51
  mathematical applications, through 61–6
  mathematical modelling, through 66–74
  meaning 46
  measurement 271–2
  middle years 47–8
  real world context 67–8
  teachers and 46
  technology, use of 79–80
  triadic dialogue 52
mathematical language 236
mathematical literacy 6, 17, 170, 381, 396, 399
mathematical modelling 66–74
  contextualised problems 67
  example of 70–1
  formulation skills of students 73
  meaning of term 66–7
  measurement 271–2
  metacognitive activity 71
  process of 68–72
  real-world problems 69–70
  sub-skills 72–4
  validation 71
mathematical problems
  awareness 39
  definition 38
  factors contributing to successful problem-
    solving 39
  heuristics 39
  knowledge base 39

measurement 271–2
metacognition 39
problem-solving component of learning 38–41
procedural complexity 12
processes for solving 13
regulation 39
relationship among 12–3
mathematical reasoning 37–8, 241–3
mathematical symbols 170–1, 199
mathematical talent, students with
  accelerated programs 365–6
  assessment tools 362–3
  defining and identifying 360–2
  enrichment programs 366–7
  teaching 364–5
  teaching strategies for 360–7
mathematical thinking
  categories of student thinking 34–5
  complexity of 35
  development in students 35–7
  open questions 51
  problem-solving 38–41
  reasoning 37–8
mathematics
  definition 5
  numeracy 6–7
  parental and community attitudes towards 417–9
  school mathematics 27–30
  student beliefs on nature of 8–9, 39–40
  understanding of *see* mathematics,
    understanding of
mathematics curriculum 17
  cross-curricular connections 48–9
  culturally response-able 383–6
  differentiated 346
  mathematical applications 61, 67–8
  mathematical modelling 66–8
  models 110–35
  numeracy 126–30
  process aspects of mathematics in 35
  technology, use of 75–6, 90–5
mathematics curriculum models
  attained 111
  authority-innovation-decision-making model
    117–8

classroom level 112
content 113–23
curriculum choices and orientations 115
curriculum concepts 111–2
decision-makers 115–8
democratic access to powerful mathematical
 ideas 113–4
equal encouragement for success 114
equal participation 114
essential components 112
historical overview of development in Australia
 118–23
implemented 112
inclusivity and rights 114
intended 111
intent or reality 111
national curriculum 121, 122–3
1980s initiatives 120–1
numeracy 126–32
organisation of 123–35
problem-solving 114, 128
process-driven 124–6
social justice 132–5
stakeholders in development 116
state and territory determination of 110
system-level documents 112
teachers' role 118
mathematics learning, assessment of 137–65
alternative methods 146–7
aspects of 137
assessment data, use of 137
assessment items for tests 153–5
assessment tasks 146–7
collecting and interpreting evidence of student
 learning 146–55
comparison studies 141
competency-based assessment 143
consistency of teacher judgments 157–62
criterion-referenced judgment
 models 142–3
development of rubric 161
developmental-based assessment (DBA) 145–6
everyday rubric grading 159
evidence-based judgment models 142–5
exemplars 144

flowchart for preparing assessment task for
 students 158
formative assessment 139
guidelines for assessment quality and equity
 156–7
IMPACT procedure 149
moderation of school-based assessment systems
 159
multiple analytic rubrics 161
norm-referenced judgment model 142
open-ended questions 149–50
performance assessment 140, 151
performance tasks, examples of 151
preparing assessment tasks for students 156–7
purposes of assessment 138–40
recording students' progress 163
reporting 163–5
rich assessment tasks (RATs) 150
rubrics 159–62
self-assessment tasks 148–9
standards-referenced judgment models 143
Structure of the Observed Learning Outcome
 (SOLO) 145
summative assessment 140
task-specific rubrics 160
teacher autonomy 157
teacher classroom questions 147–8
tests and exams 146–7, 152–5
Year 12 138
mathematics teachers
becoming a reflective teacher of mathematics
 436–40
case study of learning 449–53
continuing professional learning 435–57
feedback from colleagues 442–4
feedback from students 440–2
IMPACT procedure 441
knowledge needed for teaching 14–6, 76
learning and development 448–9
perceptions of 9–11
planning for continuing professional learning
 453–6
role in developing students'
 understanding 41–4
self-analysis of lessons 437–40

technologies 75–6, 95, 101–2, 105–8
working with parents and
    communities 417–33
mathematics teaching
    current challenges 17–8
    effectiveness of 440
    factors influencing technology use in 105–8
    feedback from students 440–2
    inquiry mathematics culture 28–30
    knowledge needed for 14–6
    school mathematics culture 28–30
mathematics teaching practices
    caring 383
    characteristics of best teaching practice 454
    knowledge needed for teaching mathematics
        14–6
    mismatches in cultural values 421
    perspectives on Year 8 classrooms 11–3
    socio-cultural values 421
    technology, working with 95–101
mathematics, understanding of 18
    constructivism 30–1
    continuing process, as 26
    definition 23–7
    developing 23–44
    framework 25–6
    levels of, 26
    Pirie-Kieren theory 26–7
    role of teachers 41
    sociocultural perspectives 31–2, 64, 141
    types of 24–6
matrices 42, 230
measurement 257–72
    angle, understanding 258, 259
    area and volume 265–8
    estimating and measuring 260–2
    making connections 271–2
    metric units, relationship of 262–4
    modelling 271–2
    perimeter and circumference 264–5
    primary years 258
    problem-solving 271–2
    Pythagoras' Theorem 49, 86, 195, 204, 228, 242,
        245, 251, 256, 259, 268–9
    real-world situations 257

secondary years 259–60
    trigonometry 270–1, 228
mental computation 180–2
metric units of measurement, relationship of 262–4
middle years 47–8, 358
mixed ability classes 345–51
modelling *see* mathematical modelling
multiplication 180

national assessment program 141
national benchmarks 164
National Council of Teachers of Mathematics
    (NCTM) 77, 114, 124
national curriculum framework 121–2
network analysis 253
New Mathematics movement 120–1
non-isometric transformations 251
number 169–97
    achievement level range 173
    addition 170
    common student misconceptions 172, 175–6
    competencies and strategies 169
    concepts and skills 169–70
    decimals 170–1
    difficulties 172
    division by fractions and decimals 185–6
    division of fraction problem types 186
    errors in mental computation 180
    exponential thinking 191–3
    field laws 180–3
    fractions 170–1, 176–9
    integers 193–5
    irrational number 195–6
    mental computation 180–2
    multiplication 180, 184–5
    multiplicative thinking 183–4, 191
    percentage 190–1
    place value 173–82
    primary years, in 170–1
    proportional thinking 187–90, 251
    secondary years, in 172–3
    subtraction 170
    surds 195–6
    whole-number thinking 175–6, 178
numeracy 6–7, 17, 47, 141, 422

Australian history, in 130–2
curriculum, across 126–30
numerate person, characteristics of 128

parents and communities, working with
collaborating with the community 427
communicating 425–6
community-centred perspectives on partnership
430–2
decision-making 426–7
educational partnerships 419–33
family-centred perspectives on partnerships
428–30
government policies 422–3
learning at home 426
parental and community attitudes towards
mathematics 417–9
parenting 424–5
partnership with community, development of
432–3
school-centred perspectives on partnerships 424–8
stakeholder perspectives on partnerships 424–32
teachers' responsibilities 417
volunteering 426
pedagogical content knowledge 14–6, 106
percentage 190–1
perimeter and circumference 264–5
place value 173–82
primary to secondary school transition 47
probability and statistics 274–5, 282–4, 290–1
procedural thinkers 267
professional development and learning
becoming a reflective teacher of mathematics
436–40
collaborative professional learning 444–6
learning to teach, case study of 449–53
opportunities and resources 456–7
planning for continued professional learning
453–6
professional networks and action research 444–6
professional practice, dimensions of 435–6
teacher learning and development 448–9
understanding professional socialisation 447
working in professional communities 446–7
proficiency strands 127

Programme for International Student Assessment (PISA) 137,
153, 370–1, 375 399–401
proportional thinking 187–90
public education systems 110, 137
Pythagoras' Theorem 49, 86, 195, 204, 228, 242, 245,
251, 256, 259, 268–9

rate, concept of 215–7
ratios 187–90
reasoning 37–8, 241–3

scaffolding learning 31–2, 40–1, 55, 86
secants, slope of 310
shapes, properties of 230, 231
simultaneous equations 27
social and cultural issues
approaches to equity and social justice 381–93
democratic or critical curriculum 391–3
equitable classroom model 381–2
equity and social justice 380–2
ethnomathematics 381
geographic location 378–9
Indigenous students 375–7, 386–90
language and language background 377–8
mathematical language 385
mathematics curriculum model 132–5
mathematics problems of personal or social
relevance 392
multicultural curriculum 383–6
problem selection 371
real-world issues 392
social capital 373
sociocultural norms and teachers' practices 379
socioeconomic status and mathematics learning
370–80
student involvement in learning 392
social justice curriculum model 132–5
goals 132–3
real-world issues, example of 133–5
sociocultural perspectives 31–2, 64, 141
special learning needs 359–60
spreadsheets 77–8, 81, 83, 98
statistical literacy 17, 275, 284–5, 292
streaming 343–5
students

beliefs on nature of mathematics 8–9, 39–40
feedback from 440–2
perceptions of mathematics teachers 9–11
students with diverse learning needs
accelerated programs 365–6
achievement spectrum 341
classroom management problems 341
defining and identifying students with learning
difficulties 352–4
defining and identifying students with
mathematical talent 360–2
diagnostic tools 354–5
enrichment programs 366–7
heterogeneous classrooms 345–51
inquiry and problem-based 346–9
learning difficulties in mathematics 352–9
responding to diverse needs of students 342–5
special learning needs 359–60
streaming 343–5
students with mathematical talent 360–7
targeted teaching 349–51
teaching students with learning difficulties 356–9
tiered curriculum 349
subtraction 170
symmetry 230

teachers see mathematics teachers
teaching see mathematics teaching
teaching practices see mathematics teaching
practices
technologies
access to resources 105
algebra 210
assessment and 101–5
benefits of learning mathematics with 77–90
calculators see calculators
chance and data 274
classroom practices and 105–8
curriculum and 90–5
dependence on 95
drawbacks 95
dynamic images, working with 80–2
effect on curriculum 95
effective use of 75–108
extension of curriculum through 92

finding and sharing 87–90
functions and variables 217–20
goal of learning and use of 75
graphing software 221–3
impact of 67
instant feedback, learning from 77–8
integration into classroom practice 76
interactive whiteboards 96
internet 76, 86–7
mathematics-specific hardware 76
mathematics-specific software 76
multiple representations, making connections
between 79–80
observing patterns 78–9
online mathematical resources 86
partner in teaching, as 97–8
pedagogy and 95–101
sequencing or treatment of topics, effect
on 93–5
simulated or authentic data, exploring 82–6
sources of assistance 105–6
spreadsheets 77–8, 81, 83, 98
teaching context 105–6
technology-rich learning environments 105–8
types of 76
visualisation 86–7, 231
working with 95–101
tests and exams 152–5
tiling patterns and tessellation 250–1
transformations 231, 248–51, 254–5, 207–9
Trends in International Mathematics and Science
Study (TIMSS) 11–3, 137, 153–4, 164,
399–401, 445
trigonometry 270–1, 228, 251

understanding of mathematics see mathematics,
understanding of

variables, concept of 213–5
vectors 254–5
visual reasoning 86, 229

Year 11 mathematics lesson 42–3

zone of proximal development (ZPD) 31–2